CHEMISTRY IN CANADA

Advances in Organometallic and Inorganic Polymer Science

Advances in Organometallic and Inorganic Polymer Science

edited by

CHARLES E. CARRAHER, JR.
Department of Chemistry
Wright State University
Dayton, Ohio

JOHN E. SHEATS
Department of Chemistry
Rider College
Lawrenceville, New Jersey

CHARLES U. PITTMAN, JR.
Department of Chemistry
University of Alabama
University, Alabama

MARCEL DEKKER, INC.   New York and Basel

Library of Congress Cataloging in Publication Data

Main entry under title:

Advances in organometallic and inorganic polymer science.

"These papers originally appeared in Journal of
macromolecular science--chemistry, volume A16, number 1,
edited by George E. Ham"--Verso t.p.
    Includes index.
    1. Polymers and polymerization--Addresses, essays,
lectures. 2. Organometallic compounds--Addresses,
essays, lectures. 3. Inorganic polymers--Addresses,
essays, lectures. I. Carraher, Charles E. II. Sheats,
John E. III. Pittman, Charles U. IV. Journal of
macromolecular science. Chemistry.
QD381.7.A38   547.7   82-5074
ISBN 0-8247-1610-8   AACR2

MARCEL DEKKER, INC.
270 Madison Avenue, New York, New York 10016

Current printing (last digit):
10 9 8 7 6 5 4 3 2 1

PRINTED IN THE UNITED STATES OF AMERICA

Preface

   This book, *Advances in Organometallic and Inorganic Polymer Science*, contains chapters selected by the editors to give an overview of much of the research being conducted in the field of organometallic polymers, as it is today and where it is going. Each of the five sections--General, New Procedures and Techniques, Electrically Conducting Polymers, Polymer-bound Catalysts and Phosphonitriles--includes one or more reviews. The reviewers are both experts within the area reviewed and are actively researching within the area. Authors were encouraged to indicate not only the present "state-of-the-art" as regards the particular area, but also to note areas in need of research, projected future trends, and actual and potential applications (industrial, other areas of science) of their topic areas. Thus the reviews are more than typical serial reviews and are considerably more valuable and longer lasting in potential value to the scientific community.

   "Polymetallocenylenes - Recent Developments" by E. Neuse details recent progress in the synthesis and spectroscopic characterization of polyferrocene and polyruthenocene including potential applications. This is an update of the Neuse and Rosenberg book *Metallocene Polymers* (Marcel Dekker, Inc., N.Y., 1970). "Polymers for Controlled Release of Organotin Toxin" by Subramanian and Somasekharan describes efforts at the synthesis of polymers which simultaneously provide long-term fouling resistance and useful engineering properties. Elements affecting the reactivity and dynamic mechanical behavior are discussed and the reactivity results fitted

to mathematical models. Taylor and St. Clair reviewed the "Incorporation of Metal Related Materials into Electrically Neutral
Polymers" including addition of metals as a metal atom vapor, organometallic compounds, coordination complexes and simple hydrated
or anhydrous salts. Property - metal nature interrelationships are
discussed, including polymer flammability and electrical
conductivity.

The application of the technique of trimethylsilylation to
mineral silicates is reviewed by B. Currell and J. Parsonage. Discussed are attempts to prepare new polyorganosiloxane materials from
wholly "natural" feedstocks. The "Metal Vapor Synthesis of Organometal Polymers and Polymer-Supported Metal Clusters," an area re-
ceiving much recent coverage, is reviewed by C. Francis and G. Ozin.
A new technique for identification of thermal degradation products
using coupled thermogravimetric analysis and mass spectroscopy is
described in detail by Carraher, Molloy, Tiernan, Taylor, and
Schroeder. The procedure is applicable to a wide range of analyses
including oil shale use studies and the thermal degradation of
transportation vehicle interiors.

The enormous potentiality offered by polymer/metal complexes in
catalysis is reviewed by C. Carlini and G. Sbrana. Included are
discussions relating catalytic activity and selectivity to binding
site, nature of the metal, and correlation with mechanism.

Additional monographs are planned at regular intervals to keep
the reader abreast of new developments. Another major symposium is
planned for 1983. We anticipate continued, exciting new developments in this field.

Charles E. Carraher, Jr.
John E. Sheats
Charles U. Pittman, Jr.

Contents

CONTENTS

POLYPHOSPHAZENES

Contributors

H. R. ALLCOCK   Department of Chemistry, The Pennsylvania State University, University Park, Pennsylvania 16802

RONALD D. ARCHER   Department of Chemistry, University of Massachusetts, Amherst, Massachusetts 01003

WILLIAM H. BATSCHELET   Department of Chemistry, University of Massachusetts, Amherst, Massachusetts 01003

D. E. BERGBREITER   Department of Chemistry, Texas A&M University, College Station, Texas 77843

RICHARD K. BROWN   Chemistry Division, Argonne National Laboratory, Argonne, Illinois 60439

M. S. BURSTEN   Department of Chemistry, Texas A&M University, College Station, Texas 77843

CARLO CARLINI   Centro di Studio del C.N.R. per le Macromolecole Stereordinate ed Otticamente Attive, Instituto di Chimica Organica Industriale, Università di Pisa, 56100 Pisa, Italy

CHARLES E. CARRAHER, JR.   Department of Chemistry, Wright State University, Dayton, Ohio 45435

B. CONWAY   Neckers Laboratory, Southern Illinois University, Carbondale, Illinois 62901

B. R. CURRELL   School of Chemistry, Thames Polytechnic, Woolwich, London SE18 6PF, England

MARCO-A. DE PAOLI   Instituto de Química, Universidade Estadual de Campinas, C.P. 1170, 13.100 Campinas, SP, Brazil

C. W. DIRK   Department of Chemistry and the Materials Research Center, Northwestern University, Evanston, Illinois 60201

S. DURAJ   Neckers Laboratory, Southern Illinois University
    Carbondale, Illinois 62901

T. L. EVANS   Department of Chemistry, The Pennsylvania State
    University, University Park, Pennsylvania 16802

VAN R. FOSTER   Department of Chemistry, Wright State University,
    Dayton, Ohio 45435

COLIN G. FRANCIS   Department of Chemistry, University of Southern
    California, Los Angeles, California 90007

GARY L. HAGNAUER   Polymer Research Division, U.S. Army Materials and
    Mechanics Research Center, Watertown, Massachusetts 02172

W. P. HART   Department of Chemistry, University of Massachusetts,
    Amherst, Massachusetts 01003

M. HODGMAN   Neckers Laboratory, Southern Illinois University,
    Carbondale, Illinois 62901

MARVIN L. ILLINGSWORTH   Department of Chemistry, University of
    Massachusetts, Amherst, Massachusetts 01003

M. E. KENNEY   Department of Chemistry, Case Western Reserve
    University, Cleveland, Ohio 44106

K. KUCHEL   Neckers Laboratory, Southern Illinois University,
    Carbondale, Illinois 62901

P. M. KUZNESOF   Department of Chemistry, Agnes Scott College,
    Decatur, Georgia 30030

D. W. MACOMBER   Department of Chemistry, University of Massachusetts,
    Amherst, Massachusetts 01003

T. J. MARKS   Department of Chemistry and the Materials Research
    Center, Northwestern University, Evanston, Illinois 60201

E. A. MINTZ* Department of Chemistry and the Materials Research
    Center, Northwestern University, Evanston, Illinois 60201

H. MICHAEL MOLLOY   Department of Chemistry, Wright State University,
    Dayton, Ohio 45435

M. MORONSKI   Neckers Laboratory, Southern Illinois University,
    Carbondale, Illinois 62901

---

*Current affilliation: Department of Chemistry, University of West
    Virginia, Morgantown, West Virginia 26506

ROBERT H. NEILSON   Department of Chemistry, Texas Christian University, Fort Worth, Texas 76129

E. W. NEUSE   Department of Chemistry, University of the Witwatersrand, Johannesburg 2001, Republic of South Africa

R. NOBLE   Neckers Laboratory, Southern Illinois University, Carbondale, Illinois 62901

R. S. NOHR   Department of Chemistry, George Mason University, Fairfax, Virginia 22030

D. A. OWEN   Department of Chemistry, Murray State University, Murray, Kentucky 42071

GEOFFREY A. OZIN   Lash Miller Chemical Laboratories and Erindale College, University of Toronto, Toronto, Ontario, Canada

J. R. PARSONAGE   School of Chemistry, Thames Polytechnic, Woolwich, London, SE18 6PF, England

M. D. RAUSCH   Department of Chemistry, University of Massachusetts, Amherst, Massachusetts 01003

A. K. ST. CLAIR   NASA Langley Research Center, Hampton, Virginia 23665

GLAUCO SBRANA   Centro di Studio del C.N.R. per le Macromolecole Stereordinate ed Otticamente Attive, Instituto di Chimica Organica Industriale, Università di Pisa, 56100 Pisa, Italy

K. F. SCHOCH, JR.   Department of Chemistry and the Materials Research Center, Northwestern University, Evanston, Illinois 60201

JACK A. SCHROEDER   Department of Chemistry, Wright State University, Dayton, Ohio 45435

ARTHUR J. SCHULZ   Chemistry Division, Argonne National Laboratory, Argonne, Illinois 60439

A. SIEGEL   Department of Chemistry, Indiana State University, Terre Haute, Indiana 47809

D. W. SLOCUM*   Neckers Laboratory, Southern Illinois University, Carbondale, Illinois 62901

K. N. SOMASEKHARAN   Department of Materials Science and Engineering, Washington State University, Pullman, Washington 99164

Current affiliation: Gulf Research and Development Company, Pittsburgh, Pennsylvania 15230

R. V. SUBRAMANIAN Department of Materials Science and Engineering, Washington State University, Pullman, Washington 99164

L. T. TAYLOR Department of Chemistry, Virginia Polytechnic Institute and State University, Blacksburg, Virginia 24061

MICHAEL L. TAYLOR Department of Chemistry and the Brehm Laboratory, Wright State University, Dayton, Ohio 45435

THOMAS O. TIERNAN Department of Chemistry and the Brehm Laboratory, Wright State University, Dayton, Ohio 45435

K. WEBBER Neckers Laboratory, Southern Illinois University, Carbondale, Illinois 62901

JACK M. WILLIAMS Chemistry Division, Argonne National Laboratory, Argonne, Illinois 60439

PATTY WISIAN-NEILSON Department of Chemistry, Texas Christian University, Fort Worth, Texas 76129

K. J. WYNNE Chemistry Program, Office of Naval Research, Arlington, Virginia 22117

Advances in Organometallic and Inorganic Polymer Science

GENERAL

Polymetallocenylenes-Recent Developments

E.W. NEUSE

Department of Chemistry, University of the Witwatersrand,
Johannesburg 2001, Republic of South Africa

ABSTRACT

The polymetallocenylenes represent a class of macromolecular compounds in which units of a metallocene complex
are directly and difunctionally interconnected so as to
constitute a linear chain. This chapter represents an
account of recent progress in the synthesis and spectroscopic characterization of the only two types of polymetallocenylene known to this date, $viz.$ poly-1,1'-
ferrocenylene (obtained with $\overline{M}_n$ up to 10 000) and poly-
1,1'-ruthenocenylene (obtained in the oligomer range).
In addition, the partial oxidation of poly-1,1'-ferro-
cenylenes, which gives poly(1,1'-ferrocenylene-co-1,1'-
ferricenylene) structures possessing mixed-valence
characteristics, is treated, and a discussion is
presented of potential applications for these organometallic polymer types and expected future developments
in the polymetallocenylene field.

INTRODUCTION

Scientific challenge and technological usefulness both have
combined to place the metallocenes into a most privileged position
within the realm of organometallic research, and one finds this well
reflected in the abundance of publications in the chemical, physical
and patent literature. Although the great majority of communications deals with di-η-cyclopentadienyliron or ferrocene, many

other metallocene types less readily accessible than the iron-
organic prototype have also in recent years attracted much research
interest.

It has long been recognized that the incorporation of metallocene units into a polymeric backbone may give rise to certain bulk
properties or combinations of pertinent chemical and physical
features not found in the non-polymeric complex. Accordingly,
the problem of metallocene polymerization has received appreciable
attention, and the number of publications in the field of metallocene-containing macromolecules most likely exceeds that of all
other communications on polymeric organometallic compounds. By
far the largest proportion of polymeric metallocenes investigated
over the past twenty-odd years comprises structures containing
organic connecting segments in addition to the metallocene complex
proper, the latter either being pendent or else constituting
a component of the chain. Most of these investigations were
conducted in the 1960 - 1970 decade and have been thoroughly and
critically reviewed [1]. The evident preponderance of metallocene
polymers containing organic connecting groups is entirely traceable
to synthetic convenience. Countless reactions can be utilized in
metallocene polymer synthesis provided only that suitable
derivatization of a metallocene complex is accomplished through
attachment of reactive functional groups of the familiar types
conventionally employed for polymerization and capable of undergoing partial (in polycondensation) or total (in polyaddition)
incorporation into the growing chain. This strategy allows the
polymer chemist to prepare, and purify, the monomer(s) in a separate
operation. The subsequent polymerization will then be brought
about entirely by reaction of the functional side groups, and the
metallocene complex proper, although capable of exerting some
steric and electronic influence on these side group interactions,
will not itself actively participate in the polymerization sequence.
Whereas the synthetic problems associated with this polymerization
approach in general are reasonably manageable, the preparative task

of polycondensing metallocenes in such a fashion as to obtain
linear chains composed of metallocene units directly joined to each
other by single bonds without the interposition of other bridging
groups has proved to be a truly formidable one, requiring highly
selected strategies and specialized laboratory techniques.
A polymer thus generated by the direct interconnection of metallocene units, while correctly to be designated as a poly(metallo-
cenediyl), is commonly referred to as a polymetallocenylene.

It is well at this point to examine in some detail the causes
underlying the problems of polymetallocenylene synthesis.   In the
first place, there are severe restrictions in the number and kind
of chemical approaches that lend themselves to the task of interconnecting metallocene units directly and in high conversion, only
the most efficacious types of aryl-aryl coupling mechanisms being
applicable here.

Secondly, one should remember that in metallocene polycondensation, just as in any other step-growth polymerization process,
the number-average degree of polymerization, $\bar{X}_n$, is strongly
dependent on the extent of reaction, p, that is, the conversion
along the polymerization reaction path, and quantitative relationships derived from simple statistical considerations allow us
quite accurately to determine $\bar{X}_n$ as a function of p.   For
illustration let us consider an experiment in which the desired
synthetic step proceeds to 50% conversion.   While, by ordinary
standards, the level of conversion attained may be rated as high,
the same extent of reaction brought about in an experiment involving
a polycondensation (at balanced stoichiometry) is calculated to
give $\bar{X}_n = 2$, this rather trivial result indicating that a mere
dimerization has been achieved.   From the polymer chemist's view-
point, such an experiment would, hence, be rated a total failure.
Even a 90% conversion in the desired direction, highly respectable
by the standards of non-polymer chemistry, is calculated to be
insufficient for polymerization purposes, as, on a number-average
basis, no more than ten monomer units are assembled in one product
molecule.   The conversion along the propagation path must indeed

very appreciably exceed that of the last-named example if satisfactory polymerization is to be accomplished, and the reaction type
chosen must proceed almost exclusively in the desired direction,
the extent of concurrent side reactions being restricted to
fractions of a percent.   In addition, the side reactions that do
occur concurrently must be of such a nature as not to interfere
with polymerization nor to cause the participation of by-products
in the propagation sequence, as this would result in contamination
of the polymeric end product.

Thirdly, since a strictly linear propagation scheme requires
that exactly two interconnecting bonds be generated for every
metallocene complex involved (save the two terminal units), it is
clear that, whatever the coupling mechanism operative, it must
ensure participation of no more and no less than two sites per
complex.   It is thus imperative that experimental conditions
causing monofunctional or trifunctional behavior of a monomer unit
in the feed or in the growing chain be meticulously avoided.
Under typical conditions, the presence of a monofunctional (or
monofunctionally reacting) compound in a concentration of only one
mole percent may well reduce $\bar{x}_n$ to one-half of what it would have
been in the absence of such an impurity.   Conversely, the same
percentage of a trifunctionally reacting compound present in the
feed suffices to cause appreciable crosslinking; in a polymerization
proceeding to $\bar{x}_n > 100$, in fact, the entire batch would turn
substantially insoluble under these conditions.

Although the two requirements of strict difunctionality and
an extraordinarily high extent of reaction are not restricted to
the synthesis of polymetallocenylenes but hold for all step-growth
polymerizations, they are notoriously difficult to fulfil for
reactions involving the direct coupling of aromatic nuclei.   Such
coupling reactions typically utilize organolithium, -copper,
-nickel, and -magnesium chemistry, all of which are to a variable
extent afflicted with the evils of detrimental side reactions and
deviations from the desired functionality through loss or
alterations of sensitive substituents in intermediary stages.

In the light of the problems pointed out in the foregoing, it is not surprising to find that the number of literature reports dealing with polymetallocenylenes has until now remained conspicuously small in relation to the coverage of metallocene-containing polymers in which the metallocene units are interlinked by organic bridging groups. Well in accord with the absolute predominance of ferrocene chemistry in all metallocene research, the published topics in polymetallocenylene chemistry are almost exclusively concerned with the polyferrocenylenes. Little information is available on polymeric chain structures composed of ruthenocene units, and nothing is known about other polymetallo-cenylene types.

The following sections will cover synthetic aspects of metallocene polymerization, as well as physical and chemical features of the known polymetallocenylenes. Concluding the chapter, an attempt will be made to delineate areas of promising research potential and point out future development trends in this challenging field of macromolecular chemistry.

POLYMERIZATION REACTIONS

As all polymers to be discussed in this chapter are derived from metallocenes of the dicydopentadienylmetal type, they can summarily be represented by the structure shown below, in which M stands for a transition metal of Group VIII, specifically Fe and Ru. The centered position of the left-hand connecting bond in

this formula indicates the possibility of homoannular (i.e. 1,2- or 1,3-type) or heteroannular (1,1'-type) substituent disposition. Although, as will be seen, almost all known polymetallocenylene

structures are of the heteroannular type, several tri-, tetra-,
and pentanuclear compounds with M = Fe have been described in which
the internal units are homoannularly connected.

POLYFERROCENYLENES

Poly-1,1'-ferrocenylenes. - Following the discovery of ferrocene in
1951, the exploration of the basic substitution chemistry of this
prototype metallocene progressed at an extraordinarily rapid pace,
and the first comprehensive coverage of the subject appeared as
early as 1965 in the form of a proficiently written book [2].
However, most of the earlier studies of ferrocene substitution
behavior involved the attachment of other groups to the metallocene
complex, and little of the chemistry explored was applicable to the
problem of polymerization by propagation through direct ferrocene-
ferrocene bond formation.   Seen in this light, the first synthetic
approach toward polyferrocenylenes reported in 1960/1961 by the
groups of Korshak and Nesmeyanov[3] must be valued as a courageous
pioneering effort, triggering off, and stimulating, the great many
research efforts subsequently undertaken in the field.   This type
of ferrocene polymerization, although lastly unsuccessful, presents
an instructive example of a polycoupling reaction the outcome of
which is dramatically affected by the various aforementioned
requirements;   it will, therefore, be discussed here in some
detail despite ample coverage in previous reviews [1].

In principle the process, referred to as a polyrecombination,
involves the step-growth polymerization of ferrocene via free-
radical intermediates at elevated temperatures as summarized in
scheme 1, and under suitable experimental conditions the authors[3]

$$\text{(1)}$$

obtained in 5 - 16% yield soluble polymers to which they assigned
the structure I of poly(ferrocene-1,1'-diyl) or poly-1,1'-ferro-

cenylene. Molecular masses were up to 7000. On the surface the reaction sequence looks simple enough: ferrocene monomer in the molten state (temperature $> 200^\circ$) is exposed to a suitable free radical source, most typically t-butyl peroxide, whereupon intermediary t-butoxy and methyl free radicals, generated thermolytically from the peroxide (eqs. 2), extract a hydrogen atom from the

$$(CH_3)_3C-O-O-C(CH_3)_3 \xrightarrow{\Delta} 2(CH_3)_3C-O\bullet \qquad (2a)$$

$$(CH_3)_3C-O\bullet \xrightarrow{\Delta} \bullet CH_3 + CH_3COCH_3 \qquad (2b)$$

metallocene nucleus, leaving a ferrocenyl free radical. Two such metallocenyl radicals recombine in the first propagation step to yield biferrocene. Further hydrogen extraction from this species, followed by recombination of the resulting biferrocenyl free radical with a ferrocenyl radical, gives a trimer, and with increasingly larger oligoferrocenylenes as substrates for hydrogen extraction, and partners for recombination, the ultimate result expected is the formation of long-chain polyferrocenylenes. The overall polyrecombination reaction could thus be depicted by scheme 3 ($\sim$ = H, ferrocenyl or polyferrocenyl).

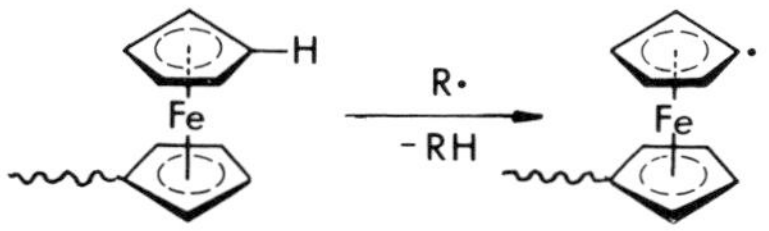

(3a)

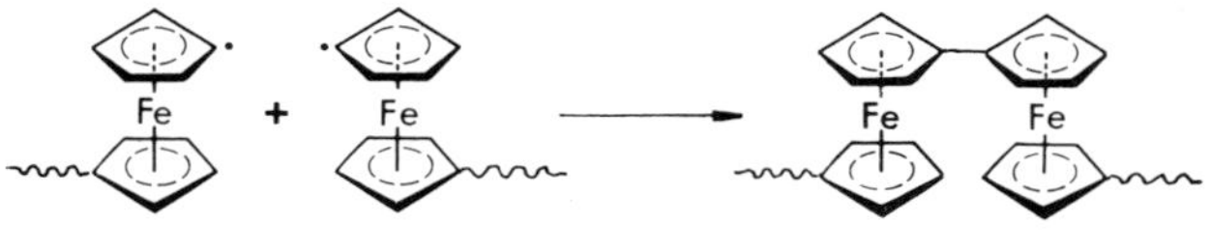

(3b)

I

However, in view of what has been said initially regarding the control of functionality and extent of reaction, doubts must arise as to the practicability of a polymerization process as implied in this scheme. Firstly, as the mono- and polynuclear substrates

offer more than two sites for reaction, one should, in an early
stage of the reaction, expect a substituent disposition not only
of the (sterically preferred) heteroannular type, but also of the
two homoannular types; in more advanced stages, one should
additionally expect tri- and polysubstitution, leading to the
generation of branches and, ultimately, crosslinks. Secondly, as
a minimum of two primary free radicals R· are required for the
formation of a single ferrocene-ferrocene bond, it stands to
reason that there should be significant competitive participation
of nonferrocene-type compounds in the propagation reaction. Not
only would one expect the primary free radicals, ·CH_3 and ·$OC(CH_3)_3$,
to engage in alkylation and alkoxylation side reactions, but also
such secondary products as acetone, t-butanol, methane, and methyl
t-butyl ether, derived from the primary radicals by hydrogen
capture or through other transfer reactions (e.g., $2(CH_3)_3C-O·\longrightarrow$
$CH_3COCH_3 + CH_3-O-C(CH_3)_3$), must be considered as potential reaction
partners. Even though their effective concentrations in the
reaction mixture will be very small under proper experimental
conditions of rapid dissipation and vaporization, hydrogen
extraction from these aliphatic compounds will be sufficiently
favored energetically over that from the aromatic ferrocene nucleus
to permit the generation of derived free radicals in instantaneous
concentrations high enough to cause alkyl and alkoxy substitution
by recombination with free radical sites on ferrocene or ferrocenylene units of the growing chain.

These various expectations were indeed borne out in later
reinvestigations[4,5] by the identification of a number of
diagnostically useful nonpolymeric by-products. Thus, the three
isomeric trimers (1,2-, 1,3-, and 1,1'-diferrocenylferrocene) were
isolated from low-molecular fractions, and large quantities of
insoluble, crosslinked material (making up the entire product batch
as conversions exceeded 90%) were collected in addition to the
9 - 27% of soluble, substantially linear products, attesting to both
the lack of regioselectivity in the primary free radical attack and
the potential polyfunctionality exhibited by the growing chain

system.   Moreover, methylferrocene and a methylated biferrocene
were detected mass-spectrometrically, and in representative
experiments both diferrocenylmethane and ferrocenylmethyl t-butyl
ether were separated and identified as by-products, giving clear
evidence for the active involvement of nonferrocene-type compounds
derived from the peroxide reactant.   In further support of this
finding, the presence of methyl, methinyl and t-butyl substituents
as well as aliphatic ether groups in the soluble oligomeric and
polymeric products was ascertained spectroscopically, and micro-
analytical results agreed with chain compositions such that, on the
average, 1 - 2 aliphatic C atoms were incorporated for every
ferrocene unit.   Taking into account all evidence presented in the
later studies, one is led to conclude that the recombination
products, far from possessing a polyferrocenylene structure, were
in fact composed of ferrocenylene and short polyferrocenylene units
bridged and substituted by a variety of aliphatic groups.   For
reasons of polyfunctional substrate behavior and massive interference by side reactions, the ferrocene polyrecombination thus
proves to be entirely unsuitable for polyferrocenylene preparation.

There has been no lack of attempts in the subsequent decade to
synthesize I by other methods designed to provide an improved
control of functionality, these efforts having in common the use or
co-use, as monomers, of ferrocene compounds with preintroduced
difunctionality.   Thus, 1,1'-dihaloferrocenes, in admixture with
monohaloferrocenes, were subjected to the conditions of Ullmann
coupling in the melt phase, giving oligomeric I in high yields [6].
Similar results were obtained in experiments comprising the selfcondensation of chloromercuriferrocenes via chloropalladated intermediates, the latter arising by palladium/mercury exchange in the
presence of lithium tetrachloropalladate [7].   In a different
approach, poly(mercuriferrocene), in which ferrocene units are
connected by mercury bridges, was demercurated thermally [8] or
with the aid of metallic silver [9], thereby yielding oligo- and
polynuclear I.   In other studies, difunctionality was introduced
in an intermediate stage, viz. through lithiation of ferrocene;

the lithiation mixture, containing both mono- and dilithioferrocene,
was allowed to undergo oligomerization by oxidative coupling in the
presence of cobaltous ion [10]. These earlier investigations, most
of which have been thoroughly reviewed and, therefore, will receive
no further attention here, all suffered from various deficiencies
with respect to 'cleanliness' of the reaction and so failed to
produce polyferrocenylenes in an acceptable molecular mass range
and/or free from structural imperfections brought about by
alkylation or introduction of cyclopentadienyl substituents.

When, several years after the first description of lithiated
ferrocene oligomerization, a method for the isolation and
purification of 1,1'-dilithioferrocene was published by Rausch's
group [11], a renewed attack was ventured on the knotty problem of
ferrocene polymerization, again utilizing the oxidative coupling
mechanism with cobaltous ion participation [12]. Use of the
prepurified bis-TMEDA complex of 1,1'-dilithioferrocene (TMEDA =
N,N,N',N'-tetramethylethylenediamine), rigorous exclusion of
moisture and oxygen, and removal of the traces of carbon dioxide
contaminating the commercial grade of argon purging gas all
combined to advantage in those experiments to give rise to linear
oligomeric and polymeric I (scheme 4). Although overall yields

$$n \quad \underset{Fe}{\overset{\text{—Li·TMEDA}}{\underset{\text{—Li·TMEDA}}{}}} \quad \xrightarrow{\text{1) CoCl}_2, \text{2)H}_3\text{O}^+} \quad \underset{I}{\left[H \underset{Fe}{\overset{}{}} H \right]_n} \qquad (4)$$

were only moderate, the average degree of polymerization attained
was about twice that of previous coupling products derived from
lithiated ferrocenes, and the polymers were devoid of the alkyl
and cyclopentadienyl substitution observed in the previous cases.

In search for ever more efficient ferrocene polymerization
approaches, a major synthetic program, prompted by these encouraging
results, was initiated three years ago in our laboratory. Latest
methods of aryl-aryl coupling made available in organic chemistry

were utilized in this program and suitably adapted to the goal of
difunctional propagation.   In the first series of experiments[13],
recourse was made again to preformed difunctional ferrocene
derivatives, such as 1,1'-dihalides, 1,1'-bis-Grignard, and
1,1'-bis(chloromercuri)ferrocenes, which were allowed to homo-couple
or cross-couple in the presence of various recently proposed
co-reactant or catalyst systems.   Table 1 summarizes reaction
variables and results for this series of experiments.

    Although, as the table shows, the results of these efforts, by
and large, turned out to be rather unpromising, they will briefly
be discussed in the following so as to high-light some of the basic
difficulties, in particular those associated with deleterious side
reactions, that one must expect to encounter in poly-coupling
experiments.

    The selfcoupling of chloromercuriarenes in polar solvents
catalyzed by chlororhodiumdicarbonyl dimer has recently been added
to the palette of the organic chemist as a highly efficacious aryl-
aryl coupling procedure, which in the originators' hands[14] gave
dimer yields as high as 80 - 90%.   Similarly high yields were
obtained in the selfcoupling of chloromercuriarenes in pyridine
with metallic copper in the presence of palladium(II) chloride
catalyst[15], this method representing an improvement over the
earlier chloropalladate process which required full stoichiometry
of the palladium salt [7].   Both methods, when applied to the
bis(chloromercuri)ferrocene, failed to give polymeric I;   only
low-molecular oligomers were isolated in addition to appreciable
quantities of unreacted bis(chloromercuri)ferrocene, chloromercuri-
ferrocene, unsubstituted ferrocene, and other coupling or degradation products (entries 1 and 2, Table 1).   Clearly, both the
observed low reactivity of the HgCl group, entailing a low extent
of reaction, and the significant degree of reductive demercuration,
with concomitant lowering of monomer functionality, are the prime
factors to be implicated in the utterly poor polymerization
efficiency realized in the two reaction types.

TABLE 1 SYNTHESIS OF LOW-MOLECULAR POLYFERROCENYLENES BY VARIOUS COUPLING REACTIONS

Entry[a]	Reagents[b] (Molarity)	Catalyst (Molarity)	Solvent[b]	Time, Temp.	Overall Coupling Yield,%[c]	$\bar{M}_n$ of product fraction with $\bar{X}_n > 5$[d]	Starting mtls. recovered or by-products isolated	Original method per ref.[e]
1	HgCl—[ferrocene]—HgCl (0.20)	[ClRh(CO)$_2$]$_2$ (2.0 × 10^{-3})	DME[f]	24 h, reflux	15 – 25	–[g]	1,1'-bis(chloromercuri)ferrocene, chloromercuriferrocene, ferrocene (50 – 60%)	14
2	HgCl—[ferrocene]—HgCl (0.25)	Cu (2.00), 2.5 × 10^{-2}	Py	22 h, reflux	35 – 40	–[g]	As above (25 – 30%). In addition, [0.0]-ferrocenophane (~5%)	15
3	MgBr[h]—[ferrocene]—MgBr (0.60)	TlBr (1.80)	THF/Bz (1:1)	12 h, reflux	60 – 65	1700(5%)	Ferrocene and Br- and Tl-containing ferrocenes (25 – 30%)	16
4	Br—[ferrocene]—Br (0.35)	Ni(COD)$_2$ (0.70)	DMF	36 h, 45°	20 – 28	–[g]	(1,1'-Dibromoferrocene), bromoferrocene, ferrocene, other ders.(45 – 55%)	17

[a]Only one representative expt. (out of three expts. conducted at varied molarity and reaction time) is listed per entry. [b]DME = 1,2-dimethoxyethane; Py = pyridine; THF = tetrahydrofuran; Bz = benzene; Ni(COD)$_2$ = bis(1,5-cyclooctadiene)-nickel(0). [c]Range of yields from three different expts. [d]Content, in percent of total coupling product, given in parentheses. [e]Original method was used in each expt.; polymeric product was separated by fractionating precipitation from original solution with hexane. [f]Replacement of DME by hexamethylphosphoramide did not improve results. [g]No fraction with $\bar{X}_n > 5$ was isolable. [h]Prepared by ref. 26.

TABLE 1   SYNTHESIS OF LOW-MOLECULAR POLYFERROCENYLENES BY VARIOUS COUPLING REACTIONS (Contd.)

Entry[a]	Reagents[b] (Molarity)	Catalyst (Molarity)	Solvent[b]	Time, Temp.	Overall Coupling Yield, %[c]	$\bar{M}_n$ of product fraction with $\bar{X}_n > 5$	Starting mtls. recovered or by-products isolated	Original method per ref.[e]
5	[diiodoferrocene] (0.03) Ni(PPh₃)₄ (0.06)	–	DMF	30 h, 55°	35 – 42	1080(8%)	Iodoferrocene, ferrocene, other ders. (35 – 40%)	18
6	[dibromoferrocene] (0.14) Ni(PPh₃)₃[f] (0.28)	–	DMF	28 h, 50°	50 – 55	1550(6%)	Bromoferrocene, ferrocene, other ders. (30 – 35%)	22
7	[dibromoferrocene] (0.25)	Ni(PPh₃)₃ (0.012)	DMF	28 h, 50°	50 – 60	1420(7%)	1,1'-Dibromoferrocene, bromoferrocene, ferrocene, other ders. (25 – 35%)	21
8	[ferrocene MgBr...Br] (0.10) [ferrocene MgBr...Br] (0.10)	NiCl₂L₂ (0.002)	THF	20 h, 0 – 50°	55 – 62	1525(9%)	Bromoferrocene, ferrocene (15 – 20%)	19,20

[a]Only one representative expt. listed in each entry. Other expts. were conducted with minor modifications.
[b]Ni(PPh₃)₃ = tris(triphenylphosphine)nickel(O); Ni(PPh₃)₄ = tetrakis(triphenylphosphine)nickel(O); NiCl₂L₂ = dichloro-1,3-propylenebis(diphenylphosphine)nickel(II); DMF = N,N-dimethylformamide.   [c]Range of yields from several expts.
[d,e]See respective footnotes on preceding page.   [f]Prepared in situ from dichlorobis(triphenylphosphine)nickel(II) and zinc powder (each 0.28 M). Triphenylphosphine (0.5M) additionally employed.

TABLE 1   SYNTHESIS OF LOW-MOLECULAR POLYFERROCENYLENES BY VARIOUS COUPLING REACTIONS (Contd.)

Entry[a]	Reagents (Molarity)	Catalyst[b] (Molarity)	Solvent[b]	Time, Temp.	Overall Coupling Yield,[c] %	$\bar{M}_n$ of product fraction with $\bar{X}_n > 5$[d]	Starting mtls. recovered or by-products isolated	Original method per ref.[e]
9	(ferrocene) MgBr / MgBr (0.21); (ferrocene) Br / Br (0.21)	PdCl₂L₂ (5.0 × 10⁻³)	THF	48 h, 0 - 50°	50	1650(7%)	(1,1'-Dibromoferrocene), bromoferrocene, ferrocene (20%)	20
10	(ferrocene) MgBr / MgBr (0.06); (ferrocene) I / I (0.06)	Ni(acac)₂ (5.0 × 10⁻⁴)	Et/Bz (1:1)	2 h, -15°; 15 h, 25°	52	1350(5%)	Iodoferrocene, bromoferrocenes, other ders. (38%)	24
11	(ferrocene) ZnCl / ZnCl (0.21); (ferrocene) I / I (0.20)	Ni(PPh₃)₄ (2.5 × 10⁻³)	Et/THF (1:6)	8 h, 25°	68	1480(8%)	Iodoferrocene, ferrocene (18%)	25
12	(ferrocene) ZnCl / ZnCl (0.22); (ferrocene) I / I (0.20)	Pd(PPh₃)₄ (3.0 × 10⁻³)	Et/THF (1:6)	9 h, 25°	60 - 73	1870(10%)	Iodoferrocene, ferrocene (15 - 20%)	25

[a] Only one experiment conducted in entries 9 - 11, three experiments in entry 12.   [b] PdCl₂L₂ = dichloro[1,1'-ferrocenylene-bis(diphenylphosphine)]palladium(II). Ni(acac)₂ = bis(acetylacetonato)nickel(II); Pd(PPh₃)₄ = tetrakis(triphenylphosphine)-palladium(0); Et = diethyl ether.   [c] In entry 12, range of yields from three experiments conducted at different molarities.   [d],[e] See respective footnotes on first page of Table 1.

An excellent method of aryl-aryl coupling, developed in the laboratories of McKillop and Taylor[16], involves heating of aryl-Grignard reagents with 1.5 molar equivalents of thallium(I) bromide in benzene.   Representative biaryl yields are in the 75 - 99% range.   This method, when applied to the 1,1'-bis-Grignard complex of ferrocene (entry 3), gave oligomer I in appreciably higher yields than determined in the two preceding entries; however, the degree of polymerization remained comparatively low despite rigorous exclusion of moisture, and, as before, large quantities of ferrocene were isolated.   This indicates that demetalation, probably as dethallation from intermediary ferrocenylthallium species, represented a major side reaction entailing a significant reduction of functionality.

Aryl halides, as was shown in the laboratories of Semmelhack[17,18], Kumada[19-21], Kende[22] and others, undergo highly efficient reductive self-coupling under mild conditions in dimethylformamide in the presence of stoichiometric or near-stoichiometric, and even catalytic, quantities of ligand-coordinated nickel(0) species.   When 1,1'-dihaloferrocenes were treated with stoichiometric amounts of either bis(1,5-cyclooctadiene)nickel(0) or tetrakis(triphenylphosphine)nickel(0) at 40 - 60° in DMF, only oligomeric I was isolated in low yields in addition to large percentages of monohaloferrocene, ferrocene, and degraded material (entries 4 and 5), the overall formation of these by-products being enhanced through use of more forcing experimental conditions, such as higher temperatures.   Slight increases in both yield and degree of polymerization were achieved (entries 6 and 7) when the nickel species was prepared in situ as tris(triphenylphosphine)nickel(0) from dichlorobis(triphenylphosphine)nickel(II) and zinc, either in stoichiometric[22] or in catalytic quantities[21], excess zinc in the latter case regenerating the catalytically active Ni(0) complex. Again, however, the presence of ferrocene and monohaloferrocene was established in the final products.   It is clear from these findings that reductive dehalogenation by low- or zero-valent nickel reduces

the monomer functionality even more drastically than was observed
in the thallium-promoted Grignard coupling reactions.

Next, attention was turned to reactions utilizing mechanisms
of cross-coupling between aryl halides and aryl-Grignard or arylzinc
compounds.   Much information published in recent years is available
on aryl-aryl coupling of this type, catalyzed or promoted by
miscellaneous transition metal complexes, and in several instances
reported coupling yields have been well in excess of 80%.   When
applied to the system 1,1'-dihaloferrocene/1,1'-bis(bromomagnesio)-
ferrocene, this approach gave rather variable results.   For
example, manganese(II) salts or catalysts[23] caused reductive
dehalogenation of the halide by the Grignard reagent to be the
principal process and so proved totally useless for the purpose.
On the other hand, dihalodiphosphinenickel(II) compounds, as
exemplified by dichloro-1,3-propylenebis(diphenylphosphine)nickel
[19,20], were found to be reasonably efficacious catalysts in this
type of reaction (entry 8), although defunctionalization at both
reactants and concomitant suppression of chain growth was still
observable.   Use of the analogous dihalodiphosphine palladium
complexes[20] as catalysts gave very similar results (entry 9).
The nickel chelate, bis(acetylacetonato)nickel(II) [24], catalyzing
the coupling of 1,1'-diiodoferrocene with the bis-Grignard
derivative of ferrocene, proved less satisfactory, the sluggishness
of the coupling reaction allowing defunctionalizing side reactions
to become predominant (entry 10).   A moderately efficient, albeit
experimentally cumbersome, coupling process proved to be the
reaction of 1,1'-bis(chlorozinc)ferrocene (from 1,1'-dilithio-
ferrocene and zinc chloride) with 1,1'-diiodoferrocene in tetrahydrofuran solution in the presence of tetrakis(triphenylphosphine)-
nickel(O) or the corresponding zero-valent palladium complex
(entries 11 and 12), using the method of Negishi[25].   Good yields
of coupling products were paired especially in the palladium complex-
catalyzed experiments with higher average degrees of polymerization
than in most of the preceding entries.

All oligo- and polyferrocenylene products prepared in this
final series of experiments (Table 1) essentially conformed to
structure I.   The higher-molecular fractions, however, were found
to be contaminated to a variable extent with non-extractable
organic (1 - 5%) and inorganic (0.01 - 0.5%) impurities stemming
from the catalyst systems, from the functional groups, or from
degraded metallocene.   Surveying the entire series of experiments
one is led to conclude that, in view of both the observed
contamination and the generally unsatisfactory degrees of
polymerization attained, these selected types of coupling reaction,
much as they should receive recognition as outstandingly useful
synthetic tools in the organic chemistry of nonpolymeric compounds,
fail to lend themselves to the preparation of polymetallocenylenes.
Neither are they sufficiently 'clean' in the sense of earlier
discussions in this chapter to preclude competition by such side
reactions as demetalation, dehalogenation, and metallocene or
catalyst degradation, nor do they proceed at rates fast enough
relative to such other reactions to prevent their active
participation in, and interference with, the propagation sequence.
One finds this failure visibly reflected in the structural imperfections of the polymeric products, the low extents of reaction, and
and the low degrees of polymerization resulting from stoichiometric
imbalances.

The second series of experiments, summarized in Table 2, was
based on organocopper chemistry.   As most copper-organic compounds
are conveniently prepared from lithium-organic precursors,
1,1'-bis(lithio-TMEDA)ferrocene (again prepared and purified in the
solid state [11]) served as the principal starting material throughout this experimental series.   While the straightforward oxidative
coupling of the lithiated ferrocene with copper(II) halide, conducted
over a wide range of temperatures, concentrations, and solvent
polarity (entry 1), did not lead to marked improvements over the
results of Table 1, the required high copper ion concentrations
causing excessive ferrocene halogenation, dimer cyclization (*vide*

TABLE 2   SYNTHESIS OF POLYFERROCENYLENES FROM DILITHIATED FERROCENE

Entry[a]	Reagents (Molarity)	Solvent[b]	Time, Temp.	Overall Coupling[c] Yield, %	$\bar{M}_n$ of product fractions with $X_n > 5$[d,e]	Starting mtls. recovered or by-products isolated	Ref.
1[f]	Li·TMEDA / Li·TMEDA (0.10) ; CuCl₂ (0.21)	DBE	3 h, -78°; 3 h, 25°; 16 h, 110°	25 - 42	1100 (26%) 2000 (6%)	Chlorinated ferrocenes and oligoferrocenylenes, ferrocene (30 - 50%); [0.0]ferrocenophane (5 - 10%)	27
2[g]	Li·TMEDA / Li·TMEDA (0.28) ; CuCl₂ (0.30)	DME/THF (1:3)	22 h, 25°	45 - 55	1300 (13%) 3400 (2%)	As above (20 - 30%); [0.0]ferrocenophane (5 - 10%)	27
3[h]	Li·TMEDA / Li·TMEDA (0.064) ; CuCl₂ (0.042)(0.020)	DIO/Hx (3:2)	3,5 h, 25°; 15 h, 50°	70 - 75	1100 (12%) 3900 (1,5%)	Ferrocene (10 - 20%)	28
4[i]	Li·TMEDA / Li·TMEDA (0.29) ; — (0.27)	DME/THF (1:1)	2 h, 0°; 20 h, 25	85 - 88	1650 (31%) 3800 (16%)	Ferrocene (1 - 4%)	30

[a]Only one representative expt (out of at least three expts conducted at varied molarity and reaction time) is listed per entry. [b]DBE = di-n-butyl ether; DME = 1,2-dimethoxyethane; THF = tetrahydrofuran; DIO = dioxan; Hx = hexane. [c]Range of yields of linear material from different experiments. [d]Content, in percent of total linear coupling product, is given in parentheses. [e]All crude products were treated with Vitride reducing agent for dehalogenation (inefficient for Cl-containing products); oligomers and polymers were separated by fractionating precipitation with hexane. [f]Similar results in DME or DME/Bz; analogous results with CuBr₂. [g]Lower yield range in DBE. [h]Similar results with CuI (Cu:Li = 0.5) or in DIO/THF. [i]Similar results in DIO/DME. Slightly higher yields with Pd(0) or Cu(I) ($\sim 10^{-3}$M) catalysts. Higher molecular masses (up to 6200 for 1st fraction), but poor composition and up to 10% ferrocene recovery, in DBE at 25 - 110°.

infra), and product contamination with copper, promising results were obtained in experiments in which the molar equivalent of copper(II) halide was halved (entry 2) [27].   This allowed for a fraction of the organolithium sites to undergo oxidative coupling with polyferrocenylene formation (eq. 5a;   Ar = ½ ferrocenylene), and for the resultant cuprous ion to suffer transmetalation with the

$$ArLi + Cu^{2+} \longrightarrow \tfrac{1}{2}Ar\text{-}Ar + Cu^{+} + Li^{+} \qquad (5a)$$

remaining organolithium functions in the reaction mixture to give ferrocenylenecopper(I) species (eq. 5b).   Raising the reaction

$$ArLi + Cu^{+} \longrightarrow ArCu + Li^{+} \qquad (5b)$$

temperature then brought about thermal coupling of the organocopper intermediates with generation of I (eq. 5c).   The oligo- and

$$ArCu \xrightarrow{\Delta} \tfrac{1}{2}Ar\text{-}Ar + Cu(0) \qquad (5c)$$

polyferrocenylenes thus prepared in 45 - 55% yield, although substantially of type I, still contained trace quantities of copper in addition to several percent of halogen present as halide end groups, effectively stopping chain growth.   Furthermore, irrespective of substrate concentration, considerable portions of dimer intermediates underwent internal cyclization to [0.0]ferrocenophane. The optimization of ferrocenophane formation (scheme 6) under

(6)

(~26%)   I
                       (~20%)

conditions of purely thermal coupling of ferrocenylenecopper(I) species [27], the latter generated in situ from 1,1'-bis(lithio-TMEDA)ferrocene and two molar equivalents of Cu(I) salt, suggests

that aggregated organocopper(I) complexes appearing as intermediates
in the polycondensation of the dilithioferrocene complex under
conditions of high relative copper ion concentrations may act as
templates.   Depending on steric factors and the degree of
association and solvation, these may decompose either *via* internal
or *via* intermolecular C-C bond formation, thus entailing either
cyclization or polymerization.

With high instantaneous copper ion concentrations established
as the principal cause of preferential ferrocenophane formation,
subsequent experiments were conducted under conditions of organo-
cuprate intermediacy[28].   Organocuprates, more copper-deficient
than organocopper compounds and in the simplest form of the general
structure $[R_2CuLi]_2$, have in recent years come to play an
increasingly significant role as reactive nucleophiles in synthetic
organic chemistry[29], their reactivity possibly deriving from the
cooperative interplay of the two copper atoms in the dimeric
complex.   They are conveniently generated *in situ* from organolithium
compound and copper(I) cation (eq. 7).   Coupling may proceed either

$$4RLi + 2Cu^+ \longrightarrow [R_2CuLi]_2 \tag{7}$$

with oxidative assistance by such other reactants as copper(II)
cation or molecular oxygen (eq. 8a) or through nucleophilic attack

$$[R_2CuLi]_2 \xrightarrow{[O]} 2R\text{-}R + 2Cu^+ + 2Li^+ \tag{8a}$$

$$[R_2CuLi]_2 + 2R'I \longrightarrow 2R\text{-}R' + 2RCu + 2LiI \tag{8b}$$

on organohalide (eq. 8b;   R,R' = alkyl,aryl).

Both reaction types were utilized for ferrocene polymerization
(R,R' = ½1,1'-ferrocenylene)[28].   Whereas the former was found to
be unsuitable because of oxidative side reactions and/or oxidant
incorporation, the latter proved highly efficaceous, affording I in
70 - 75% yield.   The reaction mixtures in these experiments were
ultimately heated to 50° in order to bring about full utilization,
through thermal coupling, of the organocopper species generated

(eq. 8b) along with coupling product.   A representative experiment,
in which the cuprous salt needed (eq. 7) was prepared *in situ* from
$CuCl_2$ (Cu:Li $\simeq$ 0.15), is summarized in entry 3.

Although, under these conditions, internal cyclization was
negligible, the products I still were not entirely free from copper
contamination.   Attention was, therefore, focussed on the problem
of direct solution condensation of the dilithioferrocene complex
with diiodoferrocene (scheme 9) [30].   This polycondensation
reaction, in the simplest mechanistic sense representing an aromatic
nucleophilic substitution process, in initial experiments gave low
p and $\bar{X}_n$ values resulting from use of insufficiently polar solvents
and inadequately purified organolithium monomer.   Later experiments, performed in di-n-butyl ether under considerably refined
experimental conditions at ultimate temperatures of 50 - 100°,
furnished polyferrocenylenes in excellent yields and degrees of
polymerization.   The polymers were marred, however, by the inclusion of (deprotonated) TMEDA chelating agent, as evidenced *inter
alia* by low (0.2 - 0.5%) nitrogen contents; in addition, they showed
an unusual aging behavior, manifested in the loss of initial
solubility after several months of storage.   The reaction conditions which, after much experimenting, were ultimately found to be
optimal for clean and efficient propagation called for the use of
strongly coordinating ether solvents, such as THF and DME, at
temperatures not substantially exceeding 25° (entry 4).   Overall
ferrocene-ferrocene coupling yields thus amounted to 85 - 88%.
The oligo- and polyferrocenylenes I were of the highest purity yet
achieved and contained no detectable quantities of [0.0] ferrocenophane.   The molecular mass of the top fraction (15 - 20% of total)
reached the 3500 - 4500 range, although it was not possible even
under the most rigorously controlled conditions entirely to suppress
demetalation and dehalogenation, evidenced by the regeneration of
up to 4% ferrocene and the comparatively large proportion of oligomeric material formed.   Use of much higher temperatures in these
experiments proved counterproductive;   while bringing about minor

further increases in $\bar{X}_n$, this invariably led to interference by
solvent and chelating agent with resultant deviations from the
ideal structure I.

$$(\frac{n}{2}+1)\ \underset{\text{Fe}}{\overset{\text{Li·TMEDA}}{\diagdown}}\text{Li·TMEDA} \ + \ (\frac{n}{2}-1)\ \underset{\text{Fe}}{\overset{\text{I}}{\diagdown}}\text{I} \ \longrightarrow$$

$$\text{TMEDA·Li}\left[\underset{\text{Fe}}{\diagdown}\text{Li·TMEDA}\right]_n \quad \xrightarrow{\text{H}_3\text{O}^+} \quad \text{H}\left[\underset{\text{Fe}}{\diagdown}\text{H}\right]_n \qquad (9)$$
                                                                              I

Other efforts further to enhance the $\bar{X}_n$ ceiling in these
organolithium-organohalide coupling reactions included catalysis
by Pd(O) and Cu(I). Although preliminary findings suggest that
minor additional molecular mass increases may be brought about by
such transition metal catalysis, the effects are not very pronounced
[30], and considerable refinement of this approach is needed if
significant improvements are to result.

    In summary, the experience emanating from the two series of
experiments covered in Tables 1 and 2 shows that the prime requirements of highest product purity and a maximal extent of reaction,
p, concomitant with highest possible $\bar{X}_n$, are not in general mutually
compatible. All efforts to increase p, and thus $\bar{X}_n$, by 'pushing'
the reaction to the right-hand side through the expediency of higher
temperatures will almost invariably result in an enhanced extent of
destructive side reactions and concomitant participation of catalyst
components or by-products in the propagation sequence with
unacceptable consequences regarding the conformance of the polymeric
products to structure I. Accordingly, a compromise solution must
be found in which the two requirements of product purity and high
molecular mass are properly balanced. In the experiment of entry

4, Table 2, the conditions of which quite obviously are the most
favorable developed to this date for the synthesis of polyferrocenylenes, this compromise is clearly evident.

Poly-1,2-ferrocenylenes.- Only a single investigation has been
communicated the aim of which was the synthesis of oligoferrocenylenes comprising internal ferrocene units joined in a 1,2-disub-
stitution pattern.   This project, studied by Rausch *et al.* [31],
involved the mixed Ullmann coupling of iodoferrocene and 1,2-diiodo-
ferrocene (or, less favorably, the homo-coupling of solely the
diiodo monomer) in the presence of activated copper in the melt at
150°.   The product mixtures, obtained in combined yields of 50 -
60° (appreciably less in the homo-coupling reactions), consisted for
the most part of biferrocene formed by selfcondensation of iodoferrocene.   As even in the homo-coupling of the diiodo compound
some 20 - 30% of total coupling product constituted biferrocene,
it is obvious that appreciable dehalogenation occurred in these
Ullmann reactions, rendering the process unsuitable for polymerization purposes.   The remaining oligomers, isolated in 0.4 - 4%
combined yield in addition to the dimer, were 1,2-terferrocene
(II, n = 3), the two (*meso-* and *d,l-*) 1,2-quaterferrocenes (II,
n = 4), and several 1,2-quinqueferrocene isomers (II, n = 5).

II

Although the organolithium-organohalide polycondensation
process and other polycoupling reactions utilized for the preparation of I could, in principle, be used as well for the synthesis
of higher-molecular II, this has not so far been attempted.

No polyferrocenylenes containing ferrocene-1,3-diyl internal
units have to this date been reported, undoubtedly for lack of

readily available starting materials, such as 1,3-dihaloferrocenes
and derived 1,3-dilithio compounds.

POLYRUTHENOCENYLENES

The ruthenocene (di-η-cyclopentadienylruthenium) and osmocene
(di-η-cyclopentadienylosmium) complexes, together with ferrocene,
represent the iron-group metallocenes, whose physical and chemical
properties show much mutual resemblance because of the similarity
of their molecular structures and valence electron systems.   As
a consequence, one should expect a similar behavior in polymerization reactions *via* the various approaches outlined above for I.
However, with ring electron density markedly decreasing in the
order ferrocene > ruthenocene > osmocene, one may predict an
increasing propensity for reaction with nucleophiles in this order
and, hence, an increasing reactivity in coupling processes in which
the metallocene is used as the dihalo derivative.   On the other
hand, decreasing nucleophilicity may be anticipated, again in the
order given, for the metallocene employed as the dilithio derivative.
although, most likely for reasons of high costs, no work has been
done in the field of osmocene polymerization, limited data are
available on ruthenocene polymerization behavior.

The first, cursory report on ruthenocene oligomerization,
dating back to 1970, and devoid of experimental details, pointed
out the feasibility of preparation of biruthenocene and higher
homologs by Ullmann coupling of mono- and 1,1'-diiodoruthenocene[32].
A thorough study of this reaction in our laboratory[13] has confirmed that the coupling of haloruthenocenes in the melt phase proceeds
indeed readily and in high overall yields as shown in scheme 10, giving bi-, ter-, and quaterruthenocene (III, n = 2-4), as well as
little higher-molecular material (entry 1, Table 3).

(10)

III
(n = 2 - 4)

TABLE 3   SYNTHESIS OF LOW-MOLECULAR POLYRUTHENOCENYLENES FROM DILITHIATED AND/OR DIIODINATED RUTHENOCENE

Entry[a]	Reagents[b] (Molarity)	Solvent[c]	Time, Temp	Overall Coupling Yield,[g,d]	$\bar{M}_n$ of product fractions with $\bar{X}_n > 5$[e,f]	Starting mtls. recovered or by-products isolated	Ref.
1	[diiodoruthenocene structure] Cu[g]	–	24 h, 130°; 2 h, 150°	68	1160 (1%)	Iodoruthenocenes, ruthenocene (25%)	13
2[h]	Li·TMEDA [structure] Li·TMEDA (0.051) (0.034)	DBE	20 h, 25°; 8 h, 70°	24 – 40	1200 (15%); 1620 (1%)	As above (20 – 50%)	33,34
3	Li·TMEDA [structure] Li·TMEDA (0.085) (0.056) (0.084) CuI	THF/DIO (2:1)	2 h, –70°; 1 h, 0 – 25°; 12 h, 50°	38	1280 (9%); 2000 (0.5%)	As above (35%)	34
4	Li·TMEDA [structure] Li·TMEDA (0.53) (0.53) CuCl₂	DBE	1 h, 0°; 19 h, 25°; 1 h, 60°	70	1250 (4%); 2400 (0.9%)	Chlorinated ruthenocenes, ruthenocene (18%)	35

[a]Only one expt. each was conducted and is listed in entries 1,3,4; only one expt. (out of four expts. conducted) is listed in entry 2.   [b]Li·TMEDA stoichiometry unproven.   [c]DBE = di-n-butyl ether; THF = tetrahydrofuran; DIO = dioxan.   [d]In entry 2, range of yields from four experiments.   [e]Content, in percent of total coupling product, is given in parentheses.   [f]All crude products were treated with Vitride reducing agent for dehalogenation (ineffective for Cl-containing products); oligomers and polymers were separated by fractionating precipitation with hexane.   [g]Cu:I = 10.   [h]Similar results in DME and with Pd(0) or Pd(II) catalysts.

In order to maintain melt conditions throughout the reaction it is
necessary to work with a mixture of mono- and dihalogenated ruthenocene, as the use of the dihalide alone leads to premature solidi-
fication of the melt and concomitant low coupling efficiency.
Another drawback is the occurrence of ample dehalogenation,
resulting in the generation of ruthenocene. The method as such is,
therefore, not readily applicable to the preparation of higher-
molecular III.

Surprisingly, the direct solution condensation of 1,1'-dilithio-
ruthenocene (chelated with TMEDA) with 1,1'-diiodoferrocene (entry
2, Table 3) under the experimental conditions so successfully
employed in the analogous ferrocene polymerization (entry 4, Table
2), turned out to be of little use, affording no more than 40% of
largely oligomeric III.   Catalysis by various transition metals
or their low-oxidation-state salts caused no improvements [33,34].
Similarly inefficient proved to be the coupling of ruthenocenylene-
cuprate with diiodoruthenocene according to eq. 8b (R = ½ ruthenocenylene), which gave predominantly oligomeric III (n = 2 - 5) in
addition to less than 5% of higher-molecular compounds (entry 3,
Table 3).   One is led to conclude from these findings that the
metalated ruthenocene complex possesses insufficient nucleophilicity
for these types of coupling reaction to proceed efficaciously.
On the other hand, the combined oxidative and thermal coupling, in
DBE solution, of TMEDA-chelated dilithioruthenocene in the presence
of copper(II) chloride (eq. 5, Ar = ½ ruthenocenylene) [35] afforded
coupling products in higher yields (70%) than realized in the
corresponding dilithioferrocene polymerization, although, again,
the degrees of polymerization attained were low, the major portion
of III being in the oligomer range.   In addition, as a result of
end group chlorination by the cupric ion reagent, the products were
found to contain up to 5% Cl resisting all attempts at removal by
reductive treatment, and only by elaborate separation techniques
was it possible to isolate the chlorine-free oligomers III,
n = 2 - 4.

As shown by the tabulated data, the results obtained to-date
in ruthenocene polymerization are by no means satisfactory, and
a healthy challenge awaits the aspiring researcher in this field.

PROPERTIES AND USES

Considering the state of infancy in which polymetallocenylene
research is found to be at this time, the paucity of property and
performance data in the literature should come as no surprise.
In fact, with just one year elapsed since publication of the more
practicable ferrocene polymerization procedures [27,28,30] , one can
hardly expect to find anything more than simple property descriptions
and most certainly no records of completed development and
application studies.   Accordingly, the reader will find the information presented in this section to be comparatively heavy on polymer charactization data but very light indeed on the question of
just what precisely these polymetallocenylenes can offer in terms
of scientific and technological use.   Furthermore, since virtually
all that is known about polymetallocenylenes refers to the ferrocene
polymers, one can only speculate at this time on the possible use of
polymers derived from other transition metal complexes.

POLYFERROCENYLENES

Physical Properties.- All oligo- and polyferrocenylenes, irrespective
of the substituent disposition at the internal recurring units, are
high-melting solids ranging in color from orange-yellow or light
orange in the low-molecular-mass region to orange-tan, tan-brown,
and, ultimately, dark brown in the region of higher molecular masses.
All individual homologs or fractions are to a varying extent soluble
in halocarbons, aromatic hydrocarbons, and a number of dipolar,
aprotic solvents.   While the low-molecular members show a distinct
saturation limit, reasonably high for the dimer and the trimers,
yet low or very low for the homologs with n = 4 - 6, the higher-

molecular fractions (of the only known polymer type I) possess excellent solubility in these solvents and may be precipitated from solution by such non-solvents as methanol or hexane. Polymorphism, manifested in differences of the X-ray powder diagrams and, occasionally, in the melting behavior, has been noticed with representative trimers and doubtlessly is associated with higher homologs as well. The individual oligoferrocenylenes are characterized by distinct melting points, with highest values shown by the members with n = 4 - 6; a tabulation of m.p. data for the known oligomers is given in Table 4.

TABLE 4   MELTING POINTS AND SELECTED ELECTRONIC ABSORPTION MAXIMA OF KNOWN OLIGOFERROCENYLENES[a]

Compound	M^b	M.p., $°C^c$	λ_{max}, nm (ϵ, M^{-1} cm^{-1})[d]			
Biferrocene (I, n = 2)	370	238 - 240	223(22900)	263(6100)	300(4800) 295(4250)	456(320) 450(300)
1,2-Terferrocene (II, n = 3)	554	202 - 204[e]	-	-	- 295(3000)	- 449(260)
1,3-Terferrocene		206 - 208	-	-	- 301(4170)	- 455(375)
1,1'-Terferrocene (I, n = 3)		229 - 230	226(20200)	267(6400)	305(5500) 302(5250)	458(480) 454(430)
1,2-Quaterferrocene, meso (II, n = 4)	738	374 - 376	-	-	-	-
1,2-Quaterferrocene, d,l (II, n = 4)		250 - 253	-	-	-	-
1,1'-Quaterferrocene (I, n = 4)		279 - 281	228(17900)	270(7900)	308(6000)	460(590)
1,2-Quinqueferrocene, meso (II, n = 5)	922	345 - 350	-	-	-	-
1,1'-Quinqueferrocene (I, n = 5)		262 - 264	-	-	-	-

[a]Data collected from refs. 5,6,7,10,27,30,31 .   [b]By mass spectrometry. [c]Only highest m.p. given.   [d]In 1,2-dichloroethane (abs. ethanol for second-row data). First column: band system X, XI in ferrocene; second column: band system VII, VIII in ferrocene; third column: band system VI in ferrocene; fourth column: band system IV in ferrocene; band system designations by McGlynn's notation (ref. 37).   [e]Polymorph melting at 187 - 189°.

The higher-molecular oligomers and polymers, available only as polydisperse fractions (i.e. mixtures of homologs) for the hetero-annular type I, are characterized by broad melting ranges (Table 5); the melting temperatures pass through a minimum for fractions with major heptamer/octomer contents and exceed the 350° level at $\bar{x}_n > 20$ [30].

The exceptional heat resistance of the ferrocene unit has prompted extensive work, mostly under government contracts, directed toward the use of ferrocene-containing polymers in high-temperature and ablative applications [1]. However, only a single study has focussed on the thermal stability behavior of polymeric I [8]. The polymer, prepared by demercuration of poly(mercuriferrocenylene), when subjected to the thermogravimetric analysis test under argon, gave a respectable thermogram, indicating relative residual weights of 98, 85, and 75%, respectively, at 400, 600, and 800°. In contrast, a sample of a polyrecombination product synthesized by Korshak's process [3] and in the same molecular mass range of 2500 - 3000 showed markedly inferior thermostability, giving residual weights of about 70% at 500° and some 65% at 600°. No thermogravimetric analysis data on more recently synthesized poly-1,1'-ferro-

TABLE 5   MELTING RANGES AND SELECTED ELECTRONIC ABSORPTION MAXIMA OF FRACTIONATED POLY-1,1'-FERROCENYLENES[a]

Fraction	$\bar{M}_n$[b]	Melting range, $^{\circ}C$	λ_{max}, nm (ε, $M^{-1}cm^{-1}$)[c]			
1	9900	> 350	228(17800)	269(6600)	308(5300)	461(650)
2	7200	> 350				
3	5800	> 350	228(18200)	270(7600)	309(5900)	460(630)
4	4700	> 350	227	270(7600)	309(5600)	460(650)
5	3200	230 - 250	228(19400)	270(8000)	310(5800)	460(610)
6	2900	215 - 230	228(18800)	269(7400)	309(5400)	461(600)
7	2100	165 - 180	229(18400)	269(7300)	309(5400)	461(580)
8	1850	140 - 150	228(18900)	269(7500)	309(5500)	460(580)
9	1300	135 - 145	227(19000)	270(7000)	308(5700)	461(620)
10	1100	205 - 215	228(18100)	268(7900)	308(6400)	461(590)

[a]Data from ref. 30.   [b]By vapor pressure osmometry.   [c]In 1,2-dichloro-ethane.   Band system notation as in Table 4.

cenylenes have as yet been reported, but there is no reason to
predict any significant further improvements over the results
obtained on the demercuration product[8], as the polymer sample
investigated was of high quality and in an acceptably high molecular
mass range.   It should be added at this point that the thermo-
oxidative stability of poly-1,1'-ferrocenylene, as determined by
thermogravimetric analysis in air, is very markedly lower than the
thermal stability in the absence of oxygen.   Catastrophic breakdown
of the molecule occurs in the vicinity of 500°, and this oxidative
degradation is probably self-catalyzed.

The two cyclopentadienyl (Cp) ring ligands in ferrocene, despite an exceedingly low barrier to internal rotation (~ 3.7 kJ mol^{-1}),
assume a preferential conformation both in the gas phase and in the
crystal.   Although, for many years, this was believed to be the
staggered (D_{5d}) type, it was more recently established[36] by electron diffraction and X-ray studies that the gas-phase equilibrium
conformation is eclipsed (D_{5h}), and even in the solid state below
the λ-point transition at 164K the relative ring orientation tends
more toward the eclipsed than the staggered conformational arrangement, whereas at room temperature rotational disorder prevails,
allowing neither a D_{5d} nor a D_{5h} symmetry assignment.   (For reasons
of tradition and convenience, the staggered conformation is still
being used in the literature, including the present chapter.)   The
dimer, in the crystalline state, likewise prefers a more eclipsed
than staggered equilibrium conformation[37], and in the 1,1'-trimer
one finds both the eclipsed (outer units) and the staggered (central
unit) conformation[38].   The directly connected Cp rings in these
two compounds are substantially coplanar, forming fulvalene-type
ligand bridges.   A lesser extent of coplanarity between directly
joined rings can be inferred from nmr data for the two homoannular
trimers, notably the sterically crowded 1,2-terferrocene (II, n = 3)
[5,31,39], although corroborating structure determinations have yet
to be performed, and the question of relative ring orientation
remains an open one.   No X-ray diffraction data are available for

the higher-molecular oligomers of I and II.   It is safe to assume,
however, that there will not be a given equilibrium conformation
for each unit, and rotational disorder may even prevail throughout
the entire chain.

As appears predictable from steric considerations, the
ferrocene units in the dimer crystal assume a mutually transoid
conformation.   The same pattern holds for the 1,1'-trimer, and
one may expect a preference for similar transoid arrangements of
adjacent units in the higher-molecular homologs of both I and II,
albeit doubtlessly associated with increased disorder in the high-
molecular-mass ranges as a consequence of greater randomness of
chain conformation.

There is no tilting of the rings in ferrocene, nor has such
tilt been observed in biferrocene, and from the insensitivity to
molecular mass changes of the chemical shift difference between α
and β proton signals in the NMR spectra of the low-molecular
oligomers of I (where such signals can still be distinguished) one
may conclude that ring tilting is insignificant as well in polymeric
I[27].   There is also no *a priori* reason for assuming tilt effects
in oligomers or polymers of the homoannularly interlinked types,
the second ring of each unit along the chain remaining unsubstituted
and unconstrained.

Various spectroscopic techniques lend themselves to the characterization of oligo- and/or polyferrocenylenes, including mass
spectrometry, solution NMR and electronic absorption spectroscopy,
as well as infrared and Mössbauer spectroscopy.

Mass spectrometry has been applied to characterize some oligoferrocenylenes of structures I and II.   While all oligomers have
in common with the parent complex, ferrocene, a pronounced resistance to fragmentation as reflected in relatively high intensities
of the parent ion peak and the corresponding doubly charged signal,
this fragmentation resistance is more pronouncedly shown by the 1,2-
type oligomers than by the 1,1'-type counterparts.   Thus, for the
trimer as well as the two tetramers of II the parent ion peak

represents the base peak, and the relative parent ion abundance
for the *meso* pentamer is still 79% of the ion at m/e 738 representing the base peak [31] .   In the analogous series of I, in contrast,
reports from the same laboratory [6b] indicate the tetramer and
pentamer to give, respectively, 61 and 6% parent ion peak intensity
(relative to m/e 368).   An interesting feature of the spectrum of
the 1,2-trimer (II, n = 3) is the isotopic abundance distribution
in the m/e 550 - 553 region characteristic of 1,2-triferrocenylenes
generated from the trimer by expulsion of 2 H and subsequent
cyclization.   An analogous process involving loss of 2 H from one
of the pentameric isomers is a strong possibility [31] .

   ^{1}H NMR spectroscopy, while a highly valued tool in the identification of lower homologs (n = 2,3,4) of both homo- and heteroannularly interlinked oligoferrocenylenes [5,10,31,39] , is of little
use for the characterization of higher-molecular fractions.   At
degrees of polymerization higher than 5 - 6, the individual proton
signals coalesce to a broad multiplet showing a maximum near
δ = 4 ppm (benzene-d_6), the latter being insensitive in position
to further molecular mass increases.   As $\bar{x}_n$ exceeds 20, the signal
becomes featureless and undergoes minor additional broadening as
a result of enhanced conformational probabilities, and no structural
information can be extracted from it [30] .

   Considerably better use for structural elucidation of both
lower- and higher-molecular polyferrocenylenes has been made of
electronic absorption spectroscopy.   The spectrum of the ferrocene
parent complex has been extensively discussed, and transition
assignments have been made, the papers by McGlynn's and Gray's
groups being representative [40] .   The maxima and extinction
coefficients for a number of individual oligoferrocenylenes and
a series of fractions of poly-1,1'-ferrocenylenes are listed in
Tables 4 and 5, the extinction coefficients having been calculated
per ferrocene unit for ready comparison.   The polymer samples
used for recording the data in Table 5 and those to follow in this
section were obtained as per entry 4, Table 2.

The marked batho- and hyperchromic shifts apparent for all
listed bands in the spectrum of biferrocene relative to ferrocene
are a reflection of the conjugation achieved between the two
directly joined, essentially coplanar Cp rings. However, on
comparing the spectrum of biferrocene with that of the next higher
heteroannular homolog, 1,1'-terferrocene, one notices but minor
further red shifts and intensity increases[10,27], and as the degree
of polymerization exceeds 4 both λ_{max} and ε remain altogether
constant within experimental error limits up to the highest $\bar{M}_n$
value ($\sim$10 000) determined. Confirming the long-known barrier
function of the metal center with respect to the transmission of
resonance from one ring to the other in the ferrocene complex,
these findings are clear evidence of the lack of significant
electronic delocalization along the polymer chain in the heteroannular polyferrocenylene and so accord well with the observed
failure of this polymer type to conduct an electrical current
(d.c. conductivity $\rho_{20} \simeq 10^{-12}$ Ohm^{-1} cm^{-1})[41]. It is informative
to compare the tabulated spectral data for the heteroannular trimer
with those for the two homoannular counterparts. Whereas 1,3-ter-
ferrocene gives virtually the same λ_{max} and ε values as shown by
1,1'-terferrocene, the pattern of 1,2-terferrocene, more closely
coincident with that of the dimer, suggests appreciable resistance
to coplanar alignment of the three directly joined Cp rings in the
molecule[5]. This in turn permits the conclusion that poly-1,2-
ferrocenylenes are unsuitable as models for the study of conjugation effects along polymer chains. The reduced steric hindrance
to coplanar alignment in poly-1,3-ferrocenylenes would seem to
suggest this polymer type to exhibit a very high extent of conjugation in chain direction as no metal barriers interrupting electronic charge delocalization along the backbone are interposed.
However, the failure of 1,3-terferrocene to show enhanced batho-
and hyperchromic shifts relative to the 1,1'-isomer indicates -
and this is supported by molecular models, as well as by NMR data -
that even a 1,3-substituent disposition does not permit unperturbed

coplanarity of the Cp rings involved in unit interconnection[39].
Accordingly, it appears doubtful whether a significantly extended
domain of conjugation can ever be brought about in a polymer
composed of 1,3-ferrocenylene units.

The infrared absorption features of homo- and heteroannular
oligoferrocenylenes have been amply described[5,6b,10,31,34,39],
and assignments of all fundamentals, skeletal vibrations, and modes
of the C-C bond connecting adjacent ferrocene units have recently
been proposed, based on the assumption that the local-symmetry
concept is valid for the sandwich-type metallocenes[30].   In going
from ferrocene to biferrocene and higher-molecular homologs of I,
one finds comparatively little change in the overall absorption
pattern, except that, as a result of changing from local point-
group symmetry C_{5v} of the unsubstituted Cp ring to C_{2v} of the mono-
substituted ring, some degeneracy of frequencies is lifted, and 21
IR- and/or Raman-active ring-vibrational modes are expected in
addition to skeletal modes and vibrational modes involving the
ferrocene-ferrocene bond.   While many of these absorb at frequencies more or less coincident with those of the pertinent ferrocene
fundamentals, others give rise to new bands, noteworthy being the
'substitution' band near 1030 cm^{-1} (assigned [30] to an a_1 CH in-
plane bending mode with admixed a_1 sym. ring-breathing component
and C-C stretching of the ferrocene-ferrocene connecting bond).
In addition, in violation of the '9,10-μ rule' of Rosenblum[42],
which states that the contribution to the two strong ferrocene bands
at 1000 and 1100 cm^{-1} vanishes for a Cp ring as it suffers substitution, biferrocene absorbs near 1000 and 1100 cm^{-1} in intensities
much higher than expected for the 'loss' of one unsubstituted ring
per unit, the reason being that new absorptions appear precisely at
these positions as a result of substitution by another ferrocene
unit.   The same observation is made with all other oligo- and
polyferrocenylenes.   As Rosenblum's rule has proved widely usable
for quantitative determinations of the contents of substituted rings
in ferrocene compounds[43], its breakdown in polyferrocenylene
chemistry represents a regrettable restriction of the diagnostic

tools available to the researcher for confirmation of the inherent
substituent patterns in homo- and heteroannularly joined polyferrocenylenes.

As the IR spectra of low- and high-molecular I are compared,
a striking insensitivity of band positions and multiplicities to
increases in chain length is noticed (Fig. 1), this being a mani-
festation of the absence of vibrational coupling between adjacent
recurring units and, hence, of the lack of significant vibrational
interaction between the pair of ring ligands in the ferrocene
complex.   The observed behavior is well in accord with experience
in the vibrational spectroscopy of non-polymeric ferrocene compounds, which are characterized by a similar lack of vibrational

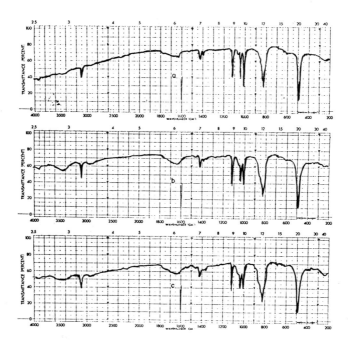

FIGURE 1   Infrared spectra (KBr pellet) of poly-1,1'-ferrocenylenes;
a: fraction of I, $\bar{M}_n$ = 550; b: fraction of I, $\bar{M}_n$ = 1600;
c: fraction of I, $\bar{M}_n$ = 4100

coupling between the two Cp rings.   It is of interest to note that
the lack of positional shifts in higher-molecular I also applies
to the skeletal infrared and Raman absorptions at 485 cm^{-1} (*asym.*
ring-iron tilting and stretching) and 310 cm^{-1} (*sym.* ring-iron
stretching), respectively, which indicates that heteroannular-type
polymerization of the ferrocene nucleus entails no constraints in
ligand tilting or metal-ligand stretching motions.

Mössbauer spectroscopy, while inconvenient or altogether
impracticable for the investigation of most other metallocenes, has
proved invaluable for the elucidation of structural and electronic
features in ferrocene compounds[44].   Both ferrocene and biferrocene give spectra virtually identical in chemical shift (δ) and
quadrupole splitting (ΔEq), and one finds this identity virtually
retained as one goes to the heteroannular trimer and, thence, to
higher-molecular I (Fig. 2).   Typical Mössbauer parameters for
selected oligomeric and polymeric fractions of I are contained in
Table 6.   The lack of $\bar{M}_n$ dependence of δ and, hence, of the s-electron density at the metal center permits the conclusion that there
is little or no d-electron interaction between adjacent ferrocene
units, as such inter-unit delocalization involving d-electron
density would certainly affect the s-electron charge distribution
at Fe because of changes in screening.

The conclusion of lacking inter-unit d-electron delocalization
in polymeric I, well in agreement with the electronic spectral
behavior, finds support in the intensitivity to $\bar{M}_n$ of the quadrupole splitting and, thus, of the electric field gradient at the
iron nucleus.   In the axially symmetric ferrocene molecule the
electric field gradient, -eq, is entirely defined by eq = V_{zz}, V_{zz}
being the z component of the electric field gradient tensor.   The
magnitude and sign of V_{zz} in ferrocene is primarily dictated by the
compounded effect of the two d_o and four $d_{\pm 2}$ electrons occupying the
the essentially non-bonding, metal-localized a_{1g} and e_{2g} orbitals,
respectively.   Therefore, in I, n > 2, the inter-unit overlap of
e_{2g} orbitals with consequent delocalization should lead to enhanced

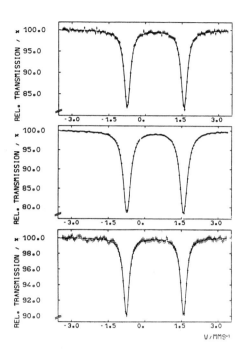

FIGURE 2. Mössbauer spectra (^{57}Co(Rh); 77K) of poly-1,1'-ferrocen-
ylenes; top: biferrocene; center: 1,1'-terferrocenyl;
bottom: fraction of I, $\bar{M}_n$ = 2500.

TABLE 6   MÖSSBAUER PARAMETERS FOR SOME
          OLIGO- AND POLYFERROCENYLENES[a]

Compound	Chemical shift δ, mm s^{-1}	Quadrupole splitting ΔEq, mm s^{-1}	Width at half height Γ, mm s^{-1}	Ref.
Biferrocenyl	0.53	2.34	0.29	45
(I, n = 2)	0.52	2.37		46
1,1'-Terferrocenyl (I, n = 3)	0.53	2.32	0.34	45
Oligo-1,1'-ferrocenylene (I, n ≈ 5)	0.53	2.33	0.29	45
Poly-1,1'-ferrocenylene (I, n ≈ 10)	0.53	2.34	0.30	45
Poly-1,1'-ferrocenylene (I, n ≈ 15)	0.53	2.34	0.26	45

[a]At 77 - 78K. All isomer shifts re-referenced to metallic iron.
^{57}Co(Rh) source in ref. 45; ^{57}Co(Cr) in ref. 46.

e_{2g} orbital diffusiveness and, hence, because of the proportionality of V_{zz} to $<r^{-3}>$ (r = distance between electron and iron nucleus), to a significant decrease of $V_{zz}(e_{2g})$ and concomitant reduction of ΔEq. Another piece of information to be gained from the observation of lacking molecular mass dependence of ΔEq concerns the distance between the central iron atom and the Cp ligands. Both δ and ΔEq have been found[47] to be critically dependent on the metal-ring distance in ferrocene compounds. The constancy of ΔEq as the degree of polymerization increases, therefore, indicates that there is no significant change in the iron-ring distance in higher-molecular I relative to the dimer or, in fact, ferrocene itself. This finding may prove important in view of the impossibility of collecting X-ray diffraction data from the polymeric material.

Concluding the discussion of physical properties, a comment is due on the magnetic behavior of poly-1,1'-ferrocenylenes. In view of the lack of inter-unit electronic interaction along the chain in this polymer type, paired with an Fe-Fe distance expected to be too large (5.1 Å in a transoid conformation) to allow for direct metal-metal bond formation and antiferromagnetic coupling, one would anticipate diamagnetic behavior in both low- and high-molecular I. For poly-1,1'-ferrocenylenes prepared by the organo-lithium/organohalide coupling route (entry 4, Table 2), suscepti-bility measurements conducted in our laboratory have indeed established diamagnetism up to the $\bar{M}_n$ range of about 5000. Higher-molecular fractions showed feeble paramagnetism; almost certainly however, this was due to paramagnetic contaminants, such as traces of adsorbed Fe^{3+} ion or ferricenium sites generated by inadvertent oxidation of ferrocene units (*vide infra*) and unremovable from the polymeric material by conventional extraction or reducing repreci-pitation treatment. It should be added that, again because of contamination, polyferrocenylenes prepared by other methods were found to tend more toward the paramagnetic side than did those of entry 4, Table 2. Thus, polymers synthesized by oxidative and thermal coupling of ferrocenylenecopper or -cuprate species proved

to be weakly paramagnetic down to $\bar{M}_n$ values of about 3000, and
polymers from other preparations, *e.g.* those summarized in Table 1,
retained feeble paramagnetism down to $\bar{M}_n$ 1500 - 2500.   The strong
paramagnetism reported in early communications[3] for Korshak's
polyrecombination products containing oligoferrocenylene segments
in the chain (*vide supra*) must likewise be traced to magnetic
impurities generated under the harsh reaction conditions[48];
indeed, recombination polymers synthesized in our laboratory and
rigorously purified were found to be diamagnetic up to $\bar{M}_n \sim 2000$,
and only faintly paramagnetic in higher molecular-mass ranges.
In summary, at the present state of knowledge in the field, it
appears safe to state that poly-1,1'-ferrocenylenes, *if* devoid of
metallic or metal ion impurities, are diamagnetic at least throughout the molecular mass range so far attained experimentally, and
there is no *a priori* reason for predicting paramagnetic behavior
for higher-molecular-mass polymers still to be synthesized.

Chemical Properties.- The chemical behavior of polyferrocenylenes
can manifest itself by reactions that may occur either at the ring
ligands or at the central metal atom.   In the former case, ring
substitution is the general result, whereas, in the latter case,
iron protonation or oxidation can typically be visualized.   In
addition, ligand exchange, which proceeds readily in ferrocene
compounds under suitable conditions, must be considered as a
possible reaction course in polyferrocenylenes.   Since a ferrocene
unit, in the process of undergoing substitution by another ferrocene
group, gains in ring basicity because of the substituent's powerful
electron donor action, the internal units in polymeric I, all
flanked by ferrocene neighbor units, should be at least as susceptible as ferrocene itself to electrophilic attack, *e.g.* in alkylation or acylation, although reduced steric accessibility of reaction sites in the polymeric substrate will by necessity have a rate-
retarding effect.   Proneness to nucleophilic attack, on the other
hand, should be lessened in the polymer relative to the monomeric

complex.   These predictions have not so far been put to the test.
Only biferrocene has been acetylated, but no polymer acylation or,
in fact, any other chemistry affecting the ring ligands in polymeric
I has to this date been investigated, although this task offers
enormous challenge with respect to both the chemical approaches
involved and the properties thereby to be gained.

    The situation is different with regard to the chemistry invol-
ving the metal center, the topic of highest current priority in
research being the change of metal oxidation state in oligo- and
poly-1,1'-ferrocenylenes brought about in suitable oxidative
environments.   The reason for the special research interest in
this topic can be traced to the mixed-valence properties expected
in a partially oxidized polyferrocenylene.   The mixed-valence
phenomenon, stated in simplified terms, can be observed in a molecular entity comprising two or more atoms of a particular element
(generally a transition metal) characterized by different oxidation
states ('valencies') and in reasonable proximity to each other to
allow a weak mutual electronic interaction.   In such a system the
valencies, although localized (trapped) in the ground state, can
undergo exchange between the two (or more) sites by an optically
induced intramolecular electron transfer requiring a comparatively
small expenditure of energy.   This exchange, referred to as intervalence transfer, manifests itself in a broad absorption band
typically in the near-infrared;   it leads to a reversal of the
oxidation states such that immediately after completed electron
transfer the donor center's coordination sphere possesses the
vibrational coordinates of the acceptor center, and *vice versa*.
The model proposed by Hush [49] has proven invaluable for the demon-
stration and understanding of intervalence transfer phenomena, and
the groups of Hush, Day, and Taube [49-51] may be cited as the foremost pioneers in the fascinating field of mixed-valence physics and
chemistry, which has in recent years become a focal point of
research activities on account of its important ramifications into
questions involving catalytic and biological electron transfer
processes.   While no details can be presented here, the reader is

referred to the many proficient articles and reviews that have since
appeared in print on pertinent aspects of the mixed-valence phenomenon [52-55].

Because of the similarity in ligand environment in both ferrocene and its one-electron oxidation product, the ferricenium cation,
it stands to reason that a system made up of two or more adjacent
ferrocene units and partially oxidized so as to give rise to a
sequence of iron centers with discrete, weakly interacting higher
(Fe(III)) and lower (Fe(II)) oxidation states should lend itself
exceptionally well to the study of the mixed-valence problem.

In the biferrocene monocation, $Fc-Fc^+$ (Fc = ferrocenyl, Fc^+ =
ferricenyl), prepared from the neutral complex by partial oxidation
with p-benzoquinone, iodine, or other oxidants (scheme 11) and

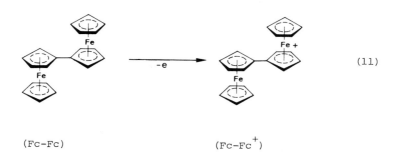

$$-e \qquad\qquad (11)$$

                    (Fc-Fc) $(Fc-Fc^+)$

physically isolated as the picrate, triiodide, tetrafluoroborate,
etc., such mixed-valence behavior has indeed been established in
experimental efforts executed most painstakingly and on a broad
front by Cowan *et al.* [52,56,57] and, subsequently, by the groups of
Hendrickson [55,58], Meyer [53,59], and others. The electrophysical
and spectroscopic data obtained on the cation are clear evidence of
a system with trapped valences, the two oxidation states being
localized, in the time scale of the characterization techniques used,
each one on a given unit. (The contrasting behavior of [0.0] ferrocenophane monocation, which represents an 'average-valence' system
with equal charge on both units, is on record [56].) The cation

possesses a (somewhat anion-dependent) magnetic moment ($\mu_{eff} \simeq 2.2$
B.M.) similar to that of mononuclear ferricenium ion, and, as in
the latter, highly anisotropic g values are derived from the ESR
spectrum.   In contrast to the ferricenium species, the dinuclear
Fc-Fc$^+$ gives a well-defined spectrum at T >> 20K because of increased
ESR relaxation time resulting from distorted symmetry in that cation
and consequent increase in the splitting between the two Kramers
doublets arising from the $^2E_{2g}$ ground-state configuration[57c].
The cation's electroconductivity, $\sigma_{20} \simeq 10^{-8}$ Ohm^{-1} cm^{-1}, is some
five orders of magnitude higher than that of ferricenium ion (excep̲t-
ing the ferricenium-TCNQ⁻ salt, which is even more highly conducting
because of anion participation) [57a].   This demonstrates convinc-
ingly that mixed valence provides a better path for electronic
conduction than would be found in a system simply composed of non-
interacting ferrocene and ferricenium units.   The Mössbauer scan
represents a superposition of the two quadrupole-split doublets
pertaining to the ferricenium (outer doublet, ΔEq = 2.1 - 2.3 mm s^{-1})
and ferrocene (inner doublet, ΔEq = 0.3 - 0.4 mm s^{-1}) parts of the
compound.   While in most cases reported the two doublets appear in
approximately equal intensity when recorded at 77 K, some anomalies
have been observed near 300 K, where the inner doublet, apparently
through contribution by a third, narrow doublet arising at this
higher temperature, possesses a higher intensity than the outer one.
The origin of the contributing resonance is not clear.   It has
been speculated that this newly appearing doublet represents an
average-valence signal and thus indicates that the electron
exchange rate between the two iron nuclei may approach 10^7 - 10^8
s^{-1} at the higher temperature, leading to partial oxidation-state
averaging within the time scale of the Mössbauer experiment[55,57b].
Mössbauer data obtained by applying an external longitudinal magnetic
field [60] seem to indicate that the lone electron at the Fe(III) site
is not metal-localized, as had previously been accepted, but rather
occupies a predominantly ligand-based orbital, e.g. a π orbital of
the fulvalene bridging ligand.   This could indeed provide a
conceivable path for thermal electron exchange and high-temperature

delocalization.   Further attention should be paid to this important
question.   The most striking spectroscopic feature of the dinuclear
cation is observed in the near-infrared absorption spectrum, recorded
on solutions in polar solvents.   The spectrum displays an intervalence transfer band at about 1900 nm for the process Fc^+-Fc $\longrightarrow$
Fc-Fc^+, and from both the frequency and width at half height of this
band one is in a position to calculate the interaction parameter α
in the simplified wave function of the mixed-valence cation[52,59].
The value derived, in the order of 10^{-1}, is in agreement with the
weak-interaction-type mixed-valence case expected for the monocation.

In addition to the mono-oxidized biferrocene, the fully oxidized
species, Fe^+-Fe^+, has also been isolated and spectroscopically
characterized[55,57b].   With each unit carrying a full positive
charge in this dication, the conditions for mixed-valence behavior
no longer obtain, and, accordingly, no intervalence transfer band
is observed.

In an excellent follow-on study, ferrocene and the three
oligomers I, n = 2 - 4, were investigated electrochemically[54].
Several partially or fully oxidized species were generated from the
neutral compounds at controlled potentials, although no products
were physically isolated.   Half-wave potentials (*versus* SCE) determined for the various oxidation stages by cyclic voltametry in
dichloromethane/acetonitrile are collected in Table 7.

It is seen that separate, discrete waves arise at increasingly
higher potentials as the ferrocene units undergo successive
oxidation, attesting to the discreteness of each oxidation state
site and to the weak interaction between them.   Also apparent is
the increasingly stronger electrostatic repulsion that must be overcome with each new Fe(III) site introduced.   Another trend of
significance is the steady shift of $E_{\frac{1}{2}}$ to more negative values as
one goes from ferrocene to the dimer, trimer, and tetramer.   Well
in accord with the stabilizing effect on a ferricenium unit exerted
by each added electron-donating ferrocene group, the trend suggests
that the propensity for mono-oxidation increases steadily, albeit

TABLE 7   HALF-WAVE POTENTIALS FOR PARTIALLY AND
           FULLY OXIDIZED OLIGOFERROCENYLENES[a]

Parent compound	Oxidation step[b]	$E_{\frac{1}{2}}$, V[c]
Ferrocene	$Fc \xrightarrow[-e]{} Fc^+$	0.40
Biferrocene	$Fc-Fc \xrightarrow[-e]{} [Fc-Fc]^+$	0.31
(I, n = 2)	$[Fc-Fc]^+ \xrightarrow[-e]{} [Fc-Fc]^{2+}$	0.65
1,1'-Terferrocene	$Fc-Fc-Fc \xrightarrow[-e]{} [Fc-Fc-Fc]^+$	0.22
(I, n = 3)	$[Fc-Fc-Fc]^+ \xrightarrow[-e]{} [Fc-Fc-Fc]^{2+}$	0.44
	$[Fc-Fc-Fc]^{2+} \xrightarrow[-e]{} [Fc-Fc-Fc]^{3+}$	0.82
1,1'-Quaterferrocene	$Fc-Fc-Fc-Fc \xrightarrow[-e]{} [Fc-Fc-Fc-Fc]^+$	0.16
(I, n = 4)	$[Fc-Fc-Fc-Fc]^+ \xrightarrow[-e]{} [Fc-Fc-Fc-Fc]^{2+}$	0.36
	$[Fc-Fc-Fc-Fc]^{2+} \xrightarrow[-e]{} [Fc-Fc-Fc-Fc]^{3+}$	0.61
	$[Fc-Fc-Fc-Fc]^{3+} \xrightarrow[-e]{} [Fc-Fc-Fc-Fc]^{4+}$	0.89

[a] Data from ref. 54.   [b] Fc = ferrocenyl (terminal) or 1,1'-ferro-
cenylene (internal).   Brackets indicate uncertainty of position(s)
of ferricenium unit(s) in the chain.   [c] *Versus* SCE at 24°, in
dichloromethane/acetonitrile (1:1), 0.1 M tetra-*n*-butylammonium
hexafluorophosphate.

in decreasing increments, as n grows from 1 to 4 and, thence, to
higher values.   A long chain of I, in fact, may suffer mono-oxidation spontaneously and with profound effects on the solubility in
nonpolar solvents.

The partial oxidation of the oligomers with n > 2 in certain
instances provides more than one possibility of positioning the
ferricenium unit(s) in the chain, and oxidation state isomerism
may thus result.   Uncertainties in the unit sequence are reflected
in the bracket denotation used in the table.   Although one may
predict certain arrangements to be energetically favored over others,
the energy differences are not large;   from electrochemical arguments,
for example, the difference in free energy between the two dicationic
isomers derived from terferrocene, $Fc^+-Fc-Fc^+$ and Fc^+-Fc^+-Fc, has
been estimated to be as small as 11.5 kJ mol^{-1} [54], the first-men-
tioned isomer being the more stable one for obvious electrostatic
reasons.

Table 8 contains the intervalence transfer band maxima measured in the same study on the electrochemically obtained salt solutions (entries 1 - 4, column A). Preliminary absorption data for the same mono- and dications obtained in the author's laboratory by chemical oxidation of the neutral parent compounds have been juxtaposed in the table for comparison (column B). Because of the broad band structure, error limits for λ_{max} are high, and this must be taken into account in any comparative evaluation. Excepting the entry 2, the tabulation shows the expected trend of intervalence transfer energies, although significant quantitative differences between the two series of results in columns A and B can be observed. Thus, the transition in the trinuclear dication (entry 3), being unsymmetrical and leading to a higher-energetic isomer, should, and

TABLE 8   INTERVALENCE TRANSFER ABSORPTION OF PARTIALLY
OXIDIZED OLIGO- AND POLYFERROCENYLENES[a]

Entry	Cation[b]	Predicted transition[c]	λ_{max}, nm A[d]	B[d]
1	$[Fc\text{-}Fc]^+$	$Fc^+\text{-}Fc \longrightarrow Fc\text{-}Fc^+$	1900	1790
2	$[Fc\text{-}Fc\text{-}Fc]^+$	$Fc\text{-}Fc^+\text{-}Fc \longrightarrow Fc^+\text{-}Fc\text{-}Fc$	1990	1780
3	$[Fc\text{-}Fc\text{-}Fc]^{2+}$	$Fc^+\text{-}Fc\text{-}Fc^+ \longrightarrow Fc^+\text{-}Fc^+\text{-}Fc$	1670	1625
4	$[Fc\text{-}Fc\text{-}Fc\text{-}Fc]^{2+}$	$Fc^+\text{-}Fc\text{-}Fc^+\text{-}Fc \longrightarrow Fc^+\text{-}Fc\text{-}Fc\text{-}Fc^{+e}$	1790	1770
5	$[I, n = 11]^{5.5+}$	$\text{-}Fc^+\text{-}Fc\text{-}Fc^+\text{-}Fc\text{-}Fc^+\text{-} \longrightarrow \text{-}Fc\text{-}Fc^+\text{-}Fc^+\text{-}Fc\text{-}Fc^+\text{-}$		1800
6	$[I, n = 24]^{12+}$	"	"	1800
7	$[I, n = 11]^{3.5+}$	$\text{-}Fc^+\text{-}Fc\text{-}Fc\text{-}Fc^+\text{-}Fc\text{-} \longrightarrow \text{-}Fc\text{-}Fc^+\text{-}Fc\text{-}Fc^+\text{-}Fc\text{-}$		1850
8	$[I, n = 24]^{8.5+}$	"	"	1840

[a]In dichloromethane/acetonitrile, 0.1 M (n-Bu)$_4$NPF$_6$ (electrochem. oxidized) or acetonitrile (chem. oxidized), under Ar.   [b]Anion is PF$_6^-$ (electrochem. oxidized; from supporting electrolyte, products not isolated) or BF$_4^-$ (chem. oxidized with p-benzoquinone/HBF$_4$).   [c]All products in vibrationally excited state. Transitions in entry 4 and higher entries speculative.

[d]Products by electrochem. oxidation (column A; from ref. 54) or chem. oxidation (column B; unpublished results, author's laboratory). Estimated uncertainty $\pm$ 10 nm for entries 1 - 4; $\pm$ 20 nm for entries 5 - 8.
[e]The less probable $Fc^+\text{-}Fc\text{-}Fc^+\text{-}Fc \longrightarrow Fc^+\text{-}Fc^+\text{-}Fc\text{-}Fc$ [54] should produce a band at higher energy observed.

does, give a maximum shifted to shorter wavelength in relation to
the monocation's band (entry 2).   The latter, perhaps for reasons
of unsymmetric transition, turns out to be at slightly higher
energy than the dinuclear monocation's maximum if column B data are
considered.   Inexplicably, however, the reverse holds for the values
of column A.   In the tetranuclear dication (entry 4) the energetic
demand for the depicted transition should be similar to that in
terferrocene monocation, and the values listed in column B indeed
are close, whereas, again, the data of column A do not conform.
With the chemical oxidation work still in progress and the findings
subject to correction, it is too early to draw significant conclusions
from these comparative observations.

    Extending the investigations discussed in the foregoing, work
is progressing in the author's laboratory towards the goal of
preparing partial oxidation products of I, n > 4.   The chemical
oxidation of polymeric I as exemplified in the segment structure
of scheme 12 for the case of an exactly half-oxidized polycation,

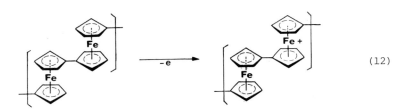

$$(12)$$

while straightforward in principle, presents extraordinary practical
difficulties arising from (i) poor solubility or absolute insolubility of most of the polysalts formed,   (ii) extreme sensitivity of
the higher-oxidized polycations in the dissolved state to impurities
capable of acting as reductants and thus causing uncontrollable
decreases in ferricenium contents, (iii) inability of most oxidizing
systems to provide rigorous stoichiometric control of the oxidation
process, and consequent variations in the degree of oxidation
attained, and (iv) polysalt resistance to combustion, causing
problems in the elemental analysis required for accurate structural

definition. For example, such conventional oxidants as H_3O^+/O_2, Ce^{4+}, tetracyanoquinodimethane, and dichlorodicyanoquinone, which have been successfully employed for the oxidation of biferrocene [55-58] or short oligoferrocenylene segments in polyrecombination products of ferrocene [61], have proved to be entirely useless for polyferrocenylene oxidation because of unreproducible degrees of oxidation achieved and/or product insolubility. The only system so far found to be moderately efficacious is $H_3O^+/$ p-benzoquinone, applied by adding HBF_4 and p-benzoquinone to the THF or benzene solution of the neutral parent polymer. While, again, giving rather unreproducible degrees of oxidation as a consequence of its propensity for undergoing both one-electron and two-electron reduction, this oxidant furnishes polysalts retaining initial solubility in rigorously purified solvents, such as acetonitrile or nitromethane. Polysalts so prepared tend to remain soluble for a period of 24-28h, permitting analytical or spectroscopic determination of desired solution properties within the limited time span. Once insoluble, the polysalts can no longer be resolubilized without concomitant destruction. As no signifi-cant differences in elemental composition and spectroscopic solid-state properties have so far been observed between products prior to, and after, the loss of solubility, one may speculate that this peculiar 'aging' characteristic is brought about by packing differences in the lattice, perhaps as a result of gradual solvent diffusion from the precipitated crude salt. Structural analyses of polysalts sufficiently low-molecular to provide a reasonable degree of crystallinity may help to shed some light on this problem.

Preliminary characterization so far performed on the poly(ferro-cenylene-co-ferricenylene) polysalts prepared in different Fc/Fc^+ ratios and at various $\bar{x}_n$ levels includes Mössbauer and electronic absorption spectroscopy. Whereas elemental analysis, for reasons pointed out above, has proved inadequate for reliable determinations of the ferricenium contents, Mössbauer spectroscopy is well suitable for this purpose, since the recoilless fractions of both the

ferrocene and the ferricenium moieties are sufficiently similar to
permit direct signal area comparison at 77K. (At much higher
temperatures, anomalies arise as earlier reported for the dinuclear
case [55,57b].) The δ and ΔEq parameters determined for a large
number of polycations possessing molecular masses of 1000 - 5000
and ferricenium contents of 35 - 60%, although showing some scatter
from sample to sample, are in the ranges of δ = 0.50 - 0.57 mm s^{-1},
ΔEq = 2.20 - 2.47 mm s^{-1}, for the outer (ferrocene) doublet and of
δ = 0.51 - 0.59 mm s^{-1}, ΔEq = 0.39 - 0.46 mm s^{-1}, for the inner
(ferricenium) doublet. They are not substantially different from
the corresponding values for the ferrocenylferricenium cation
[45,46,58] (Fig. 3). It thus appears that at least within the
range of ferricenium contents so far investigated, there is no
$\bar{x}_n$-dependent shift toward higher or lower values of either δ or

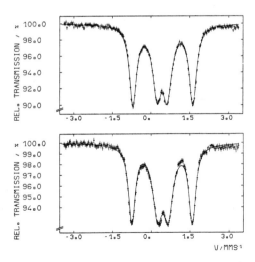

FIGURE 3. Mössbauer spectra (^{57}Co(Rh); 77K) of partially oxidized
di- and polynuclear I; top: ferrocenylferricenium
tetrafluoroborate; bottom: poly(ferrocenylene-
ferricenylene tetrafluoroborate).

ΔEq and, hence, little or no change in s-electron density and
electric field gradient at the iron nuclei of the individual Fc
and Fc$^+$ units. This suggests that the extent of electronic
interaction between adjacent units, and, thus, the mixed-valence
behavior, is not significantly altered as the ferrocenylene-
ferricenylene system becomes repetitively linked into a linear
chain of higher molecular mass.

The intervalence transfer band maxima determined for a
number of fractions of partially oxidized I are in qualitative
agreement with this inference. The bands are distinctly broader
than observed with the dinuclear cation as a consequence of the
simultaneous occurrence of many nearly isoenergetic transitions.
As seen from the exemplifying λ_{max} values listed in Table 8
(entries 5 - 8), there is no molecular-mass dependence of the
intervalence transfer energy within the range, n $\simeq$ 10 - 25, and,
in accord with expectation, the energetic demand is lowered as the
degree of oxidation decreases sufficiently below the 50%-level to
permit the random occurrence of two adjacent uncharged units in
the chain (*cf*. entries 7,8 *vs*. 5,6). However, the comparatively
high wavelength range of 1800 - 1850 nm in which these bands are
observed is somewhat puzzling. The transitions in entries 5 and
6, for example, should not be favored energetically over that in
entry 3 and, hence, should give rise to maxima closer to 1650 nm.
This apparent inconsistency requires further study, although a
possible explanation may be found simply in greater experimental
difficulties experienced in dissolving the polymer rapidly and
yet completely anaerobically, which in turn may have led to
effectively lowered ferricenium contents of the solutions investigated.

The few data presented in this section, although rather
tentative and incomplete as the experimental program continues,
suffice to permit the prediction that polysalts composed essentially of alternating Fc and Fc$^+$ units (scheme 12) and comprising
electronically balanced anions will not exhibit electrophysical

properties substantially different from those of the simple
dinuclear ferrocenylferricenium salts previously investigated
[52-58]. Specifically, increases in electroconductivity appear
to be out of the question, as do any major changes in the magnetic
properties and ESR parameters that would point to enhanced
electronic delocalization along the polycation's chain. While, to
this date, no information on structures with very much higher or
lower ferricenium contents has been made available, surprise
findings with respect to the electrophysical behavior of such
structures are quite unlikely.

Potential Uses.- In view of the limited accomplishments made until
now in the synthesis of pure polyferrocenylenes of high molecular
mass, the absence of application and performance reports in the
literature is not unexpected. In the following, an attempt will
be made, based on known applications of ferrocene itself, to point
out potential use areas for the polymer. Only some of the more
recent literature sources for ferrocene applications will generally
be cited in this context, although earlier sources, most of these
well reviewed [1,2,62], will occasionally be quoted if particularly
significant. In the majority of use areas discussed below, the
principal advantage of employing the polymer in place of ferrocene
proper lies in the firm anchoring and immobilization of the active
metal site thereby achieved. The parent complex is highly volatile
and tends to migrate and evaporate from polymeric and other substrates into which it is compounded, leading to rapid depletion of
metal contents, especially at elevated temperatures.

    The excellent heat resistance of the ferrocene complex, paired
with outstanding stability under the impact of high-energy ultraviolet radiation as encountered outside the terrestrial atmosphere,
has prompted much research on ferrocene-containing space vehicle
coatings and related compositions [63]. At the low gas densities
observed in the upper atmosphere or in outer space, the question
of retaining the metallocene in the framework of the coating is a
particularly pressing one, and use of high-molecular polyferro-

cenylenes should contribute significantly to an enhanced integrity
of the coating material.

A feature of considerable importance in connection with the
well-known functioning of the ferrocene complex as a burning rate
accelerator and combustion catalyst is the metal atom's ability to
exist in two different oxidation states and thus act as an efficient
electron transfer agent. In many areas of combustion catalysis
until now reserved for ferrocene or some of its derivatives,
e.g. in fuels (including hypergolic ones), pyrotechnic compositions,
and solid rocket propellants [64], use of polyferrocenylene should,
for the aforementioned reason, offer particular advantage. The
same argument holds for the use of polyferrocenylene in place of
ferrocene as an additive in cable insulating materials to reduce
dielectric breakdown [65], as a heat- and light-stabilizing component in numerous polymeric materials [66], as an additive exerting
a photo-sensitizing effect in polymers or in photographic compositions and other light-sensitive reproduction materials [67], and
as a constituent of epoxy resins, poly(vinyl chloride) (PVC), polyurethane, and other plastics to improve fire resistance and flame
retardance [68]. The last-named topic is of highest current
interest in materials technology, notably in aircraft and automo-
tive interior designing, as well as in high-rise building and
factory construction. Intimately connected with this question of
flame retardance is the problem of smoke suppression. Smoke and
fumes developed in an enclosed fire location tend to pose more
serious hazards to life than open fires, and successful suppression
of smoke evolution from burning organic matter, notably plastics
and rigid foam insulating materials, is therefore frequently a
vital design feature. The efficacy of ferrocene in suppressing
or inhibiting smoke development in PVC and other polymeric
materials is on record [69], and the use of polymeric I in lieu
of ferrocene is a suggested approach. A related problem concerns
the liberation of HCl during the combustion of PVC-type materials
with sometimes dramatic consequences. Ferrocene has been tested
successfully as an additive suppressing HCl evolution [70], and,

again, the polymeric product, unable to diffuse and migrate out
of the PVC composition, should show an even better performance
than the monomer.

Other applications in which the substitution of polyferrocenylenes for the monomer should offer distinct advantages include
use in radiation resist compositions and semiconductor devices
[71], in cutting fluids to enhance the durability of cutting tools
[72], in petroleum distillates for the removal of unsaturated
ingredients [73], as synergists for insecticides and acaricides
[74], as a co-initiator of the free-radical polymerization of methyl
methacrylate [75], as a curing accelerator in unsaturated polyester-
polyacrylates [76], and as a hardening agent for oligoacrylates on
cadmium surfaces [77]. In view of the ready oxidizability of ferrocene and its proneness to complexation with electron acceptors, a
particular challenge should be found in the use of polyferrocenylenes
as polymeric charge transfer complexes and, in the partially oxidized
state, as electron exchange resins and in certain electrophysical
applications where the magnetic and mixed-valence characteristics
are of interest. A case in point is the creation of surface-
attached electroactive ferrocene centers on graphite or platinum
electrodes, as well as on semiconducting, silicon-based photo-
electrodes of greatest current interest in photoelectrochemical
energy conversion projects [78]. Polyferrocenylenes can probably
be deposited readily from solutions of the neutral material. Once
deposited, the coating may be partially oxidized, whereupon it will
lose its solubility not only in nonpolar solvents but, after some
'aging' time, also in such polar solvents as nitromethane or acetonitrile; this should result in effective immobilization on the
electrode surface.

A discussion of potential polyferrocenylene applications would
be incomplete without covering the important field of medical and
medicinal use. Many ferrocene derivatives have been developed, in
some cases with considerable success, as hematinics, antibiotics,
haptens and antineoplastic agents, as specific enzyme inhibitors,
tagging agents and labels in various forms, as a pharmacon promoting

liver regeneration, and for conversion into the gamma-emitting ^{52}Fe
radionuclide to be used in nuclear medicine [79]. Hydrophilic
behavior in the oxidized state and low toxicity both combine to
render the ferrocene complex a ready substrate for medicinal
applications. However, its rapid metabolism in the living organism
presents a considerable disadvantage in most of its physiological
and pharmacological functions, and replacement by a polymeric
structure is likely to retard the metabolic process in addition to
preventing unduly fast removal from the site of action through
diffusion or resorption. This should be particularly advantageous
in hapten and other chemoimmunological, as well as labeling and
radiochemical applications.

POLYRUTHENOCENYLENE

Only the heteroannularly inter-linked compounds III will be
treated in this section, no homoannular isomers being known, and
the discussion will be restricted by necessity to the oligomeric
homologs prepared until now.

Physical Properties.- The oligo-1,1'-ruthenocenylenes III are
colorless or cream-colored solids. The dimer and trimer both
crystallize as well developed needles and, like the tetramer,
possess melting points below 300° (Table 9). Although homologs
with n > 4 have not so far been separated as pure compounds, one
may expect their melting points to show a similar trend as
observed in the ferrocene polymer series, increasing beyond 300°
at n = 4 - 6; higher fractions, as represented by entry 4 of the
table, being polydisperse, should show a dip in the melting point-
temperature relationship, followed by a steady increase beyond
350° as degrees of polymerization of 15 - 20 are exceeded. The
solubility behavior of the oligomeric ruthenocenes is similar to
that of the corresponding ferrocene analogs, although the low-
molecular members of the series are characterized by lower
saturation limits. High-molecular III, not as yet synthesized,

should possess excellent solubility in aromatic solvents and halocarbons.

While no thermostability testing has been performed owing to the unavailability of higher-molecular material, it is safe to predict, on the basis of comparative thermogravimetric analysis data reported for methylene-bridged ruthenocene polymers |80|, that high-molecular fractions of III will be superior in heat resistance even to the polyferrocenylenes.

The ruthenocene crystal is pentagonal prismatic (D_{5h}), *i.e.* the ring conformation is eclipsed, and the barrier to internal rotation is slightly higher than in ferrocene [81]. No X-ray structure data are available as yet on any crystalline oligomer; yet steric considerations suggest a transoid conformation with respect to the two metallocene units in the solid dimer, as in biferrocene, and the higher oligomers and polymers most likely are arranged in predominantly transoid conformations as well. Preliminary spectroscopic evidence (*vide infra*) agrees with an essentially coplanar alignment of the fulvalene-type bridging ligands in all oligomeric samples investigated (Table 9). There is no spectroscopic evidence for tilting of the Cp rings in oligomeric III, and no tilting is expected in compounds of higher molecular mass.

The spectroscopic data collected until now are few; with both mass spectrometry and NMR spectroscopy, as in the ferrocene series, found to be of little use in the higher polymeric range, the only major techniques employed are electronic and infrared absorption spectroscopy. The electronic spectrum of the ruthenocene parent complex has been described and analyzed, the aforementioned two papers being typical [40]; it resembles closely the ferrocene spectrum, although all pertinent transition energies are higher, and the maxima appear correspondingly shifted to lower wavelengths. The electronic spectra [13,33,35] of III, n = 2 - 4, similar to the ruthenocene spectrum, exhibit the characteristic ligand-field band near 320 nm due to the two d-d transitions, $a_1' \longrightarrow e_1''$ and $e_2' \longrightarrow e_1''$, and the two charge-transfer bands in the UV region near

250 and 265 nm, corresponding to the band systems V, IV and II in
Gray's notation [40]. The three maxima are listed in Table 9,
with ε, again, calculated per metallocene unit. Much as in the
ferrocene series, the UV bands suffer appreciable red shifts and
intensity increases as the ruthenocene molecule dimerizes, this
being the result of the introduction of the fulvalene ligand bridging the two metal atoms. The trend becomes much less pronounced as
one goes from the dimer to the trimer, and only small additional
shifts in the indicated direction can be observed in the tetramer.
The data obtained on the oligomer fraction (entry 4, Table 9) suggest
that no further shifts occur as n exceeds 4 - 5. (For the seeming
failure of the ligand-field band maximum to follow the overall trend
of bathochromic shifts, see [33].) It appears from these results
that the two Cp rings constituting each fulvalene-type ligand are
substantially coplanar. It is, furthermore, apparent that the
central metal atom in ruthenocene is not capable of transmitting
significant electronic effects from one Cp ring to the other;
accordingly, electronic interaction between adjacent 1,1'-rutheno-
cenylene units in III is essentially limited to delocalization *via*
the fulvalene bridge. Hence, no significant extent of electronic
charge delocalization along the polymer chain of III is indicated
by the UV data. The situation is thus, not unexpectedly in view
of the similar electronic structure and bonding characteristics,
quite comparable with that in the ferrocene polymer series I, and
one may safely predict that this lack of electronic inter-unit
interaction will find confirmation in the ^{99}Ru Mössbauer spectra,
the diamagnetic behavior and the electrical insulation characteristics yet to be determined on polymeric III.

The vibrational spectrum of ruthenocene has been treated
extensively [82]. As would be expected from the practically
independent vibrational behavior of each Cp ring, the ring-
vibrational fundamentals are observed at positions close to, or
coincident with, those in the ferrocene spectrum. Quite accordingly, one finds [33] the IR spectra (4000 - 500 cm^{-1}) of birutheno-

TABLE 9   MELTING POINTS AND SELECTED ELECTRONIC ABSORPTION
          MAXIMA OF KNOWN OLIGORUTHENOCENYLENES[a]

Compound	$M^{[b]}$	M.p., °C	λ_{max}, nm $(\varepsilon, M^{-1} cm^{-1})^{[c]}$		
Biruthenocene (III, n = 2)	462	238-239	249(6400)	261(5300)	322(670)
1,1'-Terruthenocene (III, n = 3)	692	276	252(9400)	265(8600)	322(1050)
1,1'-Quaterruthenocene (III, n = 4)	922	269-271	253(9400)	265(8500)	322(1300)
1,1'-Oligoruthenocenylene (III, n ≃ 6)	1420	210-230	253(9600)	264(8650)	323(1280)

[a]From refs. 13, 33, 35.   [b]By mass spectrometry (for ^{102}Ru) in first three
entries, by vapor pressure osmometry in last entry.   [c]In 1,2-dichloroethane.
Molar extinction coefficient ε calculated per metallocene unit.

cene and the higher homologs to be strikingly similar to the
spectra of the corresponding ferrocene oligomers discussed in a
previous section, showing the characteristic substitution band in
the vicinity of 1030 cm^{-1} and the absorptions near 1000 and 1100
cm^{-1}, the latter invalidating the '9,10-μ rule'. The only significant differences, relative to the ferrocene counterparts, appear in
the region of the unsymmetric skeletal vibrations below 500 cm^{-1}
where, as in the spectrum of ruthenocene itself, shifts to lower
frequencies are observed. Characteristically, and, again, as in
the case of I, there are no molecular-mass-dependent band shifts
over the entire 4000 - 250 cm^{-1} region, attesting to the absence
of significant vibrational interaction between neighboring ruthenocene units, as well as to the lack of any constraints in the ligand
tilting and metal-ligand stretching modes as n increases beyond
2 - 3.

Chemical Properties.- The ruthenocene complex, while offering
certain parallels with ferrocene in its chemical behavior [2], shows
a number of features in which it differs markedly. Thus, reduced
electronic charge in the σ carbon orbitals of the Cp rings relative
to ferrocene leads to decreased ring basicity with concomitantly

diminished reactivity towards electrophiles. As a consequence,
electrophilic ring substitution is disfavored, and nucleophilic
substitution favored, in relation to ferrocene. In contrast, the
metal basicity is greater in ruthenocene than in the iron congener;
therefore, ruthenium protonation and other electrophilic metal
substitution reactions are facilitated. None of these reaction
types have been 'tried out' as yet on oligomeric III.

   The difference between the two metallocenes is particularly
striking in the central metal's oxidation/reduction behavior.
Whereas ferrocene suffers one-step, one-electron oxidation to the
Fc^+ cation regardless of the oxidation method employed, the
ruthenium complex is capable of undergoing both one-step, one-
electron and one-step, two-electron oxidation reactions depending
on the oxidative environment [83]. Thus, reversible loss of one
electron, giving rise to the generation of the ruthenicenium
monocation, is brought about photooxidatively in the presence of
halocarbons. The same result is obtained electrochemically at the
dropping mercury electrode. The polarographically determined half-
wave potential is less positive than that of the Fc/Fc^+ system,
indicating a greater propensity for oxidation in the ruthenocene case.
Quite in contrast, if ruthenocene oxidation is accomplished
chemically (*e.g.* by iodine, $FeCl_3$, or H_3O^+/p-benzoquinone) or electrochemically at the platinum electrode, one finds a greater resistance
to oxidation than in the iron complex, the half-wave potential
having more than doubled (*i.e.* increased toward more positive values),
and the product is the ruthenicenium dication with formal Ru(IV),
generated irreversibly by a two-electron step. Whether the ruthenocene nucleus suffers oxidation more, or less, readily than the ferrocene complex thus depends critically on the oxidation method
employed. Although no work has been performed in the area of
oligoruthenocenylene oxidation, this research field should offer
high rewards in view of the unusual oxidation/reduction behavior of
the monomer. For example, the controlled, reversible one-electron
oxidation (at the dropping mercury electrode) of a fraction of the

ruthenocene units in a polynuclear compound can be expected to
furnish mixed-valence salts as in the analogous ferrocene polymer
case, although, in the light of the decreased donor power of the
ruthenocene system in relation to ferrocene, there is less likelihood of a trend of increasing ease of monooxidation in the polyruthenocenylene chain as n grows beyond the value of 2. Electrochemical treatment at the platinum electrode may lead to two-
electron oxidation as in the mononuclear compound, although perhaps
with enhanced reversibility in the polymer system. On the other
hand, the mildly electron-withdrawing effect of adjacent ruthenocene
units may raise the potential for the two-electron oxidation step
sufficiently to render one-electron oxidation more favorable, in
clear departure from the mononuclear case. Another challenging
task to explore concerns the polymer behavior under conditions of
chemical oxidation. Thus, it should be of interest to find out
whether the mode of oxidation can be changed to that of one-
electron oxidation if a ruthenocene unit tied into a chain is
treated with chemical oxidants (including the ferricenium cation)
possessing a redox potential intermediate between the potentials of
the ruthenocene/ruthenicenium monocation and ruthenocene/rutheni-
cenium dication systems. With access to the metal center likely to
be curtailed in a polymeric substrate, the simultaneous availability
at the reaction site of two one-electron-type oxidant molecules may
no longer be assured, and the mechanism of chemical oxidation could
conceivably be altered so as to facilitate the kind of one-electron
oxidation step observed at the dropping mercury electrode. This
would, of course, also result in a change of magnetic properties,
as the dication of ruthenocene is diamagnetic, whereas one-electron
oxidation entails paramagnetic behavior.

Potential Use.- As only preliminary communications regarding
oligomeric III, and no reports on polymeric III, have until now
become available, there is little reason to discuss major use areas
at this point. The high cost of the parent complex, not exactly

conducive to large-scale development and evaluation work, will
doubtlessly have a retarding effect on application research.
However, specialty fields may well call for the use of polyruthenocenylenes where costs are no primary consideration. The combined
heat and radiation resistance of ruthenocene, superior even to that
of ferrocene, should, for example, warrant the development of space
vehicle or instrument coatings containing polyruthenocenylene. The
parent complex shows low-temperature phosphorescence and is an
efficient triplet quencher, use of which could be made in polymer
applications. The most outstanding potential for investigation,
however, is offered in nuclear medicine, where excellent groundwork
involving the use of labeled ruthenocene derivatives has been
performed by Wenzel, the foremost researcher in this field [84].
The ^{103}Ru radionuclide is both a β (0.21 MeV) and a γ emitter
(0.50 MeV) with a useful half-life of 40 d. The β emission lends
itself to whole-body autoradiography, and the γ emission can be
utilized for the labeling of ruthenium compounds applied as pharmaca
or for diagnostic purposes. Accordingly, ruthenocene compounds
labeled with ^{103}Ru have found extensive use in nuclear-medical
research, $e.g.$ for the study of organ distribution and, most
recently, for the labeling of cytostatic drugs used as inhibitors
of tumor growth. These findings suggest challenging possibilities
for macromolecular, labeled ruthenocenes. The introduction of the
^{103}Ru nuclide into oligo- and polynuclear III can possibly be
accomplished with reasonable efficiency through thermal exchange
with ^{103}RuCl$_3$ by Wenzel's method developed for the labeling of
ruthenocene compounds [85].

SUMMARY AND CONCLUSIONS

Only two types of polymetallocenylene have been synthesized
until now: poly-1,1'-ferrocenylene and poly-1,1'-ruthenocenylene.
Of the many reaction paths toward polyferrocenylene investigated

in recent years, including various coupling reactions involving
organo-magnesium, -zinc, and -copper intermediates, the direct,
low-temperature polycoupling reaction of 1,1'-dilithioferrocene
with 1,1'-diiodoferrocene has proved its superiority over other
approaches, giving soluble, linear poly-1,1'-ferrocenylene in yields
of *ca* 85% and with molecular masses of up to 4000 (10 000 upon
subfractionation). The polymer gives IR and NMR spectra indicative
of the absence of molecular-mass-dependent inter-unit vibrational
and ring-current interactions, and the same insensitivity to molecular mass changes (at $\bar{x}_n$ > 4) of the electronic charge transfer and
d-d transitions is apparent from the electronic absorption spectra,
suggesting insignificant electronic charge delocalization along the
polymer chain. The Mössbauer spectra, yielding virtually identical
chemical shift and quadrupole splitting parameters irrespective of
the degree of polymerization, suggest the absence of inter-unit d_o-
and $d_{\pm 2}$-electron delocalization, thus confirming the electronic
absorption data. In accord with this spectroscopic behavior, the
polymer, within the molecular mass range investigated, does not
conduct an electrical current and in the low-molecular-mass region
is diamagnetic, although feeble paramagnetism, presumably due to
impurities, has been determined on higher-molecular samples.
Partial oxidation of oligo- and poly-1,1'-ferrocenylenes, difficult
to control by chemical means and probably very much easier to
accomplish electrochemically, gives rise to the formation of polysalts containing cationic sites in the chain; these polysalts
accordingly exhibit weak-interaction-type, mixed-valence behavior,
showing the typical intervalence transfer absorption in the near-
infrared. However, significant electronic or magnetic interaction
along the backbone of these poly(1,1'-ferrocenylene-*co*-1,1'-ferri-
cenylene) structures is not indicated by the spectral data at hand,
and electro-conduction properties are not expected.

Very much less than in the ferrocene polymer case is known
about the second polymetallocenylene type, *viz*. poly-1,1'-rutheno-
cenylene. Only oligomeric compounds or fractions have been obtained

to this date, the most successful synthetic approach involving the combined oxidative and thermal coupling of TMEDA-chelated 1,1'-dilithioruthenocene. Preliminary spectroscopic findings demonstrate the same lack of inter-unit vibrational and electronic interaction as observed in the ferrocene polymer series, and there is no reason to expect changes in spectroscopic and electronic properties as higher-molecular-mass ranges will be made available. Novel electro-physical behavior might, however, be expected in partially oxidized polyruthenocenylenes, where, depending on the oxidation method employed, monocationic or dicationic sites may appear in the chain.

The areas of potential use for both polyferrocenylene and polyruthenocenylene are quite diversified and range from radiation and heat protection, stabilization or sensitization functions, polymerization initiation and combustion catalysis to flame retar-dance and smoke suppression as well as a multitude of medical or biological applications. Whereas costs are not a major considera-tion with the polymers derived from the inexpensive iron complex, the price of ruthenocene is too high to allow for the economic utilization of the polyruthenocenylenes in many a technological field, and, hence, applications will be restricted to specialty areas where particular benefits can be achieved with polymer use in very small quantities or at low concentrations, such as in cataly-tic or certain medical applications. No major use can be foreseen for the analogous polyosmocenylenes, should these ever be synthesized, as their cost situation is even more precarious than that of the polyruthenocenylenes.

It need not be emphasized that the molecular mass ranges presently reached for polyferrocenylene, not to mention those for polyruthenocenylene, leave much to be desired, since many applica-tions require mechanical properties attainable only at a minimum molecular mass of 25000 - 30000, and even the most painstaking subfractionation of currently available bulk polymer will not pro-vide such material economically. On the other hand, it appears doubtful whether the presently available synthetic methods of

aryl-aryl coupling leave room for much improvement in the ferrocene
step-growth polymerization with respect to the overall degree of
polymerization. Although further experimental refinements can be
expected, the major drawback inherent in all the types of coupling
reaction so far utilized, *viz.* defunctionalization with concomitant
depression of p and $\bar{x}_n$, clearly sets a limit to what can be achieved
in this respect. As long as organolithium, Grignard, or organocopper
compounds are used as monomers, either directly or prepared *in situ*,
it is obvious that the problem of loss of difunctionality owing to
existing equilibria with mono- and unsubstituted (and, occasionally,
higher substituted) parent compounds must be accepted even if the
most stringent experimental conditions to ensure absence of impurities are observed. Therefore, in addition to investigations of
catalytically useful additives in the heretofore practiced coupling
reactions, in an effort to enhance the polycondensation rates relative to side reactions, it will be necessary to search for entirely
different monomers or monomer pairs, each compound having two non-
equilibrating functional substituents capable of fast reaction in
the propagation sequence and yet sufficiently unreactive with solvents, catalysts or other components to retain its difunctionality
throughout the polymerization reaction. Dihalogenated metallocenes
might be suitable as monomers of this type, to be subjected, for
example, to solution polycondensation by the Ullmann coupling
technique with the aid of a recently developed copper reagent
possessing improved activity [86], although even then some loss of
functional groups through reductive dehalogenation may not be
avoidable.

A different approach, potentially useful not only for polyferrocenylene preparation but also for the synthesis of other
polymetallocenylenes, could utilize the preformed fulvalene dianion
as the ligand-forming monomer, propagation proceeding through
double-complexation with metal cation, M^{2+}. While the method of
fulvalene dianion generation from cyclopentadienide [56,87] is
unsuitable for clean propagation because of the prominent occurrence

of side reactions, such as ligand dimerization and diene polymerization, the generation of the dianion through electrochemical reduction
of [0.0]ferrocenophane by El Murr's procedure [88] might be
feasible, provided that the solubility problem with this compound
can be overcome. At the present state of knowledge, in fact, this
procedure can be judged to provide the only conceivable synthetic
access to such other metallocene polymers as polycobaltocenylene
or polynickelocenylene, which cannot be prepared by the coupling
procedures used for polyferrocenylene synthesis as the parent
metallocenes have so far neither been successfully lithiated nor
been found themselves to suffer a clean substitution by organolithium reagents. Polycobaltocenylene is of potential interest in
the partially or fully oxidized forms. Both types of polymer should
possess properties required in certain polyelectrolyte and membrane
applications. In addition, the partially oxidized material should
provide mixed-valence features, and the fully oxidized product, to
judge from the considerable stabilization achieved in the cobaltocene system on loss of the unpaired, high-energy valence electron,
can be expected to possess extraordinarily high thermo-oxidative
(and, presumably, radiative) stability, being far superior in this
respect to any other metallocene polymer type that can be visualized.

In summary, the foregoing sections show that, while some
promising entries into polymetallocenylene chemistry have been made,
only poly-1,1'-ferrocenylene and its partial oxidation products
have been subjected to major exploration. Some rather formidable
experimental hurdles remain to be overcome if both polyferrocenylene
and polyruthenocenylene are to be obtained in a mechanically useful
molecular mass range of 25000 - 30000, and new synthetic strategies
will have to be designed for the preparation of polymers derived
from other transition metal complexes. There can be no doubt,
however, that the synthetic aspects and application goals both
provide challenge enough to render further chemical and physical
research in the field a highly rewarding task.

ACKNOWLEDGMENT

Some of the work discussed in this chapter was performed in the author's laboratory under the sponsorship of the Atomic Energy Board and with the loyal collaboration of Ladislav Bednarik. A major portion of the polymer oxidation experiments was executed by the author while on Sabbatical Leave at the Institute of Organic Chemistry, University of Mainz, and it is his pleasure to acknowledge the support provided by *Sonderforschungsbereich 41, Chemie und Physik der Makromolekūle,* the able assistantship rendered by Mrs Ursula Grimm, and, above all, the warm hospitality offered by Professor Rolf C. Schulz.

REFERENCES

[1]   E.W. Neuse, Encycl. Polym. Sci. Tech., 8, 667 (1968).
      E.W. Neuse, in Advances in Macromolecular Chemistry,
      W.M. Pasika, Ed., Vol. 1, Academic Press, New York, 1968,
      Chapt. 1.   E.W. Neuse and H. Rosenberg, Metallocene Polymers,
      Marcel Dekker, New York, 1970.

[2]   M. Rosenblum, Chemistry of the Iron Group Metallocenes,
      Part I, Wiley, New York, 1965.

[3]   V.V. Korshak, S.L. Sosin, and V.P. Alekseeva, Dokl. Akad.
      Nauk SSSR, 132, 360 (1960); Vysokomol. Soedin., 3, 1332 (1961).
      A.N. Nesmeyanov, V.V. Korshak, V.V. Voevodskii, N.S. Kochetkova,
      S.L. Sosin, R.B. Materikova, T.N. Bolotnikova, V.M. Chibrikin,
      and N.M. Bazhkin, Dokl. Akad. Nauk SSSR, 137, 1370 (1961).

[4]   H. Rosenberg and E.W. Neuse, J. Organometal. Chem., 6, 76
      (1966).   E.W. Neuse, J. Organometal. Chem., 56, 323 (1973).

[5]   E.W. Neuse, J. Organometal. Chem., 40, 387 (1972).

[6]   (a) A.N.Nesmeyanov, V.N. Drozd, V.A. Sazonova, V.I.Romanenko,
      A.K. Prokofiev, and L.A. Nikonova, Izv. Akad. Nauk SSSR,
      Otd. Khim. Nauk, 667 (1963).   (b) M.D. Rausch, P.V. Roling,
      and A. Siegel, Chem. Commun., 502 (1970);   P.V. Roling and
      M.D. Rausch, J. Org. Chem., 37, 729 (1972).

[7]   T. Izumi and A. Kasahara, Bull. Chem. Soc. Japan, 48, 1955
        (1975).

[8]   E.W. Neuse and R.K. Crossland, J. Organometal. Chem.,
        7, 344 (1967).

[9]   M.D. Rausch, J. Org. Chem., 28, 3337 (1963).

[10]   H. Watanabe, I. Motoyama, and K. Hata, Bull. Chem. Soc.
        Japan, 39, 790 (1966). I.J. Spilners and J.P. Pellegrini,
        Jr., J. Org. Chem., 30, 3800 (1965).

[11]   M.D. Rausch and D.J. Ciappenelli, J. Organometal. Chem.,
        10, 127 (1967); M.D. Rausch, G.A. Moser, and C.F. Meade,
        J. Organometal. Chem., 51, 1 (1973).

[12]   L. Bednarik, R.O. Gohdes, and E.W. Neuse, Transition Metal
        Chem., 2, 212 (1977).

[13]   E.W. Neuse and M. Loonat, unpublished results.

[14]   R.C. Larock and J.C. Bernhardt, J. Org. Chem., 42, 1680 (1977).

[15]   R.A. Kretchmer and R. Glowinski, J. Org. Chem., 41, 2661
        (1976).

[16]   A. McKillop, L.F. Elsom, and E.C. Taylor, J. Am. Chem. Soc.,
        90, 2423 (1968); Tetrahedron, 26, 4041 (1970).

[17]   M.F. Semmelhack, P.M. Helmquist, and L.D. Jones, J. Am. Chem.
        Soc., 93, 5908 (1971).

[18]   M.F. Semmelhack and L.S. Ryono, J. Am. Chem. Soc., 97, 3873
        (1975).

[19]   K. Tamao, K. Sumitani, and M. Kumada, J. Am. Chem. Soc.,
        94, 4374 (1972).

[20]   M. Kumada, Pure Appl. Chem., 52, 669 (1980).

[21]   M. Zembayashi, K. Tamao, J. Yoshida, and M. Kumada,
        Tetrahedron Lett., 4089 (1977).

[22]   A.S. Kende, L.S. Liebeskind, and D.M. Braitsch, Tetrahedron
        Lett., 3375 (1975).

[23]   G. Cahiez, D. Bernard, and J.F. Normant, J. Organometal.
        Chem., 113, 99 (1976).

[24] R.J.P. Corriu and J.P. Masse, Chem. Commun., 144 (1972).
      R.L. Clough, P. Mison, and J.D. Roberts, J. Org. Chem.,
      41, 2252 (1976).

[25] E. Negishi, A.O. King, and N. Okukado, J. Org. Chem.,
      42, 1821 (1977).

[26] H. Shechter and J.F. Helling, J. Org. Chem., 26, 1034 (1961).

[27] E.W. Neuse and L. Bednarik, Transition Metal Chem.,
      4, 87 (1979).

[28] E.W. Neuse and L. Bednarik, Transition Metal Chem.,
      4, 104 (1979).

[29] Reviews: G.H. Posner, Org. React., 19, 1 (1972); 22, 253
      (1975). H.O. House, Proc. R.A. Welch Found. Conf. Chem. Res.,
      XVII, 1974, Chpt. 4. A.E. Jukes, Advan. Organometal. Chem.,
      12, 215 (1974).

[30] E.W. Neuse and L. Bednarik, Macromolecules, 12, 187 (1979).

[31] M.D. Rausch, Pure Appl. Chem., 30, 523 (1972); P.V. Roling
      and M.D. Rausch, J. Organometal. Chem., 141, 195 (1977).

[32] H. Rosenberg and R.A. Ference, Abstr. papers, 160th Nat.
      Meeting, Am. Chem. Soc., Chicago, Ill., September 1970,
      INOR 158.

[33] L. Bednarik and E.W. Neuse, J. Org. Chem., 45, 2032 (1980).

[34] E.W. Neuse and L. Bednarik, unpublished results. See also
      L. Bednarik, Ph.D. Thesis, Johannesburg, 1979.

[35] E.W. Neuse and L. Bednarik, Org. Coat. Plast. Chem., 41,
      158 (1979).

[36] A. Haaland, Accts. Chem. Res., 11, 415 (1979), and refs.
      cited therein.

[37] A.C. MacDonald and J. Trotter, Acta Cryst., 17, 872 (1964).

[38] Z.L. Kaluski and Yu.T. Struchkov, Zhurn. Strukt. Khim.,
      6, 316 (1965).

[39] E.W. Neuse and R.K. Crossland, J. Organometal. Chem.,
      43, 385 (1972).

[40] A.T. Armstrong, F. Smith, E. Elder, and S.P. McGlynn,
      J. Chem. Phys., 46, 4321 (1967). Y.S. Sohn, D.N. Hendrickson,
      and H.B. Gray, J. Am. Chem. Soc., 93, 3603 (1971).

[41] E.W. Neuse and N.R. Byrd, unpublished results.

[42] M. Rosenblum, Ph.D. Thesis, Harvard, 1953.

[43] E.W. Neuse and D.S. Trifan, J. Am. Chem. Soc., 85, 1952 (1963).

[44] For a representative discussion, see M.L. Good, J. Buttone, and
      D. Foyt, Ann. N.Y. Acad. Sci., 239, 193 (1974).

[45] J. Ensling and E.W. Neuse, unpublished results.

[46] G.K. Wertheim and R.H. Herber, J. Chem. Phys., 38, 2106 (1963).

[47] M. Hillman and A. Nagy, J. Organometal. Chem., 184, 433 (1980).

[48] W. Flavell, Tech. Report DA-91-591-EUG 1610, on contract to
      European Research Office, U.S. Dept. of the Army, AD 268357L
      (1961).

[49] G.C. Allen and N.S. Hush, Progr. Inorg. Chem., 8, 357 (1967).
      N.S. Hush, ibid., 8, 391 (1967); Electrochim. Acta, 13, 1005
      (1968); Chem. Phys., 10, 361 (1975).

[50] M.B. Robin and P. Day, Advan. Inorg. Chem. Radiochem., 10,
      247 (1967).

[51] C. Creutz and H. Taube, J. Am. Chem. Soc., 91, 3988 (1969).
      H. Taube, Pure Appl. Chem., 44, 1 (1975).

[52] D.O. Cowan, C. LeVanda, J. Park, and F. Kaufman, Accts.
      Chem. Res., 6, 1 (1973).

[53] T.J. Meyer, Accts. Chem. Res., 11, 94 (1978).

[54] G.M. Brown, T.J. Meyer, D.O. Cowan, C. LeVanda, F. Kaufman,
      P.V. Roling, and M.D. Rausch, Inorg. Chem., 14, 506 (1975).

[55] W.H. Morrison, Jr., and D.N. Hendrickson, Inorg. Chem.,
      14, 2331 (1975).

[56] C. LeVanda, K. Bechgaard, D.O. Cowan, U.T. Mueller-
      Westerhoff, P. Eilbracht, G.A. Candela, and R.L. Collins,
      J. Am. Chem. Soc., 98, 3181 (1976).

[57] (a) F. Kaufman and D.O. Cowan, J. Am. Chem. Soc., 92, 6198
      (1970). (b) D.O. Cowan, R.L. Collins, and F. Kaufman,
      J. Phys. Chem., 75, 2025 (1971). (c) D.O. Cowan,
      G.A. Candela, and F. Kaufman, J. Am.Chem. Soc., 93, 3889,
      (1971).

[58] W.H. Morrison, Jr., and D.N. Hendrickson, J. Chem. Phys.,
      59, 380 (1973).

[59] M.J. Powers and T.J. Meyer, J. Am. Chem. Soc., 100, 4393
      (1978).

[60] A.W. Rudie, A. Davison, and R.B. Frankel, J. Am. Chem. Soc.,
      101, 1629 (1979).

[61] D.O. Cowan, J. Park, C.U. Pittman, Jr., Y. Sasaki,
      T.K. Mukherjee, and N.A. Diamond, J. Am. Chem. Soc., 94,
      5110 (1972). C.U. Pittman, Jr., Y. Sasaki, and
      T.K. Mukherjee, Chem. Letters, 383 (1975).

[62] A.N. Nesmeyanov and N.S. Kochetkova, Usp. Khim., 43, 1513
      (1974).

[63] R.M. van Vliet, U.S. Pat. 3287314, Nov. 22, 1966.

[64] C.J. Pedersen, U.S. Pat. 3038299, June 12, 1962. D.C. Sayles,
      U.S. Pat. 3447981, June 3, 1969. W.F. Avendale, U.S. Pat.
      4023994, May 17, 1977. C.I. Ashmore, C.S. Combs, Jr.,
      and W.D. Stephens, U.S. Pat. 4108696, Aug. 22, 1978.

[65] H. Kato and N. Maekawa, Japan Pat. 74119939, Nov. 15, 1974
      (Chem. Abstr., 83, 29357 (1975)); Japan Pat. 78146750,
      Dec. 20, 1978 (Chem. Abstr., 90, 169635 (1979)).

[66] H.R. Lucas, U.S. Pat. 3655606, Apr. 11, 1972. H. Kato,
      Japan Pat. 74119932, Nov. 15, 1974 (Chem. Abstr., 83, 29108
      (1975)). E.A. Kalennikov, V.S. Yuran, Ya. M. Paushkin, and
      A.E. Pereverzev, Vestsi Akad. Navuk B. SSR. Ser. Khim. Navuk,
      57 (1975) (Chem. Abstr., 83, 80184 (1975)).

[67] E.N. Matveeva, M.Z. Borodulina, M.S. Kurzhenkova,
      E.A. Kalennikov, and V.S. Yuran, U.S.S.R. Pat. 531826,
      Oct. 15, 1976 (Chem. Abstr., 86, 17745 (1977)). S. Suzuka,
      S. Maeda, and N. Endo, Japan Pat. 7829118, March 18, 1978
      (Chem. Abstr., 89, 120887 (1978)). V.A. Nefedov, U.S.S.R.Pat.
      603940, April 25, 1978 (Chem. Abstr., 89, 14852 (1978)).
      T. Shiga and M. Yoshino, Japan Pat. 7617428, Feb. 12, 1976
      (Chem. Abstr., 86, 113726 (1977)).

[68] H. Kato, Japan Pat.74119942, Nov. 15, 1974 (Chem. Abstr., 83,
      29109 (1975)). T. Tajima, Japan Pat. 7856299, May 22, 1978
      (Chem.Abstr., 89, 111523 (1978)). S. Uchida, Japan Pat.
      7872098, June 27, 1978 (Chem.Abstr., 89, 130694 (1978)).
      D.F. Lawson, J. Appl. Polym. Sci., 20, 2183 (1976).

[69] J.J. Kracklauer, Ger. Offen. 2307387, Aug. 23, 1973 (Chem. Abstr., 80, 37817 (1974)). A.F. Klarman, F. Anthony, and J.E. Horling, Int. Conf. Environ. Sensing Assess. [Proc.] 2, 41 (1976) (Chem.Abstr., 86, 77842 (1977)). L. Lecomte, M. Bert, A. Michel, and A. Guyot, J. Macromol. Sci.-Chem., A11, 1467 (1977).

[70] S. Kawawada, N. Takahata, and M. Sorimachi, Japan Pat. 74112950, Oct. 28, 1974 (Chem. Abstr., 82, 99316 (1975)).

[71] M. Miyamura, A. Miura, and O. Tada, Japan Pat. 77117616, Oct. 3, 1977 (Chem. Abstr., 88, 144343 (1978)).

[72] V.N. Lvov, P.A. Rutman, V.G. Safronov, E.B. Sokolova, and G.P. Chalykh, U.S.S.R. Pat. 649737, Feb. 28, 1979 (Chem. Abstr., 91, 60044 (1979)).

[73] E.A. Vogelfanger, T.J. Wallace, U.S. Pat. 3479163, Nov. 18, 1969.

[74] V. Mues and W. Behrenz, Ger. Offen. 2711546, Sep. 21, 1978 (Chem. Abstr., 90, 1705 (1979)).

[75] K. Kaeriyama, Polymer, 12, 422 (1971). M. Imoto, T. Ouchi, and T. Tanaka, J. Polym. Sci. Polym. Letters, 12, 21 (1974).

[76] K. Azuma, H. Kato, and H. Tatemichi, Japan Pat. 75129689, Oct. 14, 1975 (Chem. Abstr., 84, 165718 (1976)).

[77] A.A. Berlin, K.A. Frankenshtein, N.S. Gavryushenko, F.I. Dubovitskii, T. Ya. Arfell, R.V. Kronman, L.A. Konkhina, N.L. Marshavina and G.L. Popova, U.S. Pat. 4076742, Feb. 28, 1978.

[78] N. Oyama, K.B. Yap, and F.C. Anson, J. Electroanal. Chem. Interfacial Electrochem., 100, 233 (1979). M.S. Wrighton, Accts. Chem. Res., 12, 303 (1979), and references cited therein.

[79] M. Cais, S. Dani, Y. Eden, O. Gandolfi, M. Horn, E.E. Isaacs, Y. Josephy, Y. Saar, E. Slovin, and L. Snarsky, Nature, 270, 534 (1977). R.P. Hanzlik, P. Soine, and W.H. Soine, J. Med. Chem., 22, 424 (1979). L.L. Gershbein, Res. Commun. Chem. Pathol. Pharmacol., 27, 139 (1980) (Chem. Abstr., 92, 121876 (1980)). V.J. Fiorina, R.J. Dubois, and S. Brynes, J.Med. Chem., 21, 393 (1978). H. Kief, R.R. Crichton, H. Bähr, K. Engelbart, and R. Lattrell, in Proteins of Iron Metabolism, E.B. Brown, Ed., Grune and Stratton, New York, 1977, p. 107. L. Lindner and J.C. Kapteyn, Nuklearmedizin. Stand und Zukunft. 15. Intern. Jahrestagung der Ges. f. Nuklearmedizin, Groningen, Sept. 1977, F.K. Schattauer Verlag, Stuttgart,1978, p. 819.

[80] E.W. Neuse, J. Organometal. Chem., 6, 92 (1966).

[81] G.L. Hardgrove and D.H. Templeton, Acta Cryst., 12, 28 (1959).
      C.H. Holm and J.A. Ibers, J. Chem. Phys., 30, 885 (1959).

[82] See, for example, D.M. Adams and W.S. Fernando, J. Chem. Soc.
      Dalton Tr., 2507 (1972), and refs. cited therein.

[83] P. Borrell and E. Henderson, J. Chem. Soc. Dalton Tr., 432
      (1975). L.I. Denisovich, N.V. Zakurin, A.A. Bezrukova, and
      S.P. Gubin, J. Organometal. Chem., 81, 207 (1974). Y.S. Sohn,
      A.W. Schlueter, D.N. Hendrickson, and H.B. Gray, Inorg. Chem.,
      13, 301 (1974); and refs. cited in these papers.

[84] M. Wenzel, M. Schneider, J. Bier, P. Benders, and G. Schach-
      schneider, J. Cancer Res. Clin. Oncol., 95, 147 (1979), and
      refs. cited therein.

[85] D. Langheim, M. Wenzel, and E. Nipper, Chem. Ber., 108, 146
      (1975).

[86] R.D. Rieke and L.D. Rhyne, J. Org. Chem., 44, 3445 (1979).

[87] A. Davison and J.C. Smart, J. Organometal. Chem., 49, C43 (1973);
      J.C. Smart and C.J. Curtis, Inorg. Chem., 16, 1788 (1977).
      U.T. Mueller-Westerhoff and P. Eilbracht, J. Am. Chem.Soc.,
      94, 9272 (1972).

[88] N. El Murr, A. Chaloyard, and E. Laviron, Nouv. Chim., 2,
      15 (1978).

Polymers for Controlled Release of Organotin Toxin

R. V. SUBRAMANIAN AND K. N. SOMASEKHARAN

Department of Materials Science and Engineering
Washington State University, Pullman, Washington 99164

ABSTRACT

An extensive investigation of organotin polymers capable of
simultaneously providing long-term fouling resistance and useful
engineering properties is discussed. New synthetic routes have
been developed for organotin epoxy polymers utilizing (i) the
crosslinking reaction of diepoxides with the free carboxyl groups
present on a base polymer partially esterified with tributyltin
oxide (TBTO), (ii) the polymerization of TBT acrylate and TBT
methacrylate with vinyl monomers carrying functional groups
capable of crosslinking, and (iii) simultaneous vinyl polymerization and epoxide crosslinking reactions. Flexible polymers
curing under ambient conditions have also been developed (iv) by
first preparing epoxy-terminated prepolymers by reacting TBT
esters of ω-amino acids with diepoxides, and then crosslinking
with diethylenetriamine. Room-temperature-curable organotin
polymers developed also include (v) urethanes, (vi) aziridines
and (vii) polyesters. Finally, (viii) an ablating polymer is
developed by copolymerizing methyl methacrylate with TBT
methacrylate.

The network structure is varied, and the average separation,
length and type of crosslinks or pendant organotin groups are
altered by appropriate changes in monomers and synthetic routes.
Resultant changes in measured strength, fracture toughness and
dynamic mechanical behavior of the polymer systems have been
correlated with the structural variables employed.

The structure and reactivity of TBT carboxylate group have come
under careful investigation. The bioactive species released from

these polymers has been identified as tributyltin chloride by
spectroscopic and chromatographic techniques. The release rate
of tin has been determined in the laboratory, and the results
fitted to mathematical models corresponding to the bulk abiotic
bond cleavage.

An important aspect of organotin compounds that has drawn
close attention in recent years is their biological effects which
have led to their widespread application in biocidal compositions
[1]. Investigations in this area have clearly shown the potential
of organotin polymers for use as antifouling coatings. In this
context, not only the synthesis and characterization of organotin
polymers, but also the understanding of the mechanism of release
of organotin toxin from the polymers assume importance. Current
research leading to the progressive understanding of structure-
property relationships, including the biocidal properties of
organotin polymers, is therefore stressed in this review.

Antifouling Coatings

Fouling, i.e., the growth of marine organisms on submerged
surfaces, is one of the worst problems in marine environments;
and prevention of fouling is next in importance only to corrosion
prevention in the protection of ships' bottoms [2]. Fouling of
the ship hull leads to a large increase in frictional resistance
which, in turn, results in a wastage of fuel. Fouling causes an
unpredictable distortion in signals from immersed acoustic and
electronic navigational devices. In many cases, the fouling
organisms also destroy the anticorrosion coating on marine
equipment, leading to serious corrosion damage to such surfaces.

While many techniques have been tried in the prevention of
fouling such as ultraviolet irradiation [3], ultrasonic vibration
[4] or covering by non-sticky silicone rubber coatings [5], up to
now, the use of antifouling paints has been the only economically
and technically feasible protection against fouling [2,3]. Antifouling paints, whose mechanism of action is the leaching of
toxicants, are attractive because of easy application and main-

tenance. Their main drawback is the mechanism of action itself; it is based on the release of bioactive materials. Whereas the antifouling action is needed mainly when the ship is in port, the important fouling organisms being denizens of the sea shore, a very high percentage of bioactive material is released when the ship is moving, due to the turbulent conditions around it.

Antifouling Toxicants

Until now, the most important antifouling poison has been cuprous oxide [6-8]. The critical leaching rate, i.e., the rate required to keep off all fouling, is 10 $\mu g/cm^2/day$ for copper [9]. The poison loss rate by the $\sqrt{t}$ relation is the main problem of antifouling coatings. If the amount of poison needed were compared with that used, it would be no more than 20%; thus the efficiency of antifouling paints is very low. On the other hand, the loss due to fouling is so high that it is worthwhile wasting about 90% of the poison if the remainder serves to prevent fouling [2]. Researchers have always been looking for more active poisons [6-8,10] which are effective against the whole spectrum of foulants, because copper is somewhat deficient in its action against algal fouling.

Organoarsenic compounds are very effective; but they are a problem to apply because they irritate the eyes and nasal membranes. Tributyllead and triphenyllead compounds have similar toxicity against fouling organisms as organotin compounds, but their toxic hazards on application are greater. Tributyltin and triphenyltin radicals are very effective against all types of fouling and have been widely used. They are toxic to man, but they are not a hazard to him if some simple precautions are taken [11]. Unlike organoarsenic, organolead and organomercury compounds, organotin compounds easily degrade to nontoxic compounds in the environment. Among the other advantages claimed for the use of organotin toxins in antifouling coatings is the absence of corrosion problem, which always occurs with conventional paints containing cuprous oxide [6].

Extensive toxicological evaluations of organotin compounds
have led to the conclusion that the trialkyltin derivatives are
the most effective toxins against marine organisms [12]. A
comparison of the toxicity of a homologous series of trialkyltin
compounds has indicated that the tributyltin derivatives represent
the optimum balance between high toxicity against marine organisms
and tolerance toward mammals [6]. Field tests have given the best
fouling resistance when the tributyltin group is easily hydrolyzable [13]. Detailed investigations by Aldridge [14] and Selwyn
[15], of the influence of organotin compounds on mitochondrial
functions, support these observations. On the basis of these
considerations, tributyltin carboxylate has been the structural
feature of choice in our syntheses of polymers for potential use
in antifouling compositions.

Improving Antifouling Coatings

The practical possibilities to extend the effective lifetime
of antifouling coatings are to increase the film thickness, use
more effective bioactive materials, or to control the leaching
rate of bioactive substances.

Increasing film thickness is simple, but this does not change
the $\sqrt{t}$ dependence of leaching rate. As the rate of release of
poison has to be much higher in the initial stages, the accompanying perils like wastage and environmental hazard are aggravated.

The use of stronger poisons, an approach which involves lower
leaching rates, is a good approach to the solution of the short-
life time problem. But they will also be more toxic to people, and
this is the main limitation of this technique.

Several methods have been tried for controlling the release
rate. One of them is the use of a special top coat of water-
insoluble hydrophilic acrylic resin [16]. A more recent approach
designed to solve the excessive leaching problem is the development of organometallic polymers [17], in which biocidal organo-

metallic groups are chemically attached to their backbones. Among
these, organotin polymers are likely to have more practical application than the other organometallic polymers. It has been shown
that the organometallic polymer is optimally effective against
fouling organisms when the organic radical of the organometallic
groups is either propyl or butyl [6]. Also, studies of the biocidal action of organometallic compounds have shown that no one of
such compounds has effective action against all the sliming
bacteria [17].

   Therefore, by chemically attaching the organometallic group
to a polymer backbone, two goals are aimed. First, and most
important, it is possible to reduce the leaching rate in order to
achieve longer antifouling protection as a result of the chemical
bond between the bioactive group and the polymer. Second, by
incorporating two or more organometallic groups into a resin, a
broadening of its biocidal action is achieved. Due to the good
film characteristics generally exhibited by acrylics, the first
polymers prepared were poly(tributyltin acrylate) and poly(tributyltin methacrylate [6]. However, the film properties, mechanical properties (impact resistance, hardness, adhesion, flexibility),
weatherability, and resistance to heat and thermal shock of the
organotin acrylate homopolymers are not good enough for application in durable coatings.

Organotin Epoxide Polymers

   It is in this context that we undertook the synthesis and
characterization of thermoset polymers as a novel approach to
improve the film-forming properties and biocidal activity. The
synthetic scheme adopted for the preparation of the polymers
involves the partial esterification, with tributyltin oxide (TBTO),
of linear base polymers containing carboxylic acid or anhydride
groups as the first step. The free carboxylic acid or anhydride
groups of the prepolymers are then cured with diepoxides [18].

The properties of the network structure obtained thus have been varied over a wide range by changing the chemical structure of the base polymer as well as that of the crosslinking epoxy monomer. The tin content of the polymer is altered by controlling the degree of esterification of the base polymer, which also varies the crosslink density obtained in the second step. Greater esterification with TBTO leads to fewer available carboxyl groups for crosslinking and, consequently, to larger separation between points of crosslinking on the prepolymer. Promotion of homopoly-merization at high epoxy-to-anhydride ratios will have the general effect of extending the lengths of epoxy crosslinks; this has been achieved by appropriate choice of catalysts for the curing reaction [18]. When tertiary amines such as dimethylaniline are employed, esterification is the predominant reaction with the epoxide; under these conditions, the maximum amount of epoxide incorporated in the matrix is limited by the stoichiometry of one epoxide to one carboxylic acid unit. On the other hand, when homo-polymerization of the epoxide is facilitated by the presence of stannous octoate or uranyl nitrate, the proportion of epoxide units reacting with each carboxyl function is increased over a higher range, with attendant improvements in matrix strength and toughness.

The base polymers employed include poly(styrene-*co*-maleic anhydride) [SMA-1000A, ARCO], poly(methyl vinyl ether-*co*-maleic anhydride) [AN 139, General Aniline & Film], poly(methyl vinyl ether-*co*-maleic acid) [AT 795, General Aniline & Film], poly(1-hexene-*co*-maleic anhydride) [PA-6, Gulf] and poly(1-decene-*co*-maleic anhydride) [PA-10, Gulf]. The diepoxides used include bis(3,4-epoxy-6-methylcyclohexylmethyl) adipate [ERL 4289, Union Carbide], 3,4-epoxycyclohexylmethyl 3,4-epoxycyclohexanecarboxylate [ERL 4221, Union Carbide], 4-vinylcyclohexane dioxide [ERL 4206, Union Carbide], 2-(3,4-epoxy)cyclohexyl-5,5-spiro(3,4-epoxy)cyclo-hexane-*m*-dioxane [ERL 4234, Union Carbide] and diglycidyl ether of bisphenol-A [EPON 828, Shell]. Uranyl nitrate, stannous octoate,

dimethylaniline, 4-dimethylaminomethylphenol and 2,4,6-tris(di-
methylaminomethyl)phenol are some of the catalysts employed [18,19].

The degree of esterification of base polymers, the structure
of the epoxy monomers, and the type of catalyst used have thus
been varied in these synthetic schemes to effect changes in the
average separation between TBT groups and the length of the epoxy
crosslinks. Resultant changes in measured strength, fracture
toughness and dynamic mechanical behavior of the polymer systems
have been correlated with the structural variables employed.
Toughening by carboxyl-terminated liquid elastomers has also been
studied, and the improvement in fracture toughness correlated with
the average particle size of the dispersed elastomer phase. Of the
various catalysts studied, uranyl nitrate has caused the highest
degree of crosslinking. The formation of domains in the matrix
with independent glass transitions is indicated by variations in
loss moduli; these also reflect the structural effects of the
bulky TBT groups and of the different epoxy monomers [18].

Many variations of this scheme have been investigated,
including one which provided for simultaneous vinyl polymerization
and carboxyl-epoxide reactions [19]. Compositions of several of
these organotin-epoxy coatings, the method of preparing specimens
for antifouling tests in marine environments and procedures for
determining antifouling performance have been reported [19,20].
Fouling resistance upto 40 months has already been observed. The
results show that duration of fouling resistance is not as much
dependent on the concentration of tin as it is on matrix
characteristics [21].

Controlled Release

Performance tests in marine environments reveal that the
resistance to fouling is influenced by matrix characteristics [21].
Parallel tests in the laboratory indicate that only a very small
fraction of the available tin is released during the service life

of the coating [22]. It is thus reasonable to infer that fouling
commences when the rate of release falls below the critical rate,
and not because of the complete depletion of tin in these systems.
Matrix hydrophilicity and permeability seem to be the factors
determining the duration of fouling resistance.

Antifouling performance of these organotin carboxylate polymers indicate that their mode of action corresponds to the bulk
abiotic bond cleavage model [13]. We have carefully considered
all the controlling factors, viz.,

     (1)   diffusion of water (and possibly chloride)
           into the polymer matrix from sea water;

     (2)   hydrolysis of tributyltin carboxylates
           to produce TBTO (or TBTCl);

     (3)   diffusion, from the matrix to the surface,
           of the mobile species (TBTO or TBTCl) produced;

     (4)   phase transfer of the organotin species;

     (5)   its migration across the boundary layer; and

     (6)   possible mechanical loss of the tributyltin
           species from the surface.

NMR studies have revealed that the TBT group undergoes fast
chemical exchange, and hence a hydrolytic equilibrium is rapidly
established between TBT carboxylates and TBTO [23]. Laboratory
determination of the release rate, under laminar flow conditions,
shows that the phase transfer and migration across the boundary
layer are also relatively fast [22]. Thus we have come to the
conclusion that diffusion, from the matrix to the surface, of the
mobile tributyltin species produced, is the factor controlling
the rate of release in epoxy systems.

As the mobile species produced diffuses out, the hydrolysis
is expected to proceed at a concentration-dependent rate. The
model developed by Godbee and Joy for predicting the leachability
of radionuclides from cementitious grouts [24] closely represents
this situation. Based on their equations, the rate of release of
tin (in $g/cm^2/sec$) from the surface should be:

$$dq/dt \; = \; C_m D^{0 \cdot 5} K^{0 \cdot 5} [\mathrm{erf}(K^{0 \cdot 5} t^{0 \cdot 5}) \; + \; \frac{\exp(-Kt)}{(\pi Kt)^{0 \cdot 5}}] \qquad (1)$$

where C_m is the concentration of the mobile species in g/cm^3,
D is the effective diffusivity in the matrix in cm^2/sec,
K is the concentration-dependent hydrolysis rate in sec^{-1}, and
t is the time in sec.

The most salient feature of this model is that when Kt
becomes large, $\mathrm{erf}(Kt)^{0 \cdot 5}$ approaches unity, and the rate becomes
independent of time. This zero order rate is the coveted characteristic of controlled release systems. However, we have not been
able to realize this ideal behavior in the epoxy systems. This may
partly be due to the tight matrix in these epoxy systems.

The epoxy compositions discussed so far are highly crosslinked [18] and have glass transition temperatures around 140°C. Their
free volumes and segmental mobilities are very low. It is known
that these factors decrease the diffusivity (D) in the matrix [25].
Further, the magnitude of decrease is greater the larger the
diffusing molecule [25]; TBTO is a relatively large molecule.

Equation (1) predicts greater rate of release when C_m is
higher; the concentration of the mobile species is expected to be
higher in hydrophilic matrices.

Crosslinking of Organotin Copolymers

Since it has become apparent that matrix hydrophilicity and
permeability are the factors determining controlled release from
organotin carboxylate polymers, a new scheme for obtaining thermoset plastics that permits closer control of crosslinking sites is
pursued. An optimum balance between film properties and biocidal
action is expected by the incorporation into the polymer, by means
of copolymerization, of epoxy (or hydroxyl) groups and subsequent
crosslinking [26].

Organotin monomers are obtained by esterification of acrylic
acid and methacrylic acid with TBTO. These monomers are copoly-

merized with glycidyl acrylate and glycidyl methacrylate (which
contain epoxy groups) and with N-methylolacrylamide (that carries
a hydroxyl group). The monomer reactivity ratios of the six pairs
are determined, and the experimental values used to derive information on the distribution of the units of a particular monomer
in the copolymer chain. Thus, a suitable copolymer containing
either the blocks of organotin monomer units, or randomly distributed units of either monomer can be selected as desired. Varying
crosslink density and rigidity are achieved with crosslinking
agents like aliphatic and aromatic amines and also with catalysts
like uranyl nitrate [26]. The copolymers from the organotin monomers and N-methylolacrylamide may be crosslinked with diisocya-
nates to produce polyurethane coatings.

Information on biotoxicity is obtained by studying the inhibition of marine and soil bacteria and of a soil fungus [26]. The
nature and degree of crosslinking have a significant effect on the
size of the inhibition zone; the tighter the crosslinked network,
the smaller is the inhibition zone.

Flexible Epoxy Plastics

The extremely low diffusivity of tributyltin species in the
polymer matrix is partly due to the tightness of these epoxy matrices [22]. Past systems have only considered TBT carboxylate group
linked directly to the backbone of the polymer. Therefore, by
varying the length of the sidechain holding the TBT moiety, greater mobility of the TBT group is to be expected. To accomplish
this, diglycidyl ether of bisphenol-A (DGEBA) is first modified
by reacting with TBT esters of ω-amino acids; the resulting prepolymers are then cured with diethylenetriamine (DETA) at room
temperature. A lower T_g is actually observed as this chain is
extended, which means lower activation energy for this motion.
The decrease in matrix constraints allowing this motion may also
result in decreased resistance to the diffusion of TBTO. The synthesis and characterization of these systems have been the next

steps in pursuing the idealized controlled release from epoxy
systems [27,28].

TBT esters of glycine, 4-aminobutanoic acid, 6-aminohexanoic
acid and 11-aminoundecanoic acid are first synthesized. The prepolymer is then prepared by dissolving EPON 828 (Shell DGEBA) in
a benzene solution of the TBT ester, followed by removal of the
solvent. Epoxide assay shows the prepolymers to have epoxide equivalents ranging from 344 to 370 (Table 1). The ratio of TBT ester
to DGEBA is adjusted to give a consistent number of pendant groups
of TBT moiety per gram of the cured resin (Table 1). The prepolymers are mixed with DETA and cured at room temperature [27,28].

The loading of tin in the final cured polymer is nominally 5%
(w/w). This value varies since it is the number of pendant organotin chains which is kept constant (Table 1). The small variation
in the density of pendants is due to the slight homopolymerization
of the epoxide during drying, which changes the amount of DETA
required for curing; the variation is not great [27,28].

It is expected that the addition of the TBT esters would have
a plasticizing effect; previous work [18] with TBT esters gives
precedence for this. In fact, there is a large variation in the
plasticizing effect as the chain is lengthened (Table 2). The
glass transition varies from 92°C with TBT glycinate to 68°C with

TABLE 1.   Epoxy Equivalents and Pendant-chain Concentrations.

Polymer	Epoxy Equiv. of Prepolymer, g/equiv.	Moles of Pendant Chains per Gram of Polymer
TBT Glycinate/DGEBA/DETA	365	5.1×10^{-4}
TBT 4-Aminobutanoate/DGEBA/DETA	344	4.9×10^{-4}
TBT 6-Aminohexanoate/DGEBA/DETA	345	5.0×10^{-4}
TBT 11-Aminoundecanoate/DGEBA/DETA	370	5.0×10^{-4}

TABLE 2.   Tensile, Flexural and Glass Transition Values for the
Modified Epoxy Polymers.

Polymer*	Tensile, Kg/cm²		Flexural, Kg/cm²		T_g, °C
	E	S	E	S	
GLY/DGEBA/DETA	33,368	288	25,170	879	(50) 92
BUT/DGEBA/DETA	39,302	626	33,663	1090	(55) 90
HEX/DGEBA/DETA	30,000	619	27,497	893	(54) 80
UND/DGEBA/DETA	-- ---	---	-- ---	---	(55) 68

*GLY = TBT glycinate; BUT = TBT 4-aminobutanoate;
HEX = TBT 6-aminohexanoate; UND = TBT 11-aminoundécanoate.

TBT 11-aminoundecanoate. The reduction in strength and modulus is
also apparent, except in the case of glycinate which has a value
lower than expected [27,28],

Room-Temperature-Curable Organotin Polymers

There exists a need for more methods of preparing antifouling
coatings that can be cured at room temperature than the one described above. Our results on the incorporation of organotin groups
into room-temperature-cured urethanes, aziridines and polyesters
have been reported [21,29]. Urethanes and polyesters are also
attractive from the point of view of their greater hydrophilicity
and segmental mobility. There is evidence in the literature to
believe that vinyl and urethane polymers would release organotin
species at a greater rate [10].

Urethanes. TBT ester of tartaric acid is first synthesized
from TBTO and tartaric acid. TBT tartrate (a dihydroxy monomer) is
then reacted with excess tolylene-2,4-diisocyanate (TDI) to
produce the NCO-terminated prepolymers. The prepolymers are cured

by crosslinking with castor oil (a trihydroxy compound). The
approach guaranteed the completion of the reaction between TBT
tartrate and TDI, ensuring the complete incorporation of the
organotin moiety into the crosslinked structure. The cured polymers are flexible; they are characterized [29].

A noteworthy feature of this scheme is the fast reaction,
even at room temperature, between isocyanate and hydroxyls in the
presence of TBT carboxylates. Trialkyltin groups can, in fact, be
expected to catalyze the urethane reaction; the isocyanate group
can coordinate with tin, getting more polarized and exposing the
isocyanate carbon for nucleophilic attack by the hydroxyl oxygen.

Aziridines. Poly(styrene-comaleic acid) is prepared from
commercially available poly(styrene-co-maleic anhydride), and then
partially esterified by reacting with TBTO. The free acid groups
of these partial esters are found to react with a polyfunctional
aziridine [XAMA-2, Cordova], curing to a nontacky solid at room
temperature; the cured plastics are characterized. Because of the
low molecular weight of the base polymer used [SMA-1000A, ARCO],
a significant fraction of the prepolymer molecules does not become
incorporated in the network structure when the degree of esterification is high, and the sol fraction becomes consequently large.
However, heavy loading of tin in the crosslinked polymer is
possible in the aziridine-cured systems; with base polymers of
higher molecular weight and different compositions, it should be
possible to extend the range of modifications of the network
structure even further.

Polyesters. Organotin vinyl monomers, such as TBT acrylate
and TBT methacrylate, are mixed with unsaturated polyesters [WEP-
661 of Ashland, P-43 of Rohm & Haas] and cured with a free-radical
initiator. Benzoyl peroxide or methyl ethyl ketone peroxide is
capable of curing the mixture, with the aid of dimethylaniline and
/or cobalt naphthenate. However, the organotin group is found to
interfere with the activity of cobalt naphthenate. Hence, if
cobalt naphthenate is to be used as an accelerator, it should be

added to the reaction mixture toward the end. Gel time can be
controlled, by controlling the addition of the accelerators. The
mixtures cure at room temperature to hard, nontacky solids [29].

Ablating Organotin Polymers [30]

As pointed out earlier, the coveted characteristic of controlled release systems is the zero order delivery of the active
agent. It is acknowledged that reservoir devices, especially when
membrane encapsulated, are capable of steady-state release. Erodi-
ble devices of proper geometry can also approach a constant rate
of delivery. Matrix devices, however, are generally expected to
show $\sqrt{t}$ relation in the release profile. We have shown [31] that
a time-independent rate of release of organotin is possible from
polymer monoliths in which trialkyltin carboxylate groups are
chemically attached to the polymer network; this prediction has
yet to be realized.

As mentioned earlier, NMR studies prove that the TBT group
undergoes fast chemical exchange [23,22]. As a consequence, even
the interfacial reaction between TBT carboxylates and sodium
chloride is very fast [23,22]. Thus, if poly(methyl methacrylate-
co-TBT methacrylate) is used as a coating in marine environments,
TBT chloride would be readily released from the surface for antifouling action. As the hydrolysis proceeds, the hydrolyzed groups
become hydrophilic and the polymer molecules at the surface would
erode. Fresh surface would be exposed, and antifouling performance
would not be controlled by diffusion process.

It is known that a rate of release of 0.5 µg $Sn/cm^2/day$ is
sufficient protection against barnacles [32]. 5 µg $Sn/cm^2/day$
would be sufficient protection against any form of fouling, in any
environment. A release rate of 5 µg/cm^2/day can be expected if
the rate of ablation is about 3 x 10^{-10} cm/sec and if the copolymer containing 20% tin is used. If the coating thickness is 1 mm,
it may be expected to provide protection for about 11 years.

A detailed study of the copolymerization of methyl methacrylate and TBT methacrylate has revealed that they form an ideal
copolymer system [30,33]. In contrast to the poor film properties
of poly(TBT methacrylate) and poly(TBT acrylate), poly(methyl
methacrylate-*co*-TBT methacrylate) is found to have excellent film
characteristics, even when the tin loading is as high as 25%. The
films are clear, with non-sticky surface; they have good adhesion
to metals [30].

It has to be pointed out here that special precautions are
required in the determination of the monomer reactivity ratios,
because of the chemical exchange existing in TBT carboxylates [22].
Thus they are readily hydrolyzable, and unless anhydrous solvents
are used in the polymerization of trialkyltin carboxylate monomers,
erroneous values of r_1 and r_2 might result.

Organotin Polymers in Wood Preservation

Since some adverse effects of conventional preservatives like
pentachlorophenol, creosote, copper, chromium and arsenic have
given rise to misgivings about their continued use in the long-
term protection of wood against biodegradation, we have undertaken
the *in situ* copolymerization of organotin monomers in wood. Trialkyltin has a broad spectrum of activity; the impregnation technique minimizes the environmental hazards and also improves other
properties of wood at the same time [34].

Vinyl monomers like maleic anhydride or glycidyl methacrylate
are copolymerized *in situ* with TBT methacrylate in grand fir wood.
Wood samples are first impregnated with a solution of the monomers
and catalysts, and then heated to initiate the reaction. In addition to the vinyl polymerization, grafting of the polymer to wood
also proceeds by the reaction between the hydroxyls of wood and
the functional groups (epoxide or anhydride) in the polymer [34].

Specimens containing varying amounts of polymers are prepared
and tested. The macrodistribution of the polymer in the treated

wood is determined by scanning electron microscopy. The micro-
distribution is determined by microprobe analysis for tin atoms;
a significant amount of the polymer is located in the cell wall.

   In the longitudinal as well as in the transverse direction,
the ultimate flexural strength, flexural modulus of elasticity and
impact strength of treated wood increase significantly, compared
to those of untreated wood. The swelling and water absorption of
treated wood are substantially less than those for untreated wood
in all cases. The resistance to degradation is established by
exposing specimens of treated wood to brown-rot, white-rot and
soft-rot fungi, as well as to a marine bacterium [34].

Summary and Conclusions

   An extensive investigation of organotin polymers capable of
simultaneously providing long-term fouling resistance and useful
engineering properties is discussed. New synthetic routes have
been developed for organotin epoxy polymers utilizing (i) the
crosslinking reaction of diepoxides with the free carboxyl groups
present on a base polymer partially esterified with TBTO, (ii) the
polymerization of TBT acrylate and TBT methacrylate with vinyl
monomers carrying functional groups capable of crosslinking and
(iii) simultaneous vinyl polymerization and epoxy crosslinking
reactions. Flexible polymers curing under ambient conditions have
also been developed (iv) by first preparing epoxy-terminated prepolymers by reacting TBT esters of ω-amino acids with diepoxides,
and then crosslinking with amine curing agents. Room-temperature-
curable organotin polymers developed also include (v) urethanes,
(vi) aziridines and (vii) polyesters. Finally, (viii) an ablating
polymer is developed by copolymerizing methyl methacrylate with
TBT methacrylate.

   The network structure is varied, and the average separation,
length and type of crosslinks or pendant organotin groups are
altered by appropriate changes in monomers and synthetic routes.
Resultant changes in measured strength, fracture toughness and

dynamic mechanical behavior of the polymer systems have been correlated with the structural variables employed. Many of the compositions show prolonged fouling resistance in marine environments;
some have excellent composite properties in fiberglass laminate
composites [35,28].

The structure and reactivity of TBT carboxylate groups have
come under careful investigation. The bioactive species released
from these polymers has been identified as tributyltin chloride by
spectroscopic and chromatographic techniques [23]; Aldridge [14]
and Selwyn [15] have already shown that TBT chloride deranges the
mitochondrial functions. The release rate of tin has been determined in the laboratory, and the results fitted to mathematical
models corresponding to bulk abiotic bond cleavage [22]. The
attainment of zero order release of organotin toxin from polymers
can be expected to be realized with further detailed studies of
the modification of polymer properties and of the mechanism of
transport of the organotin species in the polymer matrix.

In conclusion, reference must be made to various other types
of organotin polymers reported in recent years [36] whose bioche-
mical activity and biocidal properties are under various stages of
investigation and confirmation. Particularly noteworthy among
these are the novel classes of organotin polymers, incorporating
tin atoms in the backbone of the polymer chain, which have been
synthesized by Carraher and coworkers [37-46] and many of which
would prove to have diverse and interesting applications.

ACKNOWLEDGEMENTS

It is a pleasure to acknowledge the contributions of our co-
workers, Madhu Anand, Jaime Corredor, Brij Garg, Jim Jakubowski,
Jorge Mendoza & R. Sam Williams, formerly graduate students in
this Department, whose work from the cited references forms a
large part of this review. Partial support for various parts of

this research was provided by the David W. Taylor Naval Ship Research & Development Center.

REFERENCES

[1] J. J. Zuckerman, Ed., *Organotin Compounds: New Chemistry and Applications*, American Chemical Society, Washington, D.C., 1976.

[2] Woods Hole Oceanographic Institution, *Marine Fouling and Its Prevention*, U.S. Naval Institute, Annapolis, MD, 1952.

[3] P. H. Benson, D. L. Brining and D. W. Perrin, *Mar. Technol.*, 10 (1), 30 (1973).

[4] D. M. James in *Treatise on Coatings*, Vol. 4, R. R. Myers and J. S. Long, Eds., M. Dekker, New York, 1975, Chapter 8.

[5] A. M. van Londen, S. Johnsen and G. J. Govers, *J. Paint Technol.*, 47 (600), 62 (1975).

[6] A. T. Phillip, *Prog. Org. Coat.*, 2, 159 (1973).

[7] A. M. van Londen, *A Study of Ship Bottom Paints, in Particular Pertaining to the Behavior and Action of Anti-fouling Paints*, Report No. 54C, Netherlands' Research Centre T.N.O. for Shipbuilding and Navigation, Delft, September 1963.

[8] F. H. De La Court and H. J. De Vries, *Prog. Org. Coat.*, 1, 375 (1973).

[9] B. H. Ketchum, J. D. Ferry, A. C. Redfield and A. E. Burns, *Ind. Eng. Chem.*, 37, 456 (1945).

[10] R. F. Bennett and R. J. Zedler, *J. Oil Colour Chem. Assoc.*, 49, 928 (1966).

[11] C. J. Evans, *Tin Its Uses*, 84, 7 (1970).

[12] M. W. Cooksley and D. N. Parham, *Surf. Coat.*, 2, 280 (1966).

[13] V. J. Castelli and W. L. Yeager in *Controlled Release Polymeric Formulations*, D. R. Paul and F. W. Harris, Eds., American Chemical Society, Washington, D.C., 1976, p. 239.

[14] W. N. Aldridge in *Organotin Compounds: New Chemistry and Applications*, J. J. Zuckerman, Ed., American Chemical Society, Washington, D.C., 1976, p. 186.

[15] M. J. Selwyn in *Organotin Compounds: New Chemistry and Applications*, J. J. Zuckerman, Ed., American Chemical Society, Washington, D.C., 1976, p. 204.

[16] A. M. van Londen, "A Study on the Importance of the Ships' Hull Condition: An Approach to Improving the Economy of Shipping," 3rd International Congress on Marine Corrosion and Fouling, Gaithersburg, MD, 1972.

[17] E. J. Dyckman and J. A. Montemarano, *Am. Paint J.*, 58 (5), 66 (1973).

[18] R. V. Subramanian and M. Anand in *Chemistry and Properties of Crosslinked Polymers*, S. S. Labana, Ed., Academic Press, N.Y., 1977, p. 1.

[19] R. V. Subramanian and B. K. Garg in *Proc. Int. Controlled Release Pestic. Symp.*, R. L. Goulding, Ed., Oregon State University, Corvallis, Oregon, Aug. 22-24, 1977, p. IV-154.

[20] R. V. Subramanian, B. K. Garg, J. J. Jakubowski, J. Corredor, J. A. Montemarano and E. C. Fischer, *Am. Chem. Soc., Div. Org. Coat. Plast. Chem., Pap.*, 36 (2), 660 (1976).

[21] R. V. Subramanian, B. K. Garg and K. N. Somasekharan, *Am. Chem.Soc., Div.Org.Coat.Plast.Chem., Pap.*, 39, 572 (1978).

[22] K. N. Somasekharan and R. V. Subramanian in *Modification of Polymers*, C. Carraher and M. Tsuda, Eds., American Chemical Society, Washington, D.C., 1980, p. 165.

[23] K. N. Somasekharan and R. V. Subramanian, *Am. Chem. Soc., Div. Org. Coat. Plast. Chem., Pap.*, 40, 167 (1979).

[24] H. W. Godbee and D. S. Joy, *Assessment of the Loss of Radioactive Isotopes from Waste Solids to the Environment, Part 1: Background and Theory*, Oak Ridge National Laboratory, Oak Ridge, Tennessee, 1974.

[25] V. Stannett in *Diffusion in Polymers*, J. Crank and G. S. Park, Eds., Academic Press, N.Y., 1968, Chapter 2.

[26] R. V. Subramanian, B. K. Garg and J. Corredor in *Organometallic Polymers*, C. E. Carraher, J. E. Sheats and C. U. Pittman, Eds., Academic Press, N.Y., 1978, p. 181.

[27] R. V. Subramanian, R. S. Williams and K. N. Somasekharan, *Am. Chem. Soc., Div. Org. Coat. Plast. Chem., Pap.*, 41, 38 (1979).

[28] R. S. Williams, Ph.D. Thesis, Washington State University, Pullman, WA, 1980.

[29] R. V. Subramanian and K. N. Somasekharan in *Proc. 7th Int. Symp. Controlled Release of Bioactive Materials*, D. H. Lewis, Ed., Ft. Lauderdale, Florida, July 28-30, 1980.

[30] K. N. Somasekharan and R. V. Subramanian, unpublished results.

[31] K. N. Somasekharan and R. V. Subramanian in *Controlled Release of Bioactive Materials*, R. W. Baker, Ed., Academic Press, N.Y., 1980 (in press).

[32] S. M. Miller, *Ind. Eng. Chem. Prod. Res. Dev.*, $\underline{3}$ (3), 226 (1964).

[33] N. A. Ghanem, N. N. Messiha, N. E. Ikaldious and A. F. Shaaban, *Eur. Polym. J.*, $\underline{15}$, 823 (1979).

[34] R. V. Subramanian, J. A. Mendoza and B. K. Garg in *Proc. 5th Int. Symp. Controlled Release of Bioactive Materials*, Gaithersburg, MD, Aug. 14-16, 1978, p. 6.8.

[35] R. V. Subramanian, J. J. Jakubowski and R. S. Williams, *SPE Annual Technical Conference: Technical Papers*, Seattle, WA, Aug. 10-12, 1976, p. 111.

[36] R. V. Subramanian and B. K. Garg, *Polym. Plast. Technol. Eng.*, $\underline{11}$ (1), 81 (1978).

[37] C. E. Carraher and R. L. Dammeier, *Makromol. Chem.*, $\underline{135}$, 107 (1970).

[38] C. E. Carraher and G. Scherubel, *J. Polym. Sci.*, A-1, $\underline{9}$, 983 (1971).

[39] C. E. Carraher and D. O. Winter, *Makromol. Chem.*, $\underline{141}$, 237 (1971).

[40] C. E. Carraher and G. Scherubel, *Makromol. Chem.*, $\underline{152}$, 61 (1972).

[41] C. E. Carraher and R. L. Dammeier, *J. Polym. Sci.*, A-1, $\underline{10}$, 413 (1972).

[42] C. E. Carraher, *Inorg. Makromol. Rev.*, $\underline{1}$, 271 (1972).

[43] C. E. Carraher and D. O. Winter, *Makromol. Chem.*, $\underline{152}$, 55 (1972).

[44] C. E. Carraher and G. A. Scherubel, *Makromol. Chem.*, 160, 259 (1972).

[45] C. E. Carraher, S. Jorgensen and P. J. Lessek, *J. Appl. Polym. Sci.*, 2255 (1976).

[46] C. E. Carraher, D. J. Giron, W. K. Woelk, J. A. Schroeder and M. F. Feddersen, *J. Appl. Polym. Sci.*, 23, 1501 (1979).

Incorporation of Metal Related Materials into Electrically Neutral Polymers

A. K. St. Clair
NASA Langley Research Center
Hampton, VA   23665

L. T. Taylor*
Department of Chemistry
Virginia Polytechnic Institute
   and State University
Blacksburg, VA   24061

ABSTRACT

Metal containing polymers are numerous and highly diverse. Such systems include ionomers, polymer-bound coordinating ligands, organometallic polymers wherein the metal may be a part of the polymer "backbone", metal-polymer composites and metal incorporated neutral polymers. The latter system, which is reviewed herein, is normally concerned with the modification of polymer properties which may range from electrical conductivity to polymer flammability. Metallic additives may be a metal atom vapor, organometallic compounds, coordination complexes and simple hydrated or anhydrous salts. Representative polymer systems include polyamides, polyimides, polyalcohols, polyesters, polyacetylene, polyethylene and polysiloxane. Each metal is almost a case unto itself regarding a specific polymer property (i.e. different metals produced different properties). The chemical state of the metal and the ligand environment of the metallic species are extremely important concerning polymer and metal compatibility. The literature cited, while not exhaustive, is representative of this rapidly expanding area of polymer research.

SCOPE

The incorporation of metallic species into polymers is an extremely active area of research. As evidence for this ob-

servation, a search of Chemical Abstracts during the period
1977–1979 revealed over a thousand references to polymers containing metals. Obviously this review cannot attempt to deal
with the entire area of polymer-metal systems. It, however,
will prove useful to outline the various research areas which
we perceive to compose the general field of metal-containing
polymers.

These studies may be roughly subdivided into five categories.
One group would be represented by polymers that have an ionic
group covalently attached which, in most cases, is neutralized
by a metallic counterion.[1] Incorporation of the metal ion may
occur after formation of the ionic polymer as in the binding of
copper(II) to poly(methacrylic acid)[2] or a metal salt monomer
may be employed for conducting the polymerization. The latter
situation is best illustrated by the use of magnesium or calcium
salts of mono(hydroxyethyl)phthalate as an ionic monomer[3] in the
production of metal-containing polyesters via reaction with
various anhydrides and bisepoxides. The reaction scheme I is
outlined below where the divalent metal cross-links the adjacent

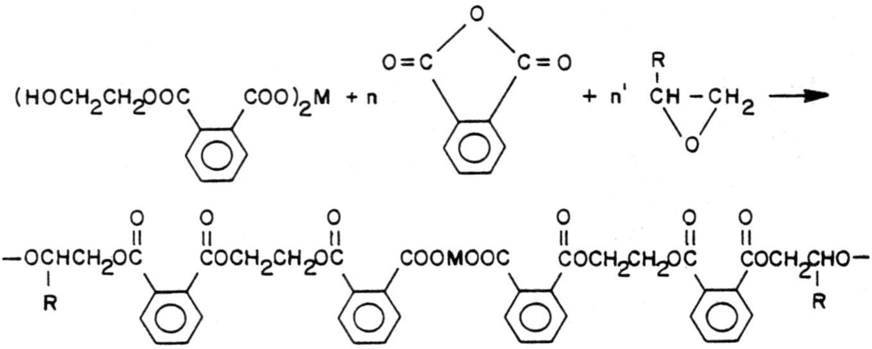

SCHEME I

carboxyl groups. The field of ionomers has reached an advanced
level of maturity; yet, it continues to develop at a rather fast
rate.

A second very large area of metal-polymer studies is
represented by those polymers which have organically bonded
(i.e. through carbon) metals. These metals are generally part
of the monomer and appear in the polymer "backbone". In the
truest sense of the word these would be organometallic polymers.
Several groups[4-6] have been quite active in this area during the
past decade. Common commercial monomers would include vinylferrocene, ferrocenylmethylmethacrylate and tricarbonylbenzyl-
acrylate chromium. Reaction scheme II outlines a typical
organometallic polymer. For a much more expanded discussion of
this field the reader is referred to several other chapters of
this monograph.

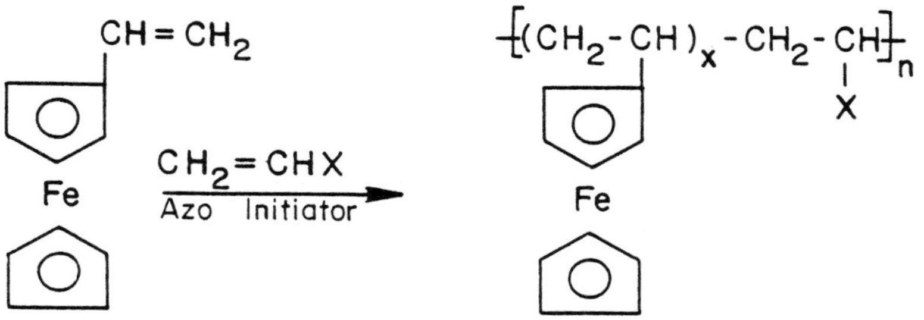

SCHEME II

A somewhat related group of polymers which would constitute
a third area of metal-polymer systems is represented by
neutral chelate resins or neutral polymer ligands. This field
has received much attention due to the wide variety of uses for
such polymer complexes. Applications include (1) sequestering
of metal ions[7,8] in general or sequestering of one of several

specific metals, (2) models for enzymic and other biological reactions such as the decomposition of hydrogen peroxide[9] and (3) homogeneous catalytic reactions whereby the precious-metal catalyst is "heterogenized" by being anchored to a polymer.[10] This class of polymer normally is synthesized by taking strong coordinating ligand monomers such as 4-vinylpyridine[11] and 4'-vinylbenzo-18-crown-6[12] (Structures I and II) or utilizing a more classical polymer system which has been chemically modifed

I   II

by appending known ligands to the polymer matrix. Reaction schemes III and IV are known examples of the latter approach.[13,14] Studies such as mentioned above have focused on the extent of metal binding to the polymer (i.e. stability constants) or the reactivity of the metal bound polymer. Many fine reveiws already exist on this topic.[15]

Other less explored polymer-metal areas include polymer film laminates and metal-filled polymers. The former[16] would be represented by the vapor deposition of metals onto polyester films for the purpose of producing electroconductive films. The laminate includes the polymer substrate and a metal layer bonded to the substrate by a polymeric coupler. No doubt some metal ion could be incorporated into the polymer. The use of "so-called" inorganic fillers with preformed polymers has received some attention. Adhesive studies with polyimides have been

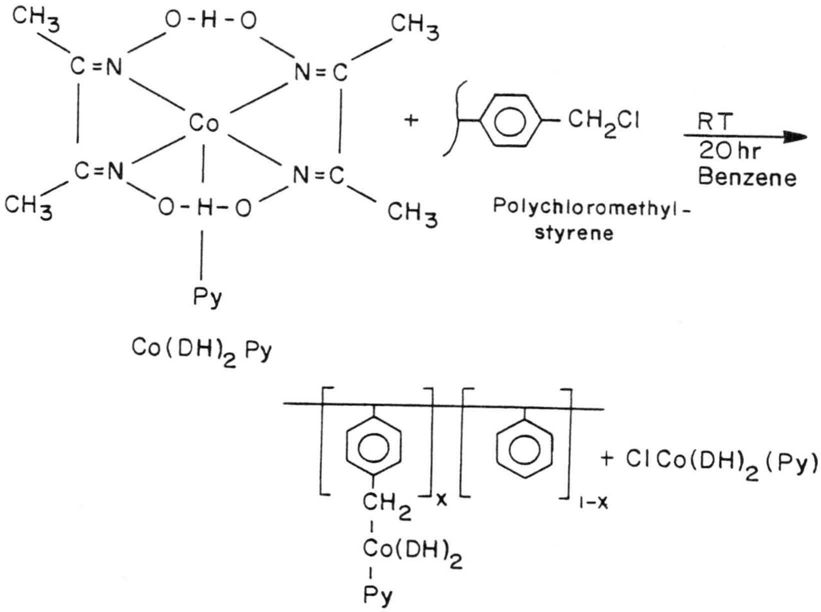

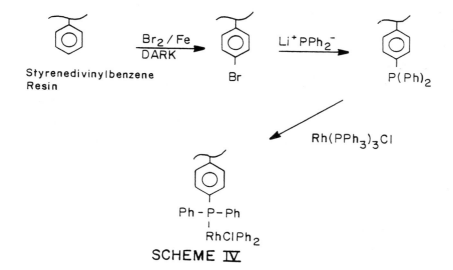

SCHEME III

SCHEME IV

conducted wherein the resin contained 5-40% metal or metal oxide
filler. Enhanced adhesive properties with titanium adherands
have been noted with aluminum powder in this regard.[17] Another
study has examined the mechanical properties of polyurethane
rubber filled with various fractions (0-50%, v/v) of sodium
chloride.[18]

   The final area of metal-polymer systems would consist of
neutral polymers to which have been added dissolved metal salts,
metal complexes, organometallic compounds and metals. This area
has probably received the least study and constitutes the focus
of this review. While this division of metal-polymer studies
may seem quite arbitrary, we believe it is readily identifiable,
rapidly expanding and (as this review hopefully will attest) highly
fascinating. The uniqueness of this area of metal-containing
polymers rests on two observations: (1) the polymer systems
employed are electrically neutral and have no readily ionizable
groups; and (2) investigations are normally concerned with adding
metals for the purpose of modifying polymer properties such as
increasing the glass transition temperature (T_g), decreasing the
melt temperature (T_m), enhancing adhesive properties, or increasing
the electrical conductivity. Numerous reports have appeared
which describe the incorporation of "non-metal" species such as
oxygen[19], iodine[20], arsenic pentafluoride[21], and tetracyanoquinodimethane salts[22] to modify electrical properties of polymers.
However, this review shall deal exclusively with "metal" related
dopants. As will become evident, a number of polymer properties
have been measured for a large variety of neutral polymers containing metal dopants. The discussion which follows has been
arranged into categories of neutral polymers.

REVIEW

Polyamides

   Polyamides constitute one of the most common polymer systems
to which have been added a large variety of metal salts. Poly-

(caproamide), Structure III, has received considerable study in
several Italian Laboratories. A decade ago it was observed that

$$-\!\!\left[(CH_2)_6-NH-\overset{\overset{\displaystyle O}{\|}}{C}-(CH_2)_4-\overset{\overset{\displaystyle O}{\|}}{C}-NH\right]_x$$

III

aqueous solutions of inorganic salts altered the crystalline
structure of biological macromolecules. Subsequently, the effect
of added KCl, LiCl, LiBr and $CaCl_2$ on the melting behavior of
Nylon 6 as determined by differential scanning calorimetry was
reported. Small amounts of these salts in the complete absence
of water caused a melting temperature depression in excess of the
theoretically predicted temperature depression.[23] Intimate mixing
of the polymer and dopant at approximately 260°C was crucial in
order to obtain large temperature depressions. The extent of
depression depended on the type of metal salt. Potassium chloride
(a salting-out agent) showed no effect; while, the depression noted
with LiCl (a salting-in agent) increased with increasing concentration. Available data at the time suggested a strong direct
interaction between Nylon 6 and salting-in agent in the absence
of water. A serious problem arises in this approach in that
thermal polymer degradative effects occur during metal ion incorporation. Nevertheless, the authors believe the investigation
has certain implications for the processing of high melting polymers.

The search for further proof regarding the occurrence of a
direct interaction between salt and Nylon 6 was detailed several
years later.[24] The T_m of Nylon 6 was decreased upon the addition
of metal salts in the order KCl < LiCl < LiBr in the composition
range 0–12%, w/w. X-ray analysis indicated that LiCl and LiBr
tend to favor the crystallization of the γ form of Nylon 6 over
the more common α form. The crystallization rate of Nylon 6
is drastically lowered by LiCl and LiBr; and the melt viscosity

at 250°C is considerably increased in the presence of LiCl. The
nature of the salt polymer interaction was postulated to be
between the amide group of the amorphous polymer and the salt.
Several questions exist in this study regarding which part of
the amide unit (i.e. nitrogen or oxygen) is involved. Also,
it has yet to be decided whether the lithium ion alone or the
ion pair interacts with the polymer.

Stabilized Nylon 6 has been substituted[25] in this work in
an effort to minimize thermal degradation. Depression of the T_m
was comparable to data obtained on unstabilized Nylon 6. Added
lithium salts, however, had no effect on polymer morphology, but
there was a reduction in the rate of crystallization. Further
investigation[26] surprisingly revealed the T_g ($\sim50°$) remained unaffected by salt additions or concentration. All three salts
caused an increase in melt viscosity. The increase with KCl was
believed to be due to its insolubility in Nylon 6 in which the
KCl acts as an inert filler. Lithium bromide resulted in higher
viscosities than LiCl.

The T_g independence of added salt to poly(caproamide) has
more recently been shown[27] to be incorrect because the effect of
water absorption. When strict precautions to exclude moisture
are taken, the addition of 4% LiCl has been shown to raise the
T_g by 25°C. T_g's of salted and unsalted Nylon 6 are lowered in
the presence of moisture. The amount of moisture absorbed at
equilibrium by Nylon 6 doped with 4% LiCl is double. The bene-
fical effect in T_g is lost when samples equilibrate with moisture.
It was also observed that the elastic modulus of oriented samples
increases with added salt provided all moisture is absent.
Severe restrictions are thus placed on the practical use of this
system.

A somewhat earlier study was conducted with various nylons
regarding the effect of metal salts on the stress cracking of
polyamides.[28] Halides were found to be most effective; while,
acetates and sulfates did not promote cracking. Later work

revealed that thiocyanates and nitrates[29] were suitable sub-
stitutes. The preparation of specimens for stress cracking was
performed in several ways. These were (1) treatment of the
stretched film with a concentrated solution of metal salts, (2)
immersion of polymer films in solutions of the active salts or
(3) cast film of the nylon-metal salt mixture. The mechanism
of rupture does not appear to be one of simple hydrolysis or
of metal catalyzed hydrolysis. Metal haldies which were found
to be active stress cracking agents induce characteristic changes
in the IR and NMR spectra of nylons. Two types of changes were
observed[30] depending on the metal halide involved which were
attributed to complex formation between the metal and amide
group. Type I metal halides such as Zn(II), Co(II), Cu(II), and
Mn(II) form complexes in which the metal atom is bonded to the
carbonyl oxygen (Structure IV). These agents are believed to

IV

cause stress cracking by interference with the hydrogen bonding
in the polyamide. Type II halides such as Li(I), Ca(II) and Mg(II)
are believed to form proton donating, solvated, species which act
as direct solvents for nylon in a manner similar to phenols and
formic acid. In other words, Type II halides cause simple solvent
cracking. Additional work in this area is not available.

Polyacrylamides

        Polyacrylamides (PAA) (Structure V) are another group of
polymers which has received significant study. Tough films of
PAA with less than 50% (w/w) $CuCl_2$ have been produced from aqueous
solution.[31] The Cu(II) doped PAA was not conductive (surface
resistivity = 10^9 ohms). The PAA-Cu(II) film when combined with
iodine, however, exhibited good conductivity (surface resistivity
$\sim 10^3$ ohms). The iodine treated film was prepared first by mixing

$$-\left[CH_2-CH-CH_2-CH \right]_n$$

(Structure with $C=O$ and NH_2 groups on each CH)

$$\underline{\underline{V}}$$

aqueous solutions of PAA and $CuCl_2$, casting a film of the resulting
viscous solution, drying at 100°C followed by dipping the film in
an iodine-acetone solution and drying further at 100°C to remove
solvent and excess solvent. Optimum surface conductivity was
achieved for 20-40% (w/w) $CuCl_2$ and more than 1% of iodine. Higher
concentrations of Cu(II) ruined the film flexibility and other
properties. Although the highest conductivity was obtained with
$CuCl_2$, similar results were produced using $Cu(NO_3)_2$, $CuSO_4$, and
$Cu(C_2H_3O_2)_2$. Films prepared under both non-oxygenated and
oxygenated conditions exhibited the same surface conductivities.

        Treatment of the PAA-Cu(II)-I_2 polymer with KOH significantly
increased the electrical conductivity[32] of films, and a maximum
conductivity was achieved using an equimolar amount of hydroxide
to the Cu(II) salt. No conduction was observed when the chelate
solution was treated with twice the equimolar amount of hydroxide.
The authors claim that the conductivity of the PAA-Cu(II) film
modified with iodine was dependent on the structures of the

chelate in solution. Structures VI and VII were believed to be
particularly effective; while, Structure VIII was thought not to
participate in the conduction. A very limited ESR study of these
films has been performed[33] in an effort to ascertain the specific

PAA-Cu(II) interaction. A mononuclear Cu(II) coordinated with
two water molecules and chelated with two oxygen or nitrogen
atoms attached to the polymer has been suggested. Optical and
scanning electron microscopy have been performed[34] on films of
PAA-Cu(II)-I_2 in hopes of discovering the reason why semi-conduct-
ive behavior is produced after treatment with I_2. Highly coagu-
lated states of PAA-Cu(II)-I_2 appear to favor the higher surface
conductivity. SEM shows a new reaction product after treatment
with I_2. Iodine is believed to attack those parts of the polymer which are rich in Cu(II). Optical absorption spectra proved
the existence of γ-CuI on the film surface which is believed
to be responsible for the enhanced conductivity.

Although our discussion is focused along polymer types, it
seems appropriate to mention here a very closely related
effort[33-35] to the PAA work involving poly(vinylalchol) (PVA).
A PVA-Cu(II) film had surface resistivity of 10^{13} ohm, but
after dipping the film in iodine/acetone and drying at 80°C
surface resistivity decreased to 10^3 ohm. For comparison PVA-I_2

with no Cu(II) yielded 10^9 ohm indicating the need for Cu(II).
Satisfactory results were given by more than 15% (w/w) of
copper(II) salt and about 4% (w/w) of iodine. Again KOH was
observed to be effective with the maximum conductivity being
reached at a 1:1 mole ratio. Several structures of PVA-Cu(II)
have been postulated such as deprotonated, $[Cu(RO_2)_2]^{-2}$ and
$[Cu(RO_2)R(OH)_2]°$, and non-ionized species where $R(OH)_2 \equiv PVA$. No
mention of γ-CuI was made. Absorption spectra of the PVA-Cu(II)-
I_2 after washing the film to remove adsorbed I_2 revealed the
presence of I_3^- and multilayer iodine. This observation suggested
that iodine in the chelate film participated in the conduction
in forms of I_3^- and/or miltilayer iodine. Other metal salts
capable of forming chelates with PVA such as Fe(III), Zn(II),
Sn(IV) and Cr(II) salts did not give favorable results, the
resistivities being greater than 10^9 ohm.

Polyimides

Approximately twenty years ago Angelo[36] patented the synthetic
procedure for the addition of metal ions to numerous types of
polyimides (Structure IX depicts an aromatic-type polyimide) along
with the results of selected physical measurements on these
mixtures. The object of the invention was a process for forming
particle-containing (< 1μ) transparent polyimide shaped structures.
Unlike the work discussed previously all of the metals were added
in the form of coordination complexes (e.g. β-diketone and
β-ketoester chelates) rather than as simple anhydrous or hydrated

IX

salts. This offers the advantage that the coordination complex
may be more soluble in the polymer solution, more compatible
with the bulk polymer and more evenly distributed throughout
the polymer.

The process employed in this study involved (1) forming the
polyamic acid, (2) adding the metal to the polyamic acid, (3)
shaping the polymer-metal ion mixture and (4) converting to the
metal-containing polyimide. The starting materials for polymer
formation were aromatic or aliphatic diamines and aromatic or
aliphatic tetracarboxylic acid dianhydrides. The major goal of
the invention was the production of polyimide films for use as
decorative or electrically conductive tapes and packaging
materials. Few properties of the films are available from the
patent and no comparison with films cast from the neat polymer
solutions was discussed. The room temperature properties
of a film cast from an N,N-dimethylformamide (DMF) solution of
4,4'-diaminodiphenyl methane, pyromellitic dianhydride and bis-
(acetylacetonato)copper(II) are given below: Percent copper,
3.0%; dielectric constant, 3.6; dissipation factor, 0.004-0.01;
and volume resistivity, 8×10^{12} ohm-cm. No data was given regarding
the glass transition temperature, thermal stability, adhesive
properties, mode of interaction between metal and polyimide, state
of the metal, thermal conductivity, surface resistivity, etc.
Unfortunately further patents or published work in this area are
not available.[37]

More recent work[38] with polyimides and metal compounds concerns polymers derived from 3,3', 4,4'-benzophenone tetracarboxylic acid dianhydride (BTDA) (Structure X), and 3,3'-diaminobenzo-
phenone (m,m'-DABP) and 4,4'-diaminobenzophenone (p,p'-DABP)
(Structure XI) in DMF, N,N-dimethylacetamide (DMAC) or diethylene-
glycol dimethylether. Approximately twenty metals in a variety
of forms were added to polyamic acid solutions. In many cases
films could then be cast and thermally converted to the corresponding polyimide. Non-brittle polyimide films were subjected

X   XI

to thermomechanical analysis, thermogravimetric analysis, weight
loss on prolonged heating and infrared analysis. Relatively large
positive and negative changes in thermal properties of the metal
ion containing polymers have been noted with different metals.
Predictive trends with certain polymers, however, were not possible with specific metals.

The best system studied in this regard involves tris(acetylacetonato)aluminum(III), Al(acac)$_3$, addition to the above two
polymer systems. Table I compares a few physical measurements on
the BTDA + m,m'-DABP polyimide alone and the Al(acac)$_3$/polyimide
mixture. An inspection of the data reveals that the T_g of the
Al(acac)$_3$/polyimide is increased without sacrificing any polymer
thermal stability. These effects work together to improve the
high temperature adhesive properties of the polyimide with
respect to bonding titanium surfaces.

The choice of metal, its chemical state, and the counterion
were critical in this investigation as evidenced by the typical
data shown in Table I regarding NiCl$_2$·6H$_2$O incorporation. As
further proof of this, experiments with AlCl$_3$, AlCl$_3$·6H$_2$O and
Al(NO$_3$)$_3$ apparently caused immediate crosslinking (rubber-like
formation) of the polyamic acid such that no film could be cast.
The T_g's were more dramatically increased with the p,p'-DABP
polyimide but at the expense of considerable loss in thermal
stability (Table I). No changes in chemical functionality in
the polyimide were apparent as judged by infrared spectral

TABLE I

COMPARISON OF PURE POLYIMIDE AND METAL

ION-INCORPORATED POLYIMIDES

Polyimide Film	$T_g{}^a$ (°C)	TGA^b (°C)	Adhesive Lap Shear Strength at 275°C^c (psi)
BTDA + m,m'-DABP	251	570	440
BTDA + m,m'-DABP + Al(acac)$_3{}^d$	271	555	1640
BTDA + m,m'-DABP + NiCl$_2$·6H$_2$O^d	279	495	610
BTDA + p,p'-DABP	283	525	–
BTDA + p,p'-DABP + CaCl$_2{}^d$	360	475	–

^{a}Apparent glass transition temperature from thermomechanical analysis

^{b}Thermogravimetric analysis polymer decomposition temperature

^{c}Performed on 50 mil titanium adherends

^{d}Metal complex to polyamic acid mole ratio = 1:4 in DMAc

comparisons of polyimide and polyimide-metal films regardless of the metal employed.

Incorporation of Al(acac)$_3$ into the polyimide disappointingly showed no significant reduction in volume resistivity relative to the polymer alone. Similar results were obtained on NiCl$_2$·6H$_2$O filled polyimides. Numerous efforts to prepare high quality films incorporating other metal ions into BTDA + m,m'-DABP were not satisfactory because metal ion addition resulted in a decreased solution viscosity. These results suggest that during the film curing process, the non-conducting Al(acac)$_3$ and NiCl$_2$ maintain their integrity rather than being converted to the more con- ducting aluminum or nickel metal as originally envisioned. X-ray photoelectron spectroscopic (XPS) measurements have confirmed this hypothesis.

Simultaneous with the above work, superior antistaic properties were reported[39] for newly available soluble polyimide (DAPI-Polyimide) film loaded with either $LiNO_3$ or LiCl. Physical properties and film smoothness remained unchanged except that electrical resistance was sharply lowered. Conductivity was increased about 20-fold or 2000% over the standard unfilled polyimide. Additional tests showed that the films were very slightly hygroscopic in the presence of lithium ions. This phenomenon may, in fact, account for the lowered resistivity. It also was not clear from the NASA brief whether the enhancement was in surface or volume conductivity.

A more recent effort to incorporate metal species into polyimides for the purpose of decreasing polymer resistivity has appeared.[40,41] Palladium-filled polyimide films have been prepared using the following dianhydride-diamine pairs: BTDA + DABP, BTDA + 4,4'-oxydianiline (ODA) (Structure XII), and BTDA + 3,3'-diaminodiphenylcarbinol (DADPC) (Structure XIII). A number of palladium additives were screened, many of which proved unacceptable because of insufficient solubility in DMAc or in the polymer-DMAc solution. While good quality films containing evenly dispersed palladium could be produced with slightly soluble $PdCl_2$

XII   XIII

and Na_2PdCl_4, only minor modifications were realized in polymer properties. Best results have been obtained with Li_2PdCl_4 and

$Pd(S(CH_3)_2)_2Cl_2$ as additives. Flexible, dark red-brown films
have been fabricated for the three monomer pairs noted above.

The synthetic procedure involves formation of the polyamic
acid (Structure XIV) in DMAc, intimate mixing of the palladium
complex and polyamic acid, and thermal imidization in air to the
palladium-filled polyimide film. An alternate _in situ_ method
whereby polymerization to the polyamic acid was performed in the

XIV

presence of the palladium complex also proved satisfactory. Several
films possessed noticeably different surfaces depending upon
whether the film had been exposed to the glass or to air during
the imidization procedure. This difference was very noticeable
for the two films containing $Pd(S(CH_3)_2)_2Cl_2$ (BTDA + ODA and
BTDA + DABP). While the glass side had a dark red-brown appearance, the air-side possessed a definite silvery metallic appearance.
The presence of oxygen during the imidization process appears
crucial, since BTDA + ODA and BTDA + DABP doped with
$Pd(S(CH_3)_2)_2Cl_2$ do not give rise to metallic surfaces when
cured in either a dry N_2, Ar, N_2/H_2 or moist Ar atmosphere.

The primary purpose for this study was to ascertain whether
or not the addition of palladium would lower the electrical
resistivities of the polyimides. Table II outlines the results.
Four different combinations of dianhydrides, diamines, and metal

TABLE II

SURFACE AND VOLUME RESISTIVITIES OF CONDUCTIVE
PALLADIUM-CONTAINING BTDA-DERIVED POLYIMIDES[a,b]

METAL COMPLEX	ODA	DABP	DADPC	CURING ATMOSPHERE
Li_2PdCl_4	9.5×10^5 ohm[c] 2.0×10^6 ohm-cm[d]	NC[e]	1.3×10^7 ohm 1.0×10^7 ohm-cm	Air
$Pd(S(CH_3)_2)_2Cl_2$	$<10^5$ ohm $<10^5$ ohm-cm	$<10^5$ ohm $<10^5$ ohm-cm	NC	Air
Li_2PdCl_4	5.1×10^7 ohm 8.9×10^7 ohm-cm	NC	2.1×10^{10} ohm 1.4×10^{11} ohm-cm	N_2
$Pd(S(CH_3)_2)_2Cl_2$	NC	NC	NC	N_2

[a]Resistivity values are quoted for the best quality films. Values for replicate films do not differ by more than one order of magnitude.

[b]Polymer alone surface and volume resistivities are 10^{17} ohm and 10^{17} ohm-cm, respectively.

[c]Surface resistivity

[d]Volume resistivity

[e]NC = No change in resistivity relative to the polymer alone.

additives yielded polyimides with dramatically lowered resistivities when cured in an air atmosphere. Conductive BTDA + ODA films were produced using both palladium additives. Surprisingly Li_2PdCl_4 gave lowered resistivity values with BTDA + DADPC; while $Pd(S(CH_3)_2)_2Cl_2$ with the same monomer pair exhibited values equivalent with the polymer alone. The results with BTDA + DABP, however, were reversed. The metallic surface on the air side displayed by the two conductive $Pd(S(CH_3)_2)_2Cl_2$ films, no doubt, lowers resistivity ($<10^5$ ohm-cm). This metallic surface is apparently not a necessity, since conductive Li_2PdCl_4 films do not display a metallic surface. However, as Table II attests, electrical resistivities are higher for the Li_2PdCl_4 film (with non-metallic surface).

The results on curing the films in a non-oxygenated atmosphere are equally interesting, Table II. No metallic surfaces are produced with $Pd(S(CH_3)_2)_2Cl_2$ as an additive and no resistivity lowering is observed. Moist argon and forming gas (N_2/H_2) give the same unchanged results. A nitrogen curing atmosphere, however, does not change the resistivity results appreciably for the air-cured, conductive Li_2PdCl_4 films. It is significant that in each case, with Li_2PdCl_4, the resistivity values are always one to three orders of magnitude higher for nitrogen cured films.

X-ray photoelectron spectroscopy (XPS) has proven valuable in studying some of these palladium-containing polyimides. Measured XPS binding energies (Pd $3d_{5/2,3/2}$) indicate that an appreciable amount of palladium has been reduced to the elemental state in each film. In other words, during the imidization process reduction of palladium has occurred. Those films which have lowered resistivities exhibit the most reduced palladium. Differences between conductive film surfaces exposed to air versus glass are very apparent here again, since a larger quantity of reduced palladium always accompanies the side of the film exposed to the air.

XPS for non-conductive films differs in several respects
from the above. First of all, the palladium signals are relatively weak and the measured binding energies fall between those for
Pd metal and $PdCl_2$. Secondly, an additional pair of photopeaks
are observed at approximately 10 eV higher energy (349.8 and
346.1 eV). The relative intensity of the two pair of peaks has
been found to vary depending upon what place on the film was
being sampled. Conductive films do not show this extra higher
energy pair of photopeaks. This phenomenon is attributed to
differential charging of the surface palladium. Some islands of
palladium are more insulated than other islands by the non-conductive polyimide and cannot dissipate the photo-charge produced
by the x-ray beam. High resistivity films, therefore, appear to
be highly heterogeneous insofar as palladium is concerned. If
palladium is the charge-carrier as we expect, an unequal distribution of palladium can result in an interruption of the charge
transfer mechanism. The addition of higher amounts of palladium
to these films does not produce the desired results. After curing
for three hours at 300°C the XPS spectrum clearly shows evidence
for Pd(II) and Pd(0) on the surface for 2:4 metal to polymer ratios.
Longer heating times result in more complete palladium reduction
but with significantly more polymer thermal degradation.

Extension of this work to other metals is currently in progress.
Preliminary results suggest that no resistivity lowering is
achieved with those Ag, Pt, Au, Cu and Li complexes which have
been tried to date. High quality films have been produced;
and in the case of Ag, Pt and Au, reduction to the metallic state
has occurred. These results, however, have not been fully
explained.

A much earlier report[42] appeared concerning Ag incorporation
into the polyamic acid derived from pyromellitic dianhydride
(PMDA) (Structure XV) and ODA. Both Ag metal and $Ag(C_2H_3O_2)$
were added to PMDA + ODA; and films containing 0.25-1.00 gram-
atom silver per repeat unit were obtained. Electroconductivity

XV

was studied as a function of temperature and Ag content. Thermal
and electrical conductivities were increased 3-7 times for the
polyimide film containing dispersed Ag metal; but no change in
properties was noted for the film containing $Ag(C_2H_3O_2)$.

In 1963 a patent was filed[43] covering very similar work
(e.g. PMDA, 4,4'-methylenedianiline and $Ag(C_2H_3O)$). The Ag
containing polyamic acid complex was converted to the polyimide
and Ag metal by heating at 300°C in vacuo for 30 minutes. The
film was stated to be tough, flexible, opaque and metallic. At
this stage the film was not conducting. Further heating at 275°C
in air for 5-7 hours rendered it conducting although no resistivity data was reported.

A different kind of metal involvement with polyimides has
recently been reported.[44] This project was concerned with providing a room temperature bonding process for the purpose of producing
heat resistant bonds. A series of very thin polyimide films under
1 mil in thickness were vacuum metalized with various high melting
metals such as gold. These samples were put in a fixture and
successfully bonded at low pressure to test samples of indium
coated polyimide. The system is believed to make use of the
adhesion of metal films to the organic substrate by the formation
of ligands. Kapton polyimide has been the polymer of choice to
date. The process uses the principle of forming high-melting
solid solutions by the diffusion mixture of two metals where the

minor component may be a low-melting malleable oxidation resistant metal or composite.

Poly(alkylbenzimidazoles)

A series of metal salts having poly(alkylbenzimidazoles)
(Structure XVI for example) as the parent ligand have been
synthesized[45] by mixing dimethylsulfoxide solutions of the
polymer and metal salt in stoichiometric proportions. This
experimental operation gave three results depending on the metal
ion employed. The polymer-metal complex precipitated [Co(II),
Ce(III), Ni(II), Cu(II), Zn(II), Cd(II)], precipitation occurred
on dilution with acetone [Cr(III), Fe(II), Sn(II), Ca(II), Ba(II),
Mg(II), W(VI), V(VI)] or no precipitate appeared [Al(III), Na(I),
K(I)]. Values for the electrical resistivity did not significantly

XVI

change upon doping the polymer with either $CrCl_3 \cdot 6H_2O$, $CoCl_2 \cdot 6H_2O$
or $Ni(C_2H_3O_2)_2 \cdot 4H_2O$. In contrast, treatment of the polymer with
HCl thereby forming the polymer acid conjugate changed the volume
resistivity from 10^{13} ohm-cm to 10^6 ohm-cm. The nature of the
polymer metal complexes was elucidated using x-ray photoelectron
spectroscopy. Spectra of the core levels of nitrogen, chlorine
and the various metals indicated the formation of polybenzimidazole/
metal salts. In addition it was concluded that the complexes of
Co and Ni were high spin thus ruling out a planar geometry for
the Ni(II) complex.

Poly(vinyl alcohol)

Investigations into poly(vinyl alcohol) (Structure XVII) have dealt primarily with copper doping. The reader is referred to the earlier discussion regarding PVA-Cu(II)-I_2 and to "ion-omeric" type studies with PVA wherein deprotonation of the

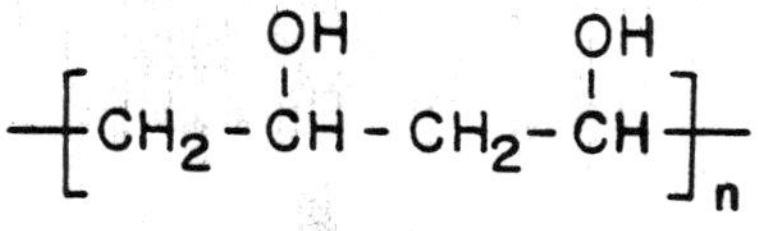

$$-\left[CH_2-\overset{\overset{\displaystyle OH}{|}}{CH}-CH_2-\overset{\overset{\displaystyle OH}{|}}{CH} \right]_n-$$

XVII

alcoholic group is followed by Cu(II) interaction.[46] One study[47] which falls into the metal-polymer area as we have defined it deals with the photoconductivity of PVA films containing Cu(II) complexes. Well-dried PVA films containing either the nitrate, sulfate, chloride or bromide salt of Cu(II) have been prepared. When the ratio $[Cu^{+2}]/[MU]$ (where [MU] denotes the concentration of PVA monomer residues) is greater than 0.015, illumination at the charge transfer UV band produces a strong photocurrent in PVA-$CuCl_2$ and PVA-$CuBr_2$ but not PVA-$Cu(NO_3)_2$ or PVA-$CuSO_4$. Drying of the films is believed to effect replacement of water molecules by two halide ions in the copper coordination sphere which is supported by optical spectra of the film. Nitrate and sulfate do not appear to enter the first coordination sphere. Temperature dependent ESR spectra indicate the existence of antiferromagnetic superexchange interaction between Cu(II) ions via intervening halide ions. The exchange interaction is greater for PVA-$CuBr_2$ than PVA-$CuCl_2$. A network structure is proposed which is believed to be responsible for photoconduction (Structure XVIII). When the PVA-$CuCl_2$ film is illuminated at the charge transfer band the following reaction is believed[46] to occur. The halogen

XVIII

$$PVA\text{-}Cu^{+2} \cdot 2Cl^{-} \xrightarrow{\text{h}\gamma} PVA\text{-}Cu^{+} \cdot Cl^{\circ} \cdot Cl^{-}$$

atom abstracts a hydrogen atom from PVA thereby producing a free
radical. The authors show some ESR evidence to support this on
irradiation. When the ratio $[Cu^{+2}]/[MU]$ is large the halogen
atoms interact more with neighboring halide ions. At this point
the reduction reaction involves the production of holes in the
network of halide ions. The photocurrent is observed when an
external electric field is applied to the film. These holes in
the network structure are believed to be responsible for the
photoconduction. A direct metal-to-metal interaction is stated
to not occur.

Polyethers

Glassy polymers have been produced from[48] solutions of $Ca(NCS)_2$
and the polymer derived from Bisphenol A (Structure XIX) and
epichlorohydrin (Structure XX). $Ca(NCS)_2$ was appreciably soluble
in the "phenoxy" polymer and films containing up to 25% dopant

XIX   XX

were clear. The anhydrous salt (25% w/w) increased the T_g from
91° to 126°, but the tetrahydrate salt increased the T_g from 162°
to 174°. There is an increased resistance to stress cracking
by polar organic liquids with the doped polymer. This may be
related to the T_g increase or to changes in the solubility
parameter, as indicated by insolubility of the salt solutions
in solvent for the pure polymer. Increased water sorption and
electrical conductivity were found to result from salt incorporation into the polymer. Surface and volume resistivities
for the polymer alone were 1×10^{15} ohm and 5×10^{13} ohm-cm; whereas,
for "phenoxy" polymer with 13% $Ca(NCS)_2$ by weight the values
were 1×10^{10} ohm and 4×10^{10} ohm-cm, respectively. The decrease
in resistivity was attributed to the high equilibrium water
content accompanying calcium ion addition.

The addition of $ZnCl_2$ and $CoCl_2$ to high (10^5) and low (2000)
molecular weight poly(propylene oxide)(PPO) has been reported.[49]
Samples were prepared by dissolving PPO and metal salt in methanol
or acetone followed by removal of the solvent in vacuo at room
temperature. $ZnCl_2$ was found to increase the T_g of both high and
low molecular weight polymers, but $CoCl_2$ increased the T_g of the
low molecular weight polymer only. A single phase system in the
$ZnCl_2$ case was indicated; while, a two phase system was apparent
in the case of the PPO/$CoCl_2$ with $CoCl_2$ acting as a filler. In
the zinc case, elevation of the T_g is believed to result from the
formation of five-membered chelate rings by coordination of two
adjacent oxygen atoms in the polymer chain with a $ZnCl_2$ molecule.

In an analogous situation, $ZnCl_2$ was added to poly(tetramethylene glycol) with similar results albeit the T_g was raised less for a given amount of metal chloride. Intermolecular coordination with ether oxygen atoms from two neighboring chains was postulated since intramolecular bonding to zinc(II) would involve the formation of a less stable seven-membered chelate ring.

Much earlier[50] the elastomeric properties of PPO mixed with $LiClO_4$ were investigated. Volume contractions were observed and viscoelastic properties were altered which were attributed to strong interaction forces between the Li^+ and polarizable ether oxygens. The glass transition temperature was raised from -70° to 40°C for 25% (w/w) $LiClO_4$ - low molecular weight PPO. Below 15% (w/w) $LiClO_4$, various low molecular weight polymer properties could be interpreted as resulting from the superposition of free polyether segments with those nearest to the $LiClO_4$. Above 15%, no free segments of PPO are believed to remain. The behavior of the high molecular weight PPO was similar, except that below 15% $LiClO_4$ the T_g observed for low molecular weight PPO was split into two transitions, T_g for PPO at -65°C and another at -10°C. The latter is believed to be a transition for polyether helices stabilized by $LiClO_4$ situated in the helix core.

A rather different type study has been performed with water soluble poly(ethylene oxide) (PEO) and various mercuric halides.[51] Fibers of PEO were exposed to saturated anhydrous ether solutions of the mercuric halides for about four weeks. Complexes were formed with a monomer unit (CH_2CH_2O) to mercuric halide ratio of 4:1 for the chloride and bromide and the monomer iodide ratio was 5:1. Examination of the fibers before and after mercuric ion doping with a polarizing microscope and by x-ray diffraction revealed that the chloride fibers remained highly oriented. The bromide fiber was much less oriented; while, the iodide fiber showed no signs of orientation. Although the polymer is water soluble, the polymer complex is not and shows no sign of swelling in contact with water. The infrared spectrum of the polymer complex was observed[52] to contain a larger number of absorption bands over that of the original polymer. This fact suggested that one of the two C-O bonds per monomeric unit changed from the trans to the gauche form.

The structures of both the 4:1 (Type I) and 1:1 (Type II) PEO polymer-HgCl$_2$ complexes have been determined[53] by x-ray diffraction. For the Type I fiber, four chains pass through the lattice and four monomeric units are contained in the fiber identity period. The conformation of PEO in the complex has been found to be $T_5GT_5\bar{G}$; that is where G and $\bar{G}$ mean the right and left-handed gauche forms, respectively. The bond length between

$$-CH_2\!\!\left(O\text{-}CH_2\text{-}CH_2\text{-}O\text{-}CH_2\text{-}CH_2\text{-}O\text{-}CH_2\text{-}CH_2\text{-}O\text{-}CH_2\text{-}CH_2\right)_n$$
$$\text{T T}\quad\text{T}\quad\text{T T}\quad\text{G}\quad\text{T T}\quad\text{T}\quad\text{T T}\quad\bar{\text{G}}$$

Hg and Cl in the complex (2.30Å) is a little longer than that of HgCl$_2$ in the crystal (2.25Å). This was consistent with the fact that the infrared absorption band associated with the antisymmetric stretching vibration of HgCl$_2$ shifts to 353 cm^{-1} in the complex from 367 cm^{-1} in the crystal. The crystal structure of the Type II complex was also determined[54] by x-ray diffraction and infrared absorption methods. For the Type II fiber, two PEO chains pass through the lattice. The conformation of PEO was near to $TG_2T\bar{G}_2$. In both cases of Type I and Type II, it was found that the HgCl$_2$ molecule is slightly distorted from its normal linear form. Coordination to the oxygen atoms of PEO may account for this. The packing of HgCl$_2$ in Type II is much closer than in Type I. The nearest Cl---Cl distance is 3.40Å which is smaller than the sum of the van der Waals radii of the chlorine atoms. The Hg-Cl distance in the Type II fiber is equal to the distances in the crystalline state. In this study it was considered that each oxygen atom of the PEO molecule coordinates to two mercury atoms with an equivalent distance and angle ($<$HgOHg$=85°$).

Polyesters

The interaction of $Cu(NO_3)_2$, $Zn(NO_3)_2$, $Ca(NO_3)_2$ and $Cd(NO_3)_2$
with various polar polymers most of which were polyesters has
been investigated.[55] Specifically cellulose acetate, poly(vinyl-
acetate), poly(methyl methacrylate) and poly(methyl acrylate)
(Structure XXI) were studied. Metal incorporation into the
polymer was achieved by mixing solutions of polymer and nitrate
followed by casting the solutions onto a KBr plate for infrared
measurements or onto a glass plate for T_g measurements. Compatibility of the additive and polymer was tested by noting the
clarity of the cast film after evaporation of the solvent. Complete clarity was indicative of solution of the salt in the
polymer. Salts were most readily soluble in cellulose acetate
with calcium and cadmium nitrates being more soluble than the
more covalent zinc and copper salts. Glass transition temperatures
varied depending on the salt and polymer system. The absolute
magnitude of the shift of T_g with salt concentration did not
always continuously increase or decrease (i.e. it maximized or
minimized).

Large shifts in the IR spectra of both nitrate and in the
polymer carbonyl and ester ether frequencies have been observed.
These observations suggested complex formation between polymer and
metal salt in the solid state. The carbonyl frequency decreased
and the ether frequency increased. The model for the polymer salt

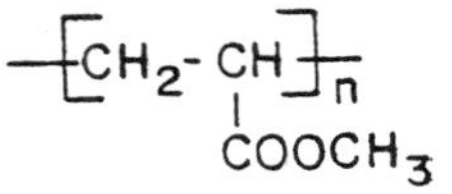

$$-\left[CH_2-CH\right]_n$$
$$\underset{COOCH_3}{|}$$

XXI

complex is shown below (Structure XXII). The metal ion is believed
to have its nitrate counterions near as well as several solvent
molecules since 4 or 6-coordination is anticipated for this metal.
A more highly hydrated system was predicted for the ionic salts.

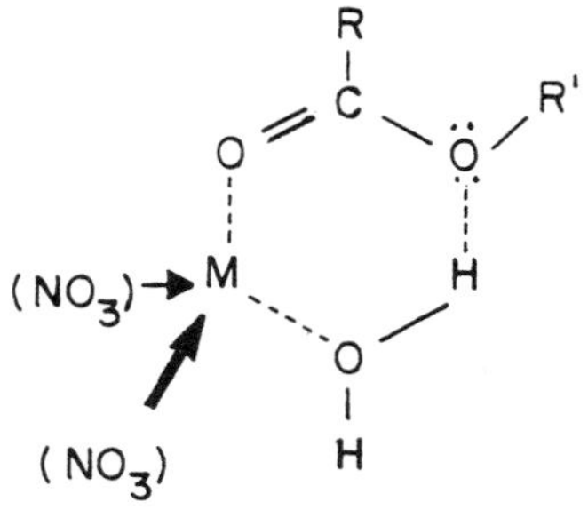

XXII

Differences in properties of the four polymer systems were stated
to be due to interaction and steric effects. The smallest T_g
effects were shown by poly(methyl methacrylate) since the salts
were least readily soluble in this polymer.

Polyacetylene

A major concentrated research effort is now being devoted
to the synthesis and properties of highly conducting derivatives
of polyacetylene (Structure XXIII). Studies have been most
extensive where nonmetal species (eletron acceptors) serve as the
dopant such as I_2, ICl, IBr, AsF_5[56], $(FSO_2O)_2$, H_2SO_4 and $HClO_4$.[57]
Overall increases in conductivity range up to eleven orders of
magnitude with the highest conducting films exhibiting room

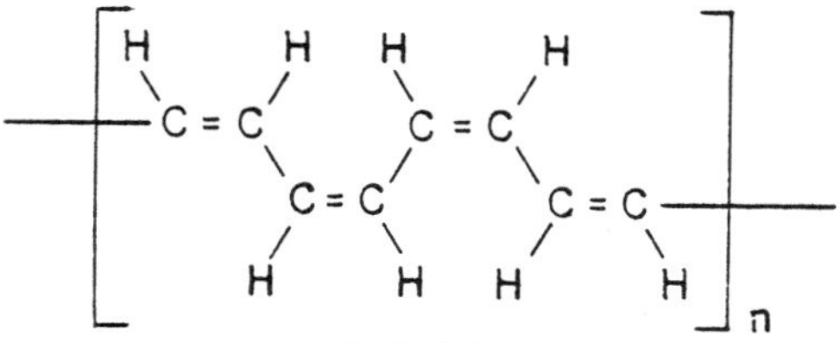

XXIII

temperature values of several hundred ohm^{-1} cm^{-1}. Since this
review is designed to cover only metal doping, the reader is
referred to several treatises[58,59] covering nonmetal species
addition to polyacetylene.

Less effort appears to have been expended toward incorporating
metal species (electron donors) into polyacetylene. Sodium doped
polyacetylene films have been prepared[60] by treating the polymer
with a solution of sodium naphthalide, $Na^+(C_{10}H_{18})^-$, in THF
whereupon electron transfer from the naphthalide radical anion
to the $(CH)_x$ was believed to occur. Employing the _trans_ form,
a film of composition $[Na_{0.28}CH]_x$ was produced with conductivity
at 25°C equal to $8x10^1$ ohm^{-1} cm^{-1}. The _trans_ polyacetylene
alone exhibited a conductivity of $4.4x10^{-5}$ ohm^{-1} cm^{-1}. Lithium
and potassium dopants under similar experimental conditions
yielded comparable results. The general features of electrical
conductivity with eiter donor or acceptor doping are the same,
although detailed differences in the saturation values and the
critical concentration for the metal-insulator transition vary.
In addition donors and acceptors can dope polyacetylene to n-type
and p-type conductors respectively.

Compensation of n-type material by acceptor. doping has been
successfully demonstrated[61] using Na(donor) and I_2 (acceptor).
The conductivity of the n-type material gradually decreases to a
minimum on exposure to an iodine atmosphere. Starting with a film
of initial composition $(CHNa_{0.27})_x$ the compensation (minimum conductivity) point occurred at $(CHNa_{0.27}I_{0.28})_x$. Continued I_2
doping led to a p-type material where the iodine was believed to
act as a polyanion in the presence of the now positively charged
donor species (D^+). The charge transfer need not be complete.
The resulting electron on the polymer chain is weakly bound to the
D^+. Compensation in the case of iodine (A) was stated to occur
through formation of (D^+A^-) leaving the polymer chain essentially
neutral and without electronic carriers.

When polyacetylene films are dipped into a solution of $AgClO_4$ or $AgBF_4$ in toluene[62], incorporation of the silver salt takes place to an extent which varies with both the concentration of the toluene solution and the exposure time. Complexes varying in stoichiometry from $(CHAg_{0.006})_x$ to $(CHAg_{0.018})_x$ have been investigated. Striking changes in the film conductivity are produced by silver salt incorporation (i.e. polymer alone, 10^{-5} ohm^{-1} cm^{-1}; $(CHAg_{0.018})_x$, 3 ohm^{-1} cm^{-1}). X-ray diffraction data suggest that (1) $(CH)_x$ remains unchanged, (2) no silver salt is observed and (3) free silver is present. Ammonia vapor causes a rapid decrease in film conductivity; yet, x-ray peaks corresponding to free silver remain unchanged. The enhanced conductivity in the doped film is concluded to not be due to silver metal. Electron microscopy and infrared data suggest that the silver ions are acting as oxidants to give free silver, immobile perchlorate anions and conducting polyolefinic cations according to the following equation:

$$(CH)_x + yAgClO_4 \rightarrow [(CH)_x^{y+}(ClO_4)^-_y] + yAg°$$

Effects similar to those described for silver are stated to be achieved using the salts of iron and copper; however, no data is available.

Poly(p-phenylene)

Results similar to those obtained with polyacetylene have been realized with poly(p-phenylene) (Structure XXIV) on doping with either electron acceptors or electron donors.[63] The advantage

<u>XXIV</u>

of the latter polymer system lies in the fact that $(C_6H_4)_x$ is more thermally stable (e.g. above 450°C in air and 550°C inert) and does not rapidly degrade in air when doped with AsF_4 which exhibits the highest conductivities.

Poly(p-phenylene) can be doped with alkali metals to provide highly conducting n-type materials which have a metallic-gold appearance. Sodium or potassium naphthalide solutions in THF increase conductivity from less than 10^{-10} ohm^{-1} cm^{-1} to a plateau value of 720 ohm^{-1} cm^{-1}. Increases in conductivity are believed to be due to the formation of charge-transfer complexes wherein the metal dopant donates a mobile electron to the polymer chain. Chemical compensation of the potassium doped poly(p-phenylene) can be accomplished with AsF_5 in a manner similar to donor doped polyacetylene.

The effect of moisture, air and halogens on poly(p-phenylene) have been investigated. The alkali metal doped polymer is much more sensitive to air exposure than the AsF_5 doped polymer.[64] Rapid tarnishing of the metallic gold appearance of the doped polymer along with a major decrease in conductivity results. While sodium and potassium dopants produce similar effects with poly(p-phenylene) and polyacetylene, halogens do not provide conducting complexes with the former.[65]

Polyethylenes

Several reports regarding polyethylene and its derivatives with various metal species have appeared, although no extensive studies have been communicated. The effect of salts of carboxylic acids (e.g. metal stearates) on the oxidation of polyethylene (Structure XXV) melts has been studied.[66] For example, zinc

$$\left[\begin{array}{cccc} H & H & H & H \\ | & | & | & | \\ -C - C - C - C - \\ | & | & | & | \\ H & H & H & H \end{array}\right]_n$$

__XXV__

stearate accelerated the oxidation. The greater the amount of
zinc stearate introduced, the greater the oxidation. Copper and
lead stearates in low concentrations accelerated polyethylene
oxidation. At higher concentrations catalysis of oxidation is
observed in the early stages of thermoprocessing, but then inhibition occurs. Active metals were found to have a similar
effect on oxidation. It was concluded that metal containing
compounds are formed which diffuse into the melt and control the
oxidation rate of the polymer.

Poly(tetrafluoroethylene) (PTFE) is by nature insoluble and
highly immiscible with most solids and liquids. This serves
as an obstacle to its development into a useful composite material. Metal-organic substances of low solubility have been
found[67] to be miscible with poly(tetrafluoroethylene) (Structure
XXVI). Iron pentacarbonyl has been absorbed into PTFE and
subsequently transformed into iron oxide applying this principle.
The PTFE samples are soaked in a 10% solution of $Fe(CO)_5$ in
ethanol at room temperature. Under these conditions the carbonyl

$$\left[\begin{array}{cccc} F & F & F & F \\ | & | & | & | \\ -C & -C & -C & -C- \\ | & | & | & | \\ F & F & F & F \end{array} \right]_n$$

XXVI

is absorbed by the polymer but ethanol is not. Irradiation of
the doped PTFE yields non-volatile $Fe_2(CO)_9$; after which, the
samples are allowed to undergo air oxidation for three days to
yield an Fe_2O_3-PTFE composite. Iron oxide contents range from
0.34 to 1.50%, (w/w). Fe_2O_3 particles are large and close to
spherical on the PTFE surface. A dramatic change in adsorbent
properties of PTFE occurred on incorporation of Fe_2O_3. The doped
polymer is stable at room temperature showing no tendency toward
phase separation or embrittlement. The authors suggest that

the oxide may be used to anchor other chemicals and incorporate
them into PTFE leading to other derivatives.

Thermal pyrolysis and oxidative pyrolysis of poly(propylene)
(PP) (Structure XXVII) with and without chromium have been
recently compared in an effort to improve flame retardancy of
PP.[68] PP was swollen in CCl_4 and then reacted with chromyl
chloride, CrO_2Cl_2, dissolved in CCl_4 at room temperature. The

$$\left[\begin{array}{ccc} CH_3 & CH_3 & CH_3 \\ | & | & | \\ CH-CH_2- & CH-CH_2- & CH-CH_2 \end{array} \right]_n$$

<u>XXVII</u>

product after drying at 60°C for 40 hours contained approximately
2% chromium. Between 388 and 438°C the <u>thermal</u> pyrolysis of PP
and chromium containing PP have comparable reaction rates.
Doping the polymer, however, lowers the thermal pyrolysis activation energy of PP from 51 to 44 KCal/mole while increasing
the temperature of the maximum pyrolysis endotherm by 15°C. The
effect of chromium on <u>oxidative</u> pyrolysis is more substantive by
comparison. Chromium suppresses the formation of all major
products although no new product was detected. It also increases
the activation energy of oxidative pyrolysis by 10 KCal/mole,
promotes char formation and becomes self-extenguishing with
about 1% of covalently bonded chromium. This behavior suggested
to the investigators that chromium possibly catalyzes termination
processes. Chromium is believed to accelerate the chain termination reaction by providing a nonradical pathway for the
destruction of the reactive hydroperoxide intermediate. The
remarkable efficacy of chromium is emphasized by the fact that
at only 1.5% level of chromium PP has a limiting oxygen index
of 26.4 and a self-ignition temperature of 400°C in air as compared to 17.4 and 250°C respectively for normal PP. The polymer

also contained about 0.6% chlorine, but its presence was not
believed to make any significant contriubtion to flame retardancy.

Poly(methylphenylsiloxane)

The above liquid polymer system provides a unique technique whereby metal species may be introduced. The employment
of metal vapor-liquid polymer synthetic techniques has been
successfully employed to control the competing processes of
metal atom diffusion, metal atom aggregation and metal atom
anchoring. Francis[69] has demonstrated that Ti, V, Cr and Mo
vapors can be deposited into a liquid poly(methylphenylsiloxane),
(Structure XXVIII) Dow Corning 510, containing a methyl:phenyl
ratio of 17:1. The reactions were carried out in a rotary
reactor in the temperature range 0 to -20°C. Organometallic
polymers in which two phenyl groups are coordinated to the
metal to give an anchored bis(arene) complex are produced.
The reaction scheme V shown below was suggested after spectroscopically monitoring the decay of the absorbance due to free

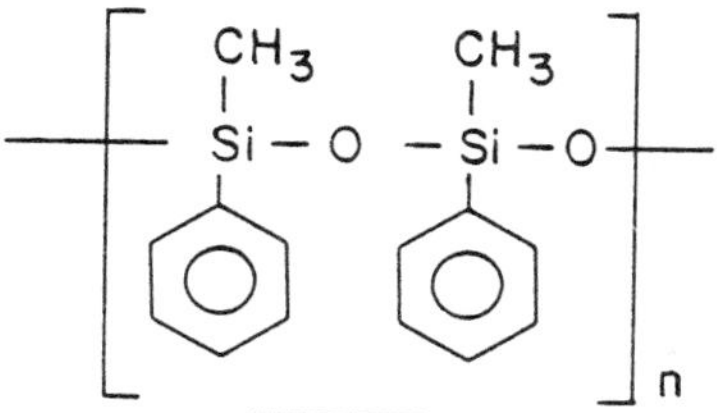

XXVIII

phenyl groups (214 nm) and the corresponding growth of the
absorbances due to products. Polymer-stabilized few-atom
clusters in this system were believed to arise from either
phenyl group solvation effects or bis(arene) metal cluster
complexation. Metal-arene interaction was suggested to
gradually diminish as the cluster nuclearity increased until

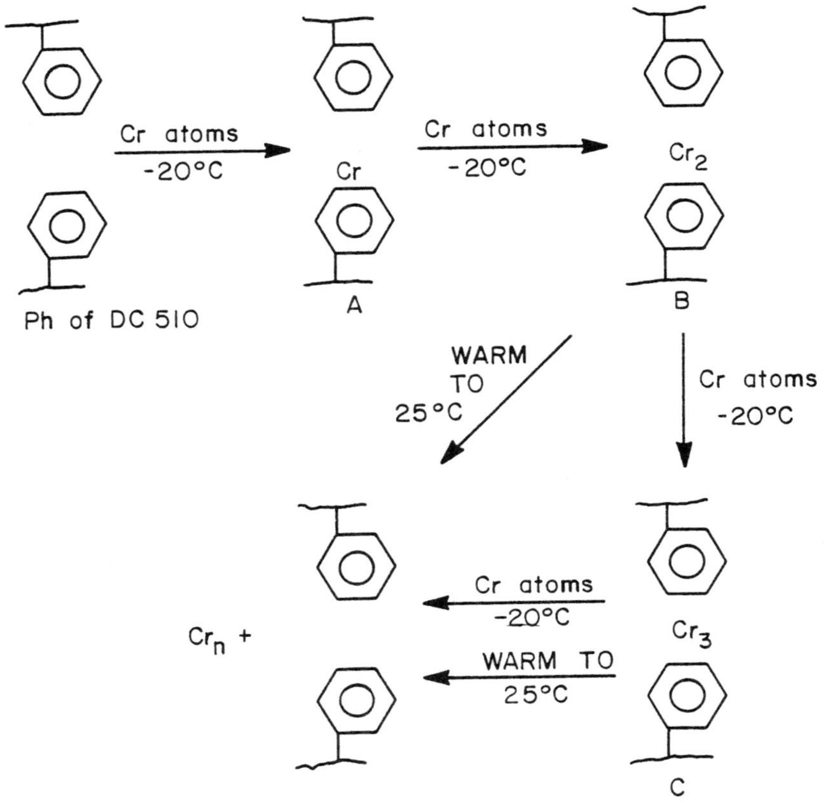

SCHEME V

at values of n greater than 3, metal cluster desolvation
occurred to release clusters containing on the order of four
to six metal atoms.

Attempts[70] to prepare systems containing two different
metals have been reported recently. The bimetallic system
was investigated by means of the following reactions: (1)
sequential Ti and Cr vapor deposition, (2) simultaneous Ti/Cr
vapor deposition, (3) saturation of the phenyl groups on
poly(methylphenylsiloxane) followed by reaction with Cr atoms

and (4) saturation with Cr vapor followed by reaction with Ti
atoms. Bimetal Ti/Cr depositions at low metal loadings led
to polymer-supported bis(arene) complexes with two different
mononuclear metal sites attached to the same polymer. At
high metal loadings binuclear sites containing polymer-stabilized
Ti_2 and Cr_2 as well as TiCr clusters are formed. Very high
Ti/Cr loadings are suggested to lead to unsolvated Cr_xTi_y bimetallic clusters approaching colloidal dimensions (Scheme

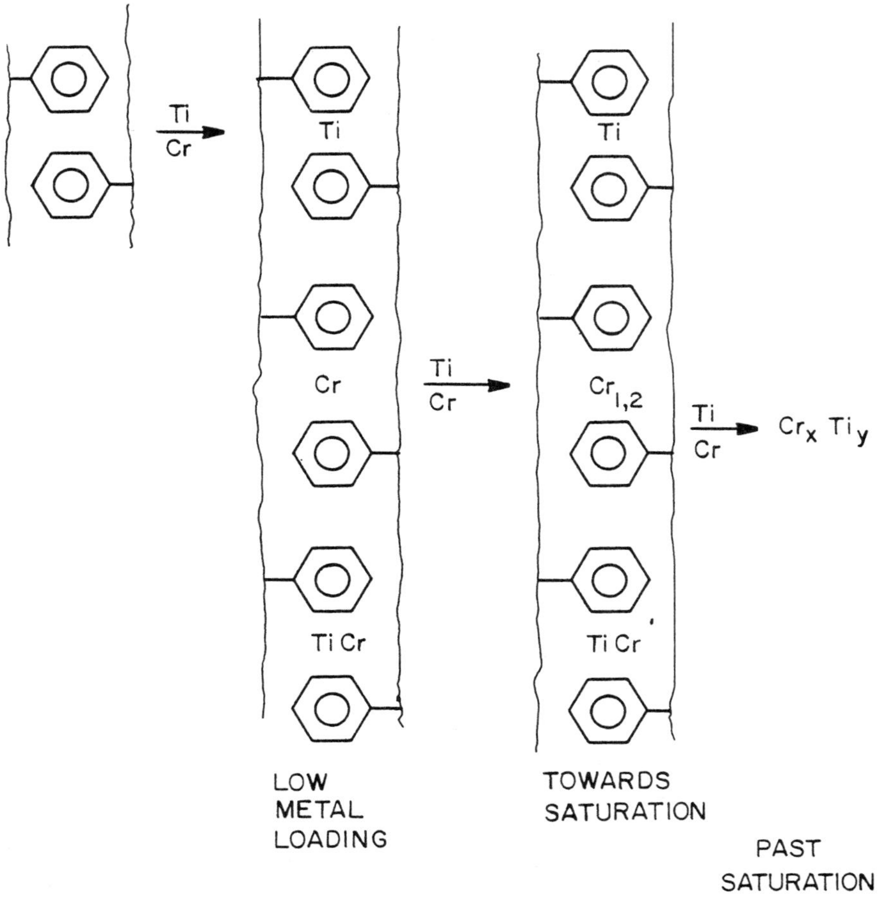

LOW
METAL
LOADING

TOWARDS
SATURATION

PAST
SATURATION

SCHEME VI

VI). Saturation by Ti followed by reaction with Cr indicated
that the Ti-containing species are kinetically unstable. For
example, the bis(arene) titanium complex appears to yield the
TiCr cluster species. Questions regarding the mechanism of
Ti displacement by Cr and vice versa were not addressed.

In conclusion, the incorporation of metal related materials
into electrically neutral polymers is an active and highly
diverse research pursuit. The metallic species and polymer
systems which have been explored todate are highly varied.
The impetus for such studies is generally the modification of
polymer properties; however, the specific properties of interest
vary greatly. Modification of adhesive properties, thermal
behavior, electrical conductivity, polymer flamability and
mechanical properties have been studied for example. The
literature cited is by no means meant to be exhaustive; but,
we believe it to be well representative of the research area.
The area no doubt will continue to expand as investigators
explore other metal related dopants and polymer systems in an
effort to understand and synthesize materials which posses
specific and highly desirable properties.

Acknowlegement - We wish to thank E. Denyszyn, T.
Wohlford and S. Ezzell for assistance in preparing this
manuscript.

REFERENCES

(1)   E. P. Otocka, J. Macromol. Sci. - Revs. Macromol. Chem.,
      C5, 275 (1971).

(2)   E. G. Kolawole and S. M. Mathieson, J. Polym. Sci. Polym.
      Chem. Ed., 15, 2291 (1977).

(3)   H. Matsuda, J. Appl. Polym. Sci., 23, 2603 (1979) and
      references therein.

(4)   C. Carraher and J. Lee, J. Macromol. Sci. Chem., A9, 191
      (1975).

(5)   C. U. Pittman, O. Ayers and S. McManus, Macromolecules,
      7, 737 (1974).

(6)   C. U. Pittman, Chem. Tech., 416 (1971).

(7)   H. Nishikawa and E. Tsuchida, J. Phys. Chem., 79, 2072
      (1975).

(8)   A. J. Varma and J. Smid, J. Polym. Sci. Polym. Chem. Ed.,
      15, 1189 (1977).

(9)   N. Hojo, H. Shirai, Y. Chujo and S. Hayashi, J. Polym. Sci.
      Polym. Chem. Ed., 16, 447 (1978).

(10) C. U. Pittman, L. R. Smith and R. M. Hanes, J. Amer. Chem.
      Soc., 97, 1742 (1975).

(11) H. Nishide, J. Deguchi and E. Tsuchida, J. Polym. Sci. Polym.
      Chem. Ed., 15, 3023 (1977).

(12) J. Pitha and J. Smid, Biochim. Biophys. Acta, 425, 287 (1976).

(13) H. Nishikawa, E. Terada and E. Tsuchida, J. Polym. Sci.
      Polym. Chem. Ed., 16, 2453 (1978).

(14) C. U. Pittman and L. R. Smith, J. Amer. Chem. Soc., 97,
      1749 (1975).

(15) E. Bayer and V. Schwrig, Chem. Tech., 212 (1976).

(16) J. C. Anderson, U. S. Patent #4,109,052, Aug. 22, 1978.

(17) D. J. Progar and T. L. St. Clair, Preprints, 7th National
      SAMPE Technical Conference, 7, 53 (1975).

(18) C. W. Van Der Waal, H. W. Bree and F. R. Schwargl, J.
      Appl. Polym. Sci., 9, 2143 (1965).

(19) J. M. Pochan, H. W. Gibson and F. C. Bailey, J. Polym.
      Sci. Polym. Lett. Ed., 18, 447 (1980).

(20) Y. W. Park, A. Denensten, C. K. Chiang, A. J. Heeger and
      A. G. MacDiarmid, Solid State Commun., 29, 745 (1979).

(21) T. C. Clarke, R. H. Geiss, W. D. Gill, P. M. Grant, J. W.
      Macklin, H. Morawitz, J. F. Rabolt, D. Sayers and G. B.
      Street, J. C. S. Chem. Comm. 332 (1979).

134   ST. CLAIR AND TAYLOR

(22) K. Mizoguchi, T. Kamiya, E. Tsuchida and I. Shinohara, J.
      Polym. Sci. Polym. Chem. Ed., 17, 649 (1979).

(23) A. Ciferri, E. Bianchi, F. Marchese and A. Tealdi,
      Makromol. Chemie, 150, 265 (1971).

(24) B. Valenti, E. Bianchi, G. Greppi, T. Tealdi and A. Ciferri,
      J. Phys. Chem., 77, 389 (1973).

(25) E. Bianchi, A. Ciferri, A. Tealdi, R. Torre and B. Valenti,
      Macromolecules, 7, 495 (1974).

(26) D. Acierno, E. Bianchi, A. Ciferri, B. DeCindio, C. Migliaresi
      and L. Nicolais, J. Polym. Sci. Symposium No. 54, 259 (1976).

(27) D. Acierno, F. P. LaMantia, G. Titamanlio and A. Ciferri,
      J. Polym. Sci. Polym. Phys. Ed., 18, 739 (1980).

(28) P. Dunn and G. F. Sansom, J. Appl. Polym. Sci., 13, 1641
      (1969).

(29) P. Dunn and G. F. Sansom, J. Appl. Polym. Sci., 14, 1799
      (1970).

(30) P. Dunn and G. F. Sansom, J. Appl. Polym. Sci., 13, 1657
      (1969).

(31) H. Kakinoki, O. Sumita, C. S. Cho and F. Higashi, J. Polym.
      Sci. Polym. Lett. Ed., 14, 407 (1976).

(32) F. Higashi, C. S. Cho, H. Kakinoki and O. Sumita, J. Polym.
      Sci. Polym. Chem. Ed., 15, 2303 (1977).

(33) O. Sumita, A. Fukuda and E. Kuze, J. Polym. Sci. Polym. Phys.
      Ed., 16, 1801 (1978).

(34) O. Sumita, A. Fukuda and E. Kuze, J. Appl. Polym. Sci.,
      23, 2279 (1979).

(35) F. Higashi, C. S. Cho, H. Kakinoki and O. Sumita, J. Polym.
      Sci. Polym. Chem. Ed., 17, 313 (1979).

(36) R. J. Angelo and E. I. DuPont DeNemours & Co., U. S.
      Patent, No. 3,073,785 (1959).

(37) R. J. Angelo, Private Communication.

(38) L. T. Taylor, V. C. Carver, T. A. Furtsch and A. K. St.
      Clair, ACS Symp. Ser., 121, 71 (1980).

(39) M. N. Sarboluki, NASA Tech. Brief, 3(2), item 36, Oct.,
      1978.

(40) A. K. St. Clair, V. C. Carver, L. T. Taylor and T. A.
      Furtsch, J. Amer. Chem. Soc., 102, 876 (1980).

(41) L. T. Taylor, V. C. Carver, T. A. Furtsch and A. K. St.
      Clair, PREPRINTS, Org. Coat. and Plas. Chem., 43, 635
      (1980).

(42) N. S. Lidorenko, L. G. Gindin, B. N. Yegorov, V. I.
      Kondratenkov, I. Y. Ravich and T. N. Toroptseva, AN SSSR,
      Doklady, 187, 581 (1969).

(43) A. L. Endrey and E. I. DuPont DeNemours & Co., U. S. Patent,
      No. 3,073,784 (1963).

(44) R. E. Frazer, NASA Tech Brief, 4(1), item 101, Aug., 1979.

(45) S. M. Aharoni and A. J. Signorelli, J. Appl. Polym. Sci.,
      23, 2653 (1979).

(46) N. Hojo, H. Shirai and S. Hayashi, J. Polym. Sci. Symp. No.
      47, 299 (1974).

(47) O. Sumita, A. Fukuda and E. Kuze, J. Polym. Sci. Polym.
      Phys. Ed., 18, 877 (1980).

(48) M. J. Hannon and K. F. Wissbrun, J. Polym. Sci. Polym.
      Phys. Ed., 13, 113 (1975).

(49) R. E. Wetton, D. B. James and W. Whiting, J. Polym. Sci.
      Polym. Lett. Ed., 14, 577 (1976).

(50) J. Moacanin and E. F. Cuddihy, J. Polym. Sci., C14, 313
      (1966).

(51) A. A. Blumberg and J. Wyatt, Polym. Lett., 4, 653 (1966).

(52) A. A. Blumberg, S. S. Pollack and C. A. J. Hoeve, J. Polym.
      Sci., A2, 2499 (1964).

(53) R. Iwamoto, Y. Saito, H. Ishihara and H. Tadokoro, J. Polym.
      Sci. Part A-2, 6, 1509 (1968).

(54) M. Yokoyama, H. Ishihara, R. Iwamoto and H. Tadokoro,
      Macromolecules, 2, 184 (1969).

(55) K. F. Wissbrun and M. J. Hannon, J. Polym. Sci. Polym.
      Phys. Ed., 13, 223 (1975).

(56) R. H. Baughman, S. L. Hsu, L. R. Anderson, G. P. Pez and
A. J. Signorelli, "Molecular Metals", NATO Conference
Series, W. E. Hatfield, Ed., Plenum Press, New York, 1979,
pp. 187.

(57) S. C. Gau, J. Milliken, A. Pron, A. G. MacDiarmid and
A. J. Heeger, J. C. S. Chem. Comm., 662 (1979).

(58) A. G. MacDiarmid and A. J. Heeger, "Molecular Metals",
NATO Conference Series, W. E. Hatfield, Ed., Plenum Press,
New York, 1979.

(59) T. C. Clarke, R. H. Geiss, W. D. Gill, P. M. Grant, H.
Morawitz, G. B. Street and D. E. Sayers, Synthetic Metals,
1, 21 (1979/80).

(60) C. K. Chiang, M. A. Druy, S. C. Gau, A. J. Heeger, E. J.
Louis, A. G. MacDiarmid, Y. W. Park and H. Shirakawa,
J. Amer. Chem. Soc., 100, 1013 (1978).

(61) C. K. Chiang, S. C. Gau, C. R. Fincher, Y. W. Park, A. G.
MacDiarmid and A. J. Heeger, Appl. Phys. Lett., 33, 18
(1978).

(62) T. C. Clarke, R. H. Geiss, J. F. Kwak and G. B. Street,
J. C. S. Chem. Comm., 489 (1978).

(63) D. M. Ivory, G. G. Miller, J. M. Sowa, L. W. Shacklette,
R. R. Chance and R. H. Baughman, J. Chem. Phys., 71, 1506
(1979).

(64) R. H. Baughman, D. M. Ivory, G. G. Miller, L. W. Shacklette
and R. R. Chance, PREPRINTS, Org. Coat. and Plas. Chem.,
41, 139 (1979).

(65) L. W. Shacklette, R. R. Chance, D. M. Ivory, G. G.
Miller and R. H. Baughman, Synthetic Metals, 1, 307 (1979/80).

(66) N. I. Egorenkov, D. G. Lin and V. A. Bely, J. Polym.
Sci. Polym. Chem. Ed., 13, 1493 (1975).

(67) F. Galembeck, C. C. Ghizoni, C. N. Inpe, C. A. Ribeiro,
H. Vargas and L. C. M. Miranda, J. Appl. Polym. Sci.,
25, 1427 (1980).

(68) J. C. W. Chien and J. K. Y. Kiang, Macromolecules, 13,
280 (1980).

(69) C. G. Francis and P. L. Timms, J. Chem. Soc. Chem. Commun.,
      466 (1977).

(70) C. G. Francis, H. Huber and G. A. Ozin, J. Amer. Chem. Soc.,
      101, 6250 (1979).

NEW PROCEDURES AND TECHNIQUES

Trimethylsilylation of Mineral Silicates

B.R. CURRELL and J.R. PARSONAGE

School of Chemistry, Thames Polytechnic,
Woolwich, London SE18 6PF

ABSTRACT

The application of the technique of trimethylsilylation
to mineral silicates is reviewed. Information can be
obtained on the molecular size distribution of the silicate
anions and also on the aluminium distribution in a silicate
backbone. Examples of the use of trimethylsilylation in
the structural determinations of silicates include examination
of sodium silicates and their solutions, calcium silicates,
various glasses, cements and concretes. Included within
the review is a discussion of the attempts to prepare new
polyorganosiloxane materials. Trimethylsilylation of certain
micas, for example, give polyorganosiloxane fluids and greases;
and attempts to physically and chemically modify the
polyorganosiloxane products are also reviewed.

INTRODUCTION

The trimethylsilylation of mineral silicates involves the
removal of ionic sites (and associated cations) and their
replacement by trimethylsilyl groups.

$$\geqslant Si - O^{\ominus}M^{\oplus} \longrightarrow \geqslant Si - O - SiMe_3$$

Interest in this subject area results from two objectives:-
(a) the determination of new information on silicate structures
    and

(b) the preparation of new polyorganosiloxane materials.

Silicate anions occur in sizes ranging from small molecules containing one silicon atom up to macromolecules of infinite size. These macromolecules are composed of building units of the general formula

$$(^-O)_n Si(O_{\frac{1}{2}})_{4-n}$$

n = 0,1,2,3 or 4

The macromolecular silicate anions may be single chains, double chains, sheets or various irregular assemblies. Successful trimethylsilylation of all these arrays gives rise to polyorgano-siloxanes composed of mainly monofunctional (M) and quadrifunctional building units (Q).

```
           O½                        Me
           |                         |
  O½ ─── Si ─── O½         Me ─── Si ─── O½
           |                         |
           O½                        Me
```

        Quadrifunctional (Q)   Monofunctional (M)

In the search for additional information on silicate structures it was hoped that the polyorganosiloxanes prepared by the trimethyl-silylation of the silicate anion would reproduce exactly the back-bone structure of the silicate used i.e. only the replacement of ionic sites occurs. If this hope were realised then the molecular size distribution of the polyorganosiloxane products would be exactly the same as the molecular size distribution of the silicate being examined. Although X-ray crystallography and related techniques give very detailed information on the relative positions of the atoms in silicate structures these techniques are unable to give information on the overall size of the silicate anions. Silicates cannot directly be dissolved in solution without molecular break-down and thus cannot be examined directly in solution by the techniques of polymer physics; however the trimethylsilylated derivatives may be quantitatively analysed by a variety of techniques including hplc, glc, mass spectrometry, nmr and thermal analysis.

Inorganic polymers have recently received much attention and books by Ray on Inorganic Polymers [1] , Wilson and Crisp entitled Organolithic Materials [2] , Holliday on Ionic Polymers [3] and Carraher on Organometallic Polymers [4] give a broad introduction to this general area. This review is the first to concentrate solely on the trimethylsilylation of mineral silicates, the equally important area of grafting organic groups to mineral silicates will be dealt with in a future review.

THE TRIMETHYLSILYLATION REACTION

The two fundamental techniques for mineral trimethylsilylation
have been called the Lentz [5] and the 'direct method' attributed
to Gotz and Masson [6] ; the "Lentz technique" consists of slurrying
the mineral with water for 1 h before adding the two-phase reaction
mixture of ice, concentrated hydrochloric acid, propan-2-ol and
hexamethyldisiloxane; the mixture is then stirred for 48 h at room
temperature and then filtered. The acid/water phase leaches the
cations from the mineral and a silicic acid is formed; this
silicic acid is capped by the trimethylsilylating agent which is
formed by the breakdown of the hexamethyldisiloxane in the presence
of hydrochloric acid. The function of the propan-2-ol is to increase
the solubility of the silylating agent in the water. The various
reaction sequences may be described as follows:-

$$
\begin{bmatrix} O^{\ominus}M^{\oplus} \\ | \\ -Si-O- \\ | \\ O^{\ominus}M^{\oplus} \end{bmatrix}_n \quad \xrightarrow[\text{Cation removal}]{H_2O/HCl} \quad \begin{bmatrix} OH \\ | \\ -Si-O- \\ | \\ OH \end{bmatrix}_n + 2nM^+
$$

$$
Me_3SiOSiMe_3 \quad \xrightarrow{H_2O/HCl} \quad 2Me_3SiOH
$$

$$
\begin{bmatrix} OH \\ | \\ -Si-O- \\ | \\ OH \end{bmatrix}_n \quad \xrightarrow{Me_3SiOH} \quad \begin{bmatrix} SiMe_3 \\ O \\ | \\ -Si-O- \\ | \\ O \\ SiMe_3 \end{bmatrix}_n + H_2O
$$

Many workers have found that trimethylsilylation does not always
proceed to completion thus in the final product some unreacted
silanol groups may be found. As a result the derivatization has
been completed by treating the initial product with Amberlyst 15
(an ion exchange resin) in the presence of hexamethyldisiloxane.
This converts the residual silanol groups to their trimethylsilyl
analogues. The application of the trimethylsilylation technique
to the determination of the molecular size of silicate anions
depends on the assumption that the value of n does not change

during the course of the reaction. Unfortunately, the polysilicic acids in acidic aqueous solutions are liable to various polymer-isation/depolymerisation equilibria as follows:-

$$\left[\begin{array}{c} OH \\ | \\ Si - O \\ | \\ OH \end{array}\right]_n \overset{H_2O}{\rightleftharpoons} HO - \begin{array}{c} OH \\ | \\ Si - OH \\ | \\ OH \end{array} + \left[\begin{array}{c} OH \\ | \\ Si - O \\ | \\ OH \end{array}\right]_{n-1}$$

The Gotz and Masson direct method was developed in an attempt to suppress this side reaction by reducing the amount of free hydrochloric acid and water available by using trimethylchlorosilane instead of hexamethyldisiloxane. Hence it is particularly applicable to those minerals containing appreciable amounts of water of crystallisation. In the initial publications [6,7,8] describing their method Gotz and Masson reported its application to, for example, olivine (an orthosilicate) and hemimorphite (a pyrosilicate); the former gave an 88% yield of QM_4 and the latter a 94% yield of Q_2M_6 in accurate reflections of the original silicate structures.
Dent Glasser and Sharma [9,10] applied trimethylsilylation to a study of sodium silicates and concluded that Lentz's method is more suitable for the study of solutions, probably because the large quantity of hydrochloric acid used rapidly reduces the pH to a point where these polymerisation/depolymerisation equilibria are relatively slow.
    Various different groups of workers have tried various reaction techniques to minimise the opportunities for silicic acid inter-mediates to undergo side-reactions. The equilibrium constants of these reactions are strongly dependent on pH and the reaction rates have a minima between pH 2 and 3; attempts to minimise these side reactions therefore involve control of pH and a reduction of the time needed for trimethylsilylation. In order to accelerate the trimethylsilylation reaction Garzo et al [11] used weakly acidified bistrimethylsilylacetamide (BSA) as the trimethylsily-lating agent and replaced propan-2-ol by acetone. Acetone was added in such an amount that the reaction mixture appeared to be a single homogeneous phase after the addition of the BSA solution. In order to demonstrate that side reactions were not occurring four main components of silicic acid solutions (QH_4, Q_2H_6, Q_3H_8 and Q_4H_8) were first determined by means of ^{29}Si nmr spectroscopy.
These results were in acceptable agreement with the glc results obtained after BSA silylation of the same sample. To provide further justification of their technique Garzo et al studied the application of their technique and that of the Lentz and the Gotz-Masson techniques to the study of mono- and disilicic acid solutions.

Freshly prepared (lifetime 1 min) mono- and disilicic acid solutions
were chosen for the silicate systems, it was assumed that hydrolysis
(leading to monosilicic acid) or condensation (leading to di-,
trimeric or higher silicic acids) takes place to only a small
extent within 1 min. Thus nearly 100% of the weighed silica is
expected to appear as QM_4 (monosilicic acid sample) and Q_2M_6
(di-silicic acid samples) and all other components indicate hydrolysis
or condensation during the silylation process. The Garzo technique
gave 77% of the silica occurring as QM_4 from the monosilicic acid
solution and 84% of the silica occurring as Q_2M_6 from the disilicic
acid solutions. Garzo et al concluded that the application of the
Gotz and Masson technique to these solutions appeared to promote
condensation reactions and the Lentz technique to favour the hydrolysis
of Q_2M_6 species. Both techniques promoted cyclisation whereas the
Garzo method seems to suppress all of these disturbing effects to
a reliable extent. They also concluded that incomplete silylation
allowed a formation of Q_2M_3H and Q_3M_7H only if Lentz or Gotz and
Masson techniques were used.

Sharma and Hoering [12] have developed anhydrous methods for
trimethylsilylation in order to reduce side reactions; 60-80% yields
of dimer, cyclotetramer and linear trimer were obtained from the
corresponding silicates i.e. hemimorphite, laumontite and
natrolite by reacting finely ground minerals with a large excess of
trimethylchlorosilane and hexamethyldisiloxane solvent; anhydrous
methanol saturated with dry hydrogen chloride and Amberlyst 15 ion
exchange resin were added to provide acidity.

After sixteen hours reaction trimethylsilylation was incomplete
and free OH groups were still present in the products. Derivatisation
was completed by stirring the reaction product with hexamethyldisiloxane
in the presence of the ion exchange resin. A second method which did
not give such good results involved stirring the mineral with a
large excess of trimethylchlorosilane and hexamethyldisilazane in
the solvent hexamethyldisiloxane. A small quantity of water was
then added slowly. Water reacts rapidly with the silylating agent
to yield hydrogen chloride and ammonium chloride which leaches the
metallic ions from the mineral. Derivatisation was again completed
by stirring the reaction product with hexamethyldisiloxane in the
presence of the ion exchange resin.

The use of Amberlyst 15 has been reported by many workers
however, it has been noted [11, 13] that some of the high molecular
weight products will redistribute on prolonged exposure to an
Amberlyst 15/hexamethyldisiloxane mixture. Masson [14] has shown
that Q_3M_6 is converted to Q_3M_8 on stirring with Amberlyst 15 for
several days at room temperature and has suggested derivatisation
using trimethylgermylchloride as an alternative. The interference
caused by the formation of isopropoxy derivatives in the trimethylsilylation of natrolite has been extensively studied by combined
gas chromatography - mass spectrometry [15]. In the absence of
side-reactions the natrolite structure should give solely octakis-

(trimethylsiloxy)trisiloxane by cleavage at the Al-O-Si sites; however siloxane cleavage and rearrangement give a wide range of linear and cyclic products. A range of various trimethyl siloxy and monoisoproxy derivatives were obtained, the most complicated product was decakis(trimethylsiloxy)tetrasiloxane and the simplest tris(trimethylsiloxy)monoisopropoxysilane.

CHARACTERISATION AND QUANTITATIVE ANALYSIS OF TRIMETHYLSILYLATED DERIVATIVES

The products of trimethylsilylation consist of QM units which may be bonded together to give a wide range of molecular possibilities. Separative methods involve the use of various chromatographic methods; gas liquid and thin layer chromatography for the lower molecular weight range and gel permeation chromatography for the higher ranges.

Early methods were based on a column of SE30 on Chemisorb W 80-100 mesh with a temperature programmed rise at 8 ^{o}C min^{-1} from 50 ^{o}C to 260 ^{o}C, with nitrogen being used as the carrier gas. These methods allowed analysis of the derivatives from QM_4 up to Q_4M_{10}. Masson [13] and Sharma [12] have both reported conditions for the detection of species up to Q_6M_{12}. The former either using SE30 on Chemisorb W or a 25m x 0.27 mm glass capillary column coated with OV 101, the latter using Dexsil 300, a silicone-carborane copolymer, for column packing. Model compounds Q_nM_{2n+2} (n = 1-4) Q_3M_6, Q_4M_8, Q_6M_{10}, Q_7M_{10}, Q_8M_8 and $Q_{10}M_{10}$ have been prepared by trimethylsilylation, the products separated by tlc and then the molecular weights and stoichiometry of the compounds were established by mass spectrometry and combustion analysis [16]. These model compounds were then chromatographed by Garzo et al [11, 16] using in separate experiments a glass capillary column and three packed columns all analyses being performed using a Perkin Elmer Model 900 gas chromatograph equipped with a flame ionization detector.

Retention indices were measured by the simultaneous injection of the sample compound, two suitable dimethylsiloxane chain oligomers $[(CH_3)_3SiO_{\frac{1}{2}}]_2$ $[(CH_3)_2SiO]_n$, and methane; the homologous series of dimethylsiloxane oligomers were used as secondary standards for the retention index determinations. Retention characteristics depend mainly on the number of M groups located on the surface of a molecule while the size and structure of the molecular skeleton have only minor effects. The relationship found between the retention indices and molecular structure was suggested to be a basis for the correct chemical and structural identification of unknown peaks.

Glasser [17] and also Cook [18] have used gel permeation chromatography (gpc) to obtain the molecular weight distributions of higher molecular weight fractions. In the case of Cook a Waters Associates Model 501hplc was used with three μ-styragel columns with permeability ranges 10^2, 5×10^2 and 10^3 Å, each having a plate count greater than 3000. Samples were injected as solutions in chloroform each containing benzene as an internal standard. Peaks were determined using refractive index and ultraviolet 254 nm detectors. The instrument was calibrated with polystyrene standards. Unfortunately no high molecular weight standards for the trimethyl-silylated esters of silicic acid exist and the only suitable commercially available standards are either polystyrene or poly-glycol, both very different chemically from the polymers under investigation. With this reservation Cook [18] reported that the highest molecular weight material derived by the trimethylsily-lation of a biotite mica was greater than 110,000 on the polystyrene scale.

Glasser et al have used gpc to study derivatives from sodium silicate solutions. The calibration curves were constructed using a combination of polydimethylsiloxane fractions and polystyrene standards. Gpc depends on the effective hydrodynamic volume of the molecules and Glasser et al realised that these calibrations were not strictly accurate. Derivatives of Si_4O_{12} and Si_8O_{20} were prepared with reasonable purity and gpc gave apparent molecular weights for these that were 10% low; indicating that their method of calibration was not grossly wrong. Recently Shimono and co-workers [19] have successfully separated QM_4, Q_2M_6 and the M_8 isomers Q_3M_8 and Q_4M_8 from the trimethylsilylation of hemimorphite using Bio-beads S-X1 as the support and chloroform isopropanol as eluent.

The simplest derivatives can of course be unambiguously characterised after separation simply by elemental analysis and molecular weight determinations, however, it is of course much more difficult to determine the structure where the molecular size allows different possibilities for the ways in which Q and M may be bonded together. To provide a basis for the study of the more complicated derivatives Harris and Newman [20] have studied the ^{29}Si and ^{13}C nmr of various derivatives. They have reported the relative ^{29}Si chemical shifts for trimethylsilylated silicate derivatives as follows:-

Compound	M - units ppm from QM_4		Q-units ppm from QM_4	
	on Q^1	on Q^2	Q^1	Q^2
Q_2M_6	0.30	-	-2.38	-
Q_3M_8	0.23	0.55	-2.64	-5.06
Q_4M_8	-	1.52	-	-3.65

These results indicate that the Q unit region of ^{29}Si nmr
spectrum is more useful than the M unit region for quantitative
analysis of a mixture because the Q unit shifts are dispersed over
a range 3 times as wide. In particular the Q units of the dimer
and trimer should definitely be resolvable where the M units
attached to these units may not be distinguished. They studied
the trimethylsilyl derivatives of wollastonite and also pseudowollastonite and concluded that the structures of the original
silicates were not retained , although the original structures
do clearly influence the product distribution; they also concluded
that ^{29}Si can be a useful technique in the analysis of such
products provided that all samples are studied at the same concentration in the same solvent; and that it is convenient to relate
chemical shifts to those of the corresponding QM_4 groups. In a
further paper [21] Harris in collaboration with Glasser,
Lachowski and Jones used these techniques to study the products
of the trimethylsilylation of tetramethylammonium silicate which
contained a compound of formula $C_{30}H_{90}O_{21}Si_{18}$ and showed that the
structure of this compound was related to a double four-membered
cage Q_8M_8 by the replacement of one Si-O-Si bridge by trimethylsiloxy units to give Q_8M_{10}.

EXAMPLES OF STRUCTURAL EXAMINATION USING TRIMETHYLSILYLATION

Study of Sodium Silicates

     As one example of the applicability of his technique Lentz
presented the results of a study of the trimethylsilylation of
sodium silicates and their solutions. In his original paper
Lentz [5] was well aware of the occurrence of side reactions but
nevertheless felt that valuable information could be obtained in
spite of this limitation. Thus for example only 43.5% of the
silicate was recovered as QM_4 from a 1 M sodium orthosilicate
solution with a Na:Si ratio of 4:1, he argued that although this
may be a true reflection of the content of the solution some of
the silicate solution is polymeric because (i) such a low yield
was never obtained with an orthosilicate mineral, (ii) a 9.6%
yield of a cyclic tetramer derivative is an improbable product of
a reaction if the solution is entirely monomeric and (iii) higher
yields of orthosilicate derivative were found as the concentration
of sodium was decreased.
     As described in earlier sections of this review the study of
sodium silicate solutions has provided a vehicle for the development of techniques which give derivatives providing a truer
reflection of the structure of the original solutions. Glasser
et al [9,17] agreed with the observations of earlier workers that
the degree of polymerisation increases with increasing concentration
and decreasing soda:silica ratio. Their results were consistent
with the view that cross-linking is a major factor as the degree

of polymerisation is increased. Of those species positively
identified only the monomer, dimer and trimer were linear, the
tetramer and hexamer were cyclic and the Si_8O_{21} derivative was
thought to be either a cage with one of the edges opened, or a
cage of four or five membered rings. Glasser and Lachowski [22]
have established that changes in pH and concentration of sodium
silicate alters the degree of polymerisation of the silicate
species as measured by trimethylsilylation. The structure of
the polymeric species formed in solution was the subject of a
second paper which established the following rules:-
(i) connectivity is maximised consistent with a lower ring
size of four tetrahedra, (ii) all tetrahedra in a given
species show as nearly as possible the same degree of connectivity
[23].

    Garzo et al[11] in their work also reported the formation
of non-linear species. They suggested the following pathways
for the formation of the more complicated derivatives formed
using either the Lentz or the Gotz and Masson techniques.

$$QM_4$$

$$\downarrow \quad +QM_2$$

$$Q_2M_6$$

$$\downarrow \quad +QM_2$$

$$Q_3M_8 \xrightarrow{-M_2} Q_3M_6 \xrightarrow{2x - M_2} Q_6M_{10} \xrightarrow{-M_2} Q_6M_8 \xrightarrow{-M_2} Q_6M_6$$

$$\downarrow$$

$$Q_4M_{10} \xrightarrow{-M_2} Q_4M_8 \xrightarrow{2x - M_2} (M>12)$$

$$\vdots \quad +QM_2$$

Although in their paper they described these molecules in their
fully trimethylsilylated form they did point out that the
condensation steps would involve the protonated forms and that
some of the trimethylsilyl groups are probably hydroxyl groups.
    The degree of polymerisation of polysilicic acids in
magnesium silicates has also been studied by the trimethylsilylation of the silicic acid produced by the acid hydrolysis
of the silicates. The results showed that some 76-82% of
the derivative consisted of insoluble polysilicic acids,
which suggested that when silicates are hydrolysed in gastric
juice most of the silicic acids produced lead to insoluble
silica gel [24] .

Analysis of glasses

The structure of glass as a three dimensional polymer has
been investigated by the trimethylsilylation method. The earliest
application of the Lentz technique [5] was in a study of the
silicate structure [25] and the effect of time and temperature
on the silica distribution in borate glasses [26] . The results
suggested a random distribution of silica tetrahedra in a glassy
boric oxide matrix; the method also enabled an estimate of the
time taken to establish the melt equilibrium. The application
of a modified Gotz and Masson technique to glass-type samples
$PbO.SiO_2$ [27] showed that glassy $2PbO . SiO_2$ and each of the
three main crystalline polymorphs are characterised by its own
specific silicate anion distribution. The agreement between
paper chromatography which can identify the higher molecular
weight derivatives and the corrected TMS results were good.
However, Nakumura in a study of $PbO.SiO_2$, and $2ZnO.SiO_2$ suggested.
that further development of the trimethylsilylation technique
was needed before its reliability could be established [28] .
The glc data obtained appeared to be dependent on the
experimental procedure adopted e.g. the amount of water added
or the method of shaking whilst temperature was found to have
no significant effect on the QM_4/Q_2M_6 ratio. Trimethylsiloxy
groups have been added on to the surface of titanium dioxide or
cobalt blue and the resulting materials were stable up to 300 $^\circ$C
[29] .
An extensive review of polymer mass distribution of silicate
glasses included trimethylsilylation as a useful additional
technique to traditional methods [30,31] . The effect of heating
lead orthosilicates at 500 $^\circ$C for different times has been
carefully investigated [32] . Initially the dimer concentration
increases reaching a maximum after 30 minutes and then decreases
continually with time, the monomer concentration also decreases
noticeably with heat treatment duration. Diffractograms showed
that the heat treated samples gradually crystallised to include
tetramers at high concentration. Lunar samples have also been
analysed by trimethylsilylation and the orthosilicate derivative
was found ,agreeing with the modal analysis which showed a
significant proportion of olivine [33,34] . Trimethylsilylation
of augite and enstatite did not yield significant amounts of
silicate ions, this showed that the ions observed did not arise
from the decomposition of pyroxenes. Discussion of the differences
between the Apollo 11 and 12 samples were given. The minor
constituent $Si_4O_{12}^{8-}$ led to speculation of yet undetected mineral
phases with discrete silicate frameworks as microlites or crypto-
crystallites in the shocked partially melted minerals and glasses
in lunar fines. Interestingly enough there seemed to be little
or no $Si_4O_{12}^{8-}$ from Apollo 14 samples compared with the Apollo 11 and
12 fines [34] .

Currell et al [35] have shown that the presence of other
metal ions in a silicate influence the yield and structure of poly-

organosiloxane derivatives from glasses. In a study of calcium
aluminosilicate glasses the yield of soluble polymer produced
was related to the % Al_2O_3 present. Evidence for   SiF bonds has
been presented by Masson [37] in a study of a glass $3PbO.PbF_2.SiO_2$
where trimethylsilylation gave evidence for $[SiO_4]^{4-}$, $[Si_2O_7]^{6-}$ and
$[Si_3O_{10}]^{8-}$ as well as a number of mono, di and trifluorosilicate
ions. The results provide direct experimental evidence for
discrete silicate and fluorosilicate ions in the glass, in line
with polymer theory predictions.

The Analysis of Dioptase

The trimethylsilyl derivative of dioptase $Cu_6Si_6O_{18}6H_2O$ was
expected to be the monocyclic hexamer $Si_6O_{18}[Si(CH_3)_3]_{12}$ but
gas chromatographic [12] analysis revealed three bicyclohexamers.
Earlier Gotz and Masson [36] had given a peak temperature of
289 °C for the monocyclic hexamer, however Hoebbel [16] suggested
that this was low and that the peaks obtained between 291 - 293 °C
correspond to bicycloisomers of Q_6M_{10} caused by intermolecular
condensation or cyclization.
The results obtained by Masson [13] are given overleaf.

Derivative	Method 1a	Method 2a	Method 1b
QM_4	0.5	8.4	0.6
Q_2M_6	0.2	Tr	0.2
$Q_4M_8 + Q_3M_8$	0.6	3.3	0.4
Unknown	0.9	2.0	0.4
$Q_6M_9(C_3H_7)$	5.9		3.9
$Q_5M_{10}Q_6M_{10}$	41.3	70.9	41.4
Unknown } Unknown }	2.9	trace	2.2
Q_6M_{12} isomer 1	3.1	9.7	9.6
Q_6M_{12} isomer 2	35.8	5.6	38.7
Q_6M_{12} isomer 3	9.0		2.6

a) 24 hours Amberlyst treatment b) 168 hours Amberlyst
treatment

Method 1   Dioptase (0.3g), HMD (9 cm^3) and IPA (1 cm^3) were
stirred together at room temperature, chlorotrimethylsilane
(2 cm^3) was added and the stirring continued for ca 17h. The HMD
layer was removed and chlorotrimethylsilane removed. The residue
was stirred with Amberlyst 15.

Method 2   Dioptase (0.3g), HMD (9 cm^3) IPA (6 cm^3) , H$_2$O(4 cm^3)
and concentrated HCl (37% 4 cm^3) were stirred together for
ca 20 h at room temperature. Separation as in 1.
The table shows that the addition of Amberlyst 15 has an effect
on the derivatives obtained, this has been discussed elsewhere
in this review. Trimethylsilylation of dioptase by method 1 gives
three isomers of both Q_6M_{12} and Q_6M_{10}. Acidification of the
reaction medium results in a decrease in the yield of Q_6M_{12} and
an increase in Q_6M_{10}. Gotz and Masson [36] and Hoebbel [16]
using more acidic conditions found the Q_6M_{10} but little or no
Q_6M_{12}.

These results suggest that the following rearrangement may occur,

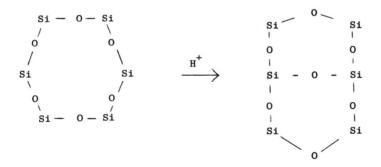

Intermineral Conversion

Masson and co-worker [38] have used the TMS technique to
look at intermineral conversion, hemimorphite $Zn_4(OH)_2Si_2O_7 \cdot H_2O$
loses water between 175-600 $^{\circ}$C and changes to a modified hemimorphite $Zn_4(OH)_2Si_2O_7$ which has $Si_2O_7^{6-}$ as the main structural

unit. Heating between 600 - 630 ^{o}C causes dehydroxylation and
yields a material with an orthosilicate structure identical to
β-ZnSiO$_4$ described by Taylor. This work confirmed by trimethylsilylation that at temperatures >835 ^{o}C βZn_2SiO_4 is converted to
α Zn$_2$SiO$_4$ without change in the orthosilicate nature of the
materials.

The Analysis of Cements, Concretes and Calcium Silicates

Lentz applied his original silylating study to cement paste
[39] tobermorite [40] and also to study the effect of carbon dioxide
on silicate structures in portland cement paste [41] . This work
complemented the acid molybdate method and thin layer chromatography [42,43] which had been pioneered by Wieker [44] , further
work on this including modifications to the tlc technique has been
reported by Tamas [45] .

The major silicate structures found in the cement samples
analysed by Lentz varied with time and consisted of an orthosilicate, a pyrosilicate and a polysilicate of unknown molecular
configuration. Small quantities of a trisilicate and a cyclic
tetrasilicate were also detected. In an attempt to further
investigate the nature of the polymerised structures formed in
aged cement pastes, tobermorite Ca$_5$(Si$_6$O$_{18}$H$_2$).4H$_2$O and wollastonite
were analysed by the trimethylsilylation technique. The results
indicate that on the basis of the trimethylsilylation technique
there is only limited similarity between the mineral tobermorite
and the calcium silicate hydrate in portland cement paste.
These original papers have been followed by investigations into
(i) the side effects that occur during the derivatisation and
(ii) attempts to improve the derivatisation procedure.
Two model compounds for the silicate structures in cement have
been investigated by Currell [46] .

Akermanite (Ca$_2$MgSi$_2$O$_7$) and β-dicalcium silicate (Ca$_2$SiO$_4$)
were trimethylsilylated under a whole variety of conditions;
the results showed that the ratio of QM$_4$ and Q$_2$M$_6$ varied with
reaction time, temperatures and concentrations of the reactants.
Frazier [47] had previously noted that the yield of QM$_4$ altered
by replacing hexamethyldisiloxane with trimethylchlorosilane as
the silylating species. Komatsu and co-workers [48] have used the
Lentz technique as a means of determining the growth of siloxane
units in calcium silicate.

Glasser and Lachowski [49] investigated the hydration of
3CaO.SiO$_2$ pastes at 25 ^{o}C by a variety of techniques and suggested
a mechanism for gradual polymerisation of silicates in cements.

The trimethylsilylation technique used in this study was
introduced by Tamas, Sarkar and Roy [50] which is the preferred
method for calcium silicates. Lachowski [51] has compared the
methods of Lentz [39] ;Gotz and Masson[6] ; and Tamas Sarkar and
Roy [50]; for the derivatisation of cement pastes.

Reagents/Author	Lentz (I)	Gotz/Masson (II)	Tamas et al (III)
Hexamethyldisiloxane	10 ml	9 ml	5 ml
Propan-2-ol	15 ml	0.8 ml	0
Hydrochloric acid	7.5 ml	0	0
Deionised water	6.3 ml	0.2 ml	0
Trimethylchlorosilane	0	2 ml	5 ml
Dimethylformamide	0	0	10 ml

Sample size of silicate 10 mg: mixtures stirred before addition
of silicate
        I (60 mins)   II (0 mins) III   (15 mins).
Post   treatment with Amberlyst 15.

    Method III has the advantages of a solvent (DMF) which
suppresses side reactions; DMF acts as a good solvent for the
calcium salts produced in the reaction. Lachowski [51] concludes
that method III gives higher, more reliable yields for those
calcium silicates which are of interest to cement chemists,
however the Ca/Si ratio needs to be fairly high. If this requirement is not met either the Lentz method [39] or Milestones variation [52] which uses t-butanol/salt as a method of preferentially
salting out the silicic acid into the organic phase are preferable.
    Milestone [52] has also suggested the use of γ-glycidoxy-
propyltrimethoxysilane as a suitable internal standard in the gas
liquid chromatographic analysis of QM_4 and Q_2M_6. Although it is
clear that in itself the QM_4/Q_2M_6 ratio is only giving general

information about the silicate polymerisation mechanism. The ratio
reflects at least two on-going mechanisms, the disappearance of
the anhydrous phase as well as the polymerisation of the hydrated
material [53] .
    The trimethylsilylation techniques of Lentz [39] Tamas [50]
and Milestone [52] have been applied to the analysis of creep and
dry shrinkage of calcium silicate pastes [54] and also autoclaved
lime-quartz materials [55] . There is still clearly a need for
better structural information on the high molecular weight derivatives and an agreed method for the analysis of the low molecular
weight species QM_4 and Q_2M_6.

TRIMETHYLSILYLATION OF COMPLEX SILICATES

Susceptibility of Minerals to Trimethylsilylation

The applicability or otherwise of the trimethylsilylation depends in a very complicated way on the structure of the silicate mineral used. The first requirement is that the cations must be leachable by hydrochloric acid. Although the susceptibility of silicate minerals to attack by acids has been discussed by Mase [56], Murata [57] and Petzold [58] the situation is complicated and is still incompletely understood. However some general points will be noted:

(a)   according to Petzold cation removal is helped by the presence of heavy metal ions e.g. Fe,
(b)   the less closely packed is the backbone the more likelihood there is of acid attack i.e. ease of access of the acid molecules.

Following cation removal the accessibility of the Si-OH sites to attack by trimethylsilanol will be the controlling factor. The size of the silicic acid molecules formed following acid removal will probably also have a major influence. In most sheet silicates which contain aluminium in the tetrahedral layer(isomorphous replacement of silicon by aluminium) the distribution of the aluminium atoms will control the size of the silicic acid molecules and is an important factor in trimethylsilylation. Currell and co-workers have investigated [59] using the Lentz technique the possibility of trimethylsilylation of a wide range of silicate minerals.

Single Chains		Group
Diopside	$CaMg(Si_2O_6)$	1
Hedenbergite	$CaFe(Si_2O_6)$	2
Hypersthene	$(Mg,Fe)_2(Si_2O_6)$	3
Bronzite	$(Mg,Fe)_2(Si_2O_6)$	1
Wollastonite	$Ca(SiO_3)$	2
Double Chains		
Anthophyllite	$(Mg,Fe)_7(Si_8O_{22})(OH)_2$	1 and 2
Tremolite	$Ca_2Mg_5(Si_8O_{22})(OH)_2$	1

Sheets

Daphnite	$(Fe,Al)_6(Si,Al)_4O_{10}(OH)_8$	3
Vermiculite	$(Mg,Fe,Al)_3(AlSi_3O_{10})(OH)_2$	2
Prehnite	$Ca_2Al_2(Si_3O_{10})(OH)_2$	1
Apophylite	$Ca_4K(Si_4O_{10})_2F.8H_2O$	2
Biotite	$K(MgFe)(AlSi_3O_{10})(OH)_2$	3
Phlogopite	$KMg_3(AlSi_3O_{10})(OH)_2$	1 and 3
Talc	$Mg_3(Si_4O_{10})(OH)_2$	1
Kaolinite	$Al_4(Si_4O_{10})(OH)_8$	1
Muscovite	$KAl_2(AlSi_3O_{10})(OH)_2$	1 and 2
Margarite	$CaAl_2(Al_2Si_2O_{10})(OH)_2$	1 and 3
Thuringite	$FeAl(Al_2Si_2O_{10}(OH)_2(MgFe)_3$ $(OH,O)_6$	1 and 3

Note:

Group 1 products whilst soluble in the reaction media were on examination shown to be unreacted minerals.

Group 2 products were insoluble in the reaction media and were on examination shown to be partially trimethyl-silylated.

Group 3 products were soluble in the hexamethyldisiloxane layer and are fully trimethylsilylated.

Note that in the case of Margarite and Thuringite Group 3 products only were obtained with reaction temperatures of 75° Group 1 products only at ambient reaction temperature ; also Phlogopite only gave Group 1 at ambient and a mixture of 1 and 3 at 75 °C.

Wollastonite was the only one of the single chain structure minerals to show some degree of trimethylsilylation. The probable reason for the relative ease of attack in the case of wollastonite is the fact that the silicate chains are less closely packed in this mineral. It should also be noted that workers at the Paint Research Association [60,61] were able to get Group 3 products by the trimethylsilylation of pseudowolloastonite and partial success in the case of Anthophyllite is possibly due to the presence of iron. The sheet silicates show an interesting variation. Kaolinite and Talc show no trimethylsilylation and the mineral is recovered i.e. they are not susceptible to the initial acid attack. Other

sheet silicates showed the formation of Group 3 or Group 2
products i.e. they were all susceptible to acid attack. Group 3
products appear to be formed when there is a certain degree of
replacement of silicon by aluminium in the silicate sheet. Thus
with very little Al in the tetrahedral sheet of Vermiculite
although the cations are removed no soluble products are
formed. Biotite, Phlogopite and Muscovite have approximate
Al:Si ratios of 1:3 while Margarite and Thuringite have
approximate Al:Si ratios of 1:1; each of these gave soluble
products. Presumably acid attack occurs at the aluminium centre
breaking the silicate sheet at these points to form fragments
small enough to allow attack by trimethylsilanol and give products
of molecular weight suitable to allow solubility in organic
solvents. To the first approximation it was reported that
average molecular weight of products could be correlated with
the distribution of aluminium in the silicate sheet. Thus
Margarite and Thuringite with high levels of aluminium in the
silicate sheet gave low molecular weight products compared with
Biotite, Phlogopite and Muscovite (with less aluminium in the
silicate sheet) which gave comparatively higher molecular weight
products. Further reports by these workers [62,63] on the
results obtained with Biotite and Phlogopite have supported this
initial conclusion but, for example, molecular weight
distributions obtained by gpc on polymers prepared from phlogopite
indicate a bimodal distribution with approximately 50% by weight
of the product in low molecular weight fragments i.e. QM_4 etc
and the rest of the product high molecular weight material with
a wide distribution. Thus the effect of aluminium distribution
on the average molecular weight is partially obscured by small
fragments breaking off from the silicic acid fragments before
complete trimethylsilylation.

Sharma [12] suggested that aluminium is released prior to
the silicate derivatisation this is in contrast to the views of
Currell and co-workers [63] who have reported that in the reaction
of phlogopite aluminium appears to be partially retained in
preference to the potassium and magnesium although throughout
the reaction the mineral retained its original structure.
Aluminium was apparently removed from the tetrahedral layer and
moved to the octahedral sites with a compensating movement of
potassium ions to counteract the shift in charge balance in the
silicate sheet. It was suggested that at least partial trimethylsilylation occurs before the final breaking of the silicon/
oxygen/aluminium bonds.

Glasser [64] has applied the trimethylsilylation technique to
melilites in order to determine the Al-Si ordering by establishing
the amount of QM_4 and/or Q_2M_6 derived from one natural and five
synthetic melilites. The synthetic gehlenite - rich melilites
are appreciably but not completely disordered. The natural melilite
is well ordered, but heating at 1050 ^{o}C introduces considerable
disorder. The Si/Al ratio has also been studied by Yokoi in
some faujasite type zeolites (Si/Al ratio : 1.20-2.65). Zeolites

have small values of Si/Al ratio (< 1.90) consist mainly of
orthosilicate and pyrosilicate ions however zeolites where Si/Al
(> 1.90) consist of three dimensional networks [65] .
     Kenney et al were the first [66,67,68] to apply the technique
of trimethylsilylation to complex silicates. They used a mixture
of isopropanol, hydrochloric acid and trimethylchlorosilane.
The reaction of chrysotile asbestos with this mixture gave a
planar organosilicon polymer in fibrous form. They concluded
that half the sites available on a given sheet were occupied by
trimethylsilyl groups and that the rest were occupied by hydroxy
groups or were linked together by oxygen bridges. It was
postulated that the first stage in the reaction is the removal of
magnesium ions from the surface of the chrysotile fibres with
hydroxyl groups substituted in their place, followed by further
substitution involving trimethylsilyl groups. The resulting
individual polymer sheets which have all the trimethylsilyl groups
on one side have internal strains which cause the   sheets to
curl up into tube like ribbons. The fibres which were characterised
by detailed electron microscopy consist of bundles of rolled up
ribbons. Kenney and co-workers also applied the same reaction
mixture to the mineral apophyllite but because of the nature of the
parent sheet trimethylsilyl groups were attached to both sides
giving balanced strains and thus flat sheets.
     Zapata and co-workers [71,72] applied a similar technique to
chrysotile and vermiculite this time using dimethylallylchloro-
silane as the silylation reagent, the grafting of the allyldimethyl-
silyl radical apparently proceeds more slowly than that of the
trimethylsilyl because of steric hindrance   effects associated with
the bulkier alkyl group. The product was found to be rubbery, this
rubbery nature was attributed to the plasticising effect of diallyl-
tetramethyldisiloxane the removal of which by extraction destroys
the rubbery character of the material. In a later paper
Mendelovici reported on the acid and heat stability of organic
derivates of chrysotile [73] . Fripiat has [69] patented a method
for grafting hydrophobic organosilicon compounds on to the anhydrous
or hydrated calcium silicate components of cement. The corrosion
and weather resistance of the cements was improved. Other calcium
silicates, including Wollastonite have been   completely derivatised
to give novel polyorganosiloxanes which on the addition of clay
prior to trimethylsilylation gave mixed Si/Al derivatives.
An earlier paper by Fripiat [47] has described the methyl
derivative of chrysotile using hexamethyldisiloxane to give
a high molecular weight two dimensional organic derivative.
Linsky has also shown the possibility of forming organosilicon sheet polymers [70] .
     Saunders, Cox and Slade in a patent [60] assigned to the
National Research and Development Corporation claimed the
preparation of a material suitable for water repellant treatment prepared using a range of possible organosilyl capping
agents. A typical example describes the reaction of synthetic
wollastonite (cyclic half α form) reacted with a mixture of

concentrated hydrochloric acid, water,propan-2-ol and hexamethyldisiloxane followed by stirring at room temperature for
1 hour. The non-aqueous fraction after drying and distillation
at low pressure gave a colourless, mobile oil. The number average
and weight average molecular weights of the oil were
determined by gel permeation chromatography as 1610
and 1730 respectively.   This product, when used
to treat ventile cotton gave a BS water repellancy spray test
rate of 5 compared to the spray test rating of 4 for a similar
sample treated with a typical commercially available silicon
water repellant composition. Saunders and Cox in a further patent
[61] also assigned to the National Research and Development
Corporation covered a process in which greatly improved yields can
be obtained from polysilicate minerals such as wollastonite. They
claimed that in this invention it is important to use the acid
in an amount sufficient to react with at least half the metal
cations of the polysilicate mineral, a water miscible organic
solvent and water in an amount such that the concentration of the
acid in the aqueous phase is not greater than 2 gm equiv.per.lit.
so that silylation of the polysilicate takes place with at least
partial retention of the silicate backbone. A typical example
quotes a reaction medium prepared by dissolving trichloroacetic
acid in a mixture of concentrated hydrochloric acid, water and
acetone. Hexamethyldisiloxane is added and the mixture stirred for
1 hour before addition of powdered wollastonite. The final product
from this reaction mixture is quoted as having a number average
molecular weight of 3000 and a weight average molecular weight
of 9000 with 17% of the polymer having a chain length in excess
of 250$\overset{o}{A}$.

     Kuroda and Kato [74] produced a glassy product by trimethyl-
silylating para-wollastonite using the Lentz technique - the
product dissolved in a range of organic solvents and its thermal
properties gave a strong sharp exothermic peak at 430 ^{o}C due to
oxidation of organic groups.
Pseudowollastonite has also been converted by treatment with
trimethylchlorosilane in t-butanol and acetone to give
Q_3M_8 (70%) [75] . One of the major drawbacks to the industrial

exploitation of the polymers derived from simple silicates has been
their low number average molecular weights. Q_3M_8 can be successfully

polymerised [64] by stirring with toluene and dmf in the presence
of BuLi in n-hexane for two days to give a polymer M.W.6000.
The recent work on complex silicates to produce high molecular
weight derivatives has centered on two research groups Currell
and co-workers in the United Kingdom and Kuroda and Kato in
Japan. Currell [76] prepared the first soluble polyorganosiloxane from an inorganic mineral silicate by trimethylsilylating
chlorite var. daphnite$\left\{(Fe, Al, Mg)_{12}\right\}\left\{(SiAl)_8O_{20}\right\}$ $(OH)_{16}$ to
give a highly viscous material $\overline{Mn}$ 1618. In a subsequent paper
the trimethylsilylation technique was applied to a range of single

chain, double chain and sheet silicate anions [59] . Kuroda
and Kato [77] have also attempted to prepare higher molecular
weight materials by the trimethylsilylation of complex silicates
such as biotite. These authors have recently reviewed a series
of their more recent papers [74, 78, 79, 80, 81] in a further
composite paper [82]. The trimethylysilyl derivative of
halloysite was organophilic and resulted from an initial attack
on the interlayer region by hydrochloric acid; the octahedral
layers are decomposed and labile tetrahedral layers, having
silanol groups on one side are generated; the silanol groups in
the tetrahedral layers are then partially trimethylsilylated [79]
Ethylene glycol was used as a linking agent in an attempt to
produce higher molecular weight trimethylsilyl derivatives from
hemimorphite [78] . Kuroda and Kato had previously investigated
the straightforward derivatisation of this mineral in a previous
study [80] and the use of an alcohol in the trimethylsilylation
reaction esterified any residual hydroxyl groups [81] . The glassy
product obtained from parawollastonite dissolved in a wide range
of organic solvents but was insoluble in water [82] .

Currell et al [83] have investigated the further chemistry
of polyorganosiloxanes prepared by the trimethylsilylation
technique, partly with the aim of providing a basis for the
development of methods for the preparation of materials with
increased molecular weight. Material with increased molecular
weight (up to $\sim$ 6000) with elimination of hexamethyldisiloxane
is obtained on heating up to 300 $^{\circ}$C [84] On heating with dimethyldichlorosilane up to 70 $^{\circ}$C for $\frac{1}{2}$ h in the presence of anhydrous
ferric chloride or the hexahydrate , pseudowollastonite polymer
gives a mixture of products including trimethylchlorosilane,
hexamethyldisiloxane, 2-chloropropane, and depending on the
reactant ratio used, polymer of increased molecular weight [85,86].
The formation of these products was rationalised in terms
of the following reactions:-

(i) $\Rightarrow$SiOSiMe$_3$ +   Me$_2$SiCl$_2$ $\rightarrow$ $\Rightarrow$SiOSiMe$_2$Cl   +   Me$_3$SiCl

(ii) $\Rightarrow$SiOSiMe$_2$Cl   +   Me$_3$SiOSi$\lessgtr$ $\rightarrow$ $\Rightarrow$SiOSiMe$_2$OSi$\lessgtr$ +   Me$_3$SiCl

(iii) $\Rightarrow$SiOSiMe$_3$ +   Me$_3$SiCl $\rightarrow$ $\Rightarrow$SiCl + Me$_3$SiOSiMe$_3$

(iv) $\Rightarrow$SiOSiMe$_2$Cl   +   ClMe$_2$SiOSi$\lessgtr$ $\rightarrow$ $\Rightarrow$SiOSiMe$_2$OSi$\lessgtr$ + Me$_2$SiCl$_2$

These polymers contain $-O_{\frac{1}{2}}SiMe_2O_{\frac{1}{2}}$ (i.e. D units) as well as the
original Q and M units. A variant of this reaction involves
the interaction of dimethyldichlorosilane with a mixture of
octamethyltetrasiloxane and the trimethylsilylated polymer.
Other methods of incorporating D units include the use of
short chain polydimethylsiloxanes (with trimethylsiloxy end
groups) as the trimethylsilylating agent in place of
hexamethyldisiloxane in the original reaction with the

mineral silicates.   After the preparation of polymer
by trimethylsilylation some residual hydroxyl groups are
present; these -OH groups show very variable reactivity
probably due to steric factors, however reactions have been
reported with compounds containing Si-Cl groups
(i.e. ($CH_3)_3SiCl$, $CH_3(CH_2 = CH)SiCl_2$ and $CH_2 = CHSiCl_3$,
and also with Ph_2PCl) [18] .

$$\rightarrow SiOH \quad + \quad ClSi \Leftarrow \rightarrow \quad \rightarrow SiOSi \Leftarrow + \quad HCl$$

$$\rightarrow SiOH \quad + \quad Ph_2PCl \rightarrow \quad \rightarrow SiOPPh_2 \quad + \quad HCl$$

Using chlorine or bromine gas in the presence of tungsten
light the trimethylsilyl methyl groups may be chlorinated or
brominated.

$$\rightarrow Si(CH_3)_3 \quad + \quad Cl_2 \quad \xrightarrow[\text{4 hours}]{\text{W light}} \quad \rightarrow Si(CH_3)_2CH_2Cl$$

In a subsequent two-step reaction sequence the chlorine atom
was then replaced by an hydroxyl group as shown below:

$$\rightarrow SiCH_2Cl \quad \xrightarrow[\text{HAc - 24h}]{\text{KAc}} \quad \rightarrow SiCH_2OCOCH_3$$

$$96h \searrow \text{dry } HCl/CH_3OH$$

$$\rightarrow SiCH_2OSi(CH=CH_2)Cl_2 \xleftarrow[CCl_4/4 \text{ h}]{CH_2=CHSiCl_3} \rightarrow SiCH_2OH$$

The reactivity of the $\rightarrow SiCH_2OH$ groups with, for example,
silicon-chlorine compounds was shown to be much more reproducible
than that of the previously mentioned Si-OH groups [87] .
The inorganic polymers listed above have a variety of properties
and molecular weights in the range from 10^3 to 10^5. They have
a number of potential industrial uses as   coatings, adhesives
and as anti-corrosive agents. This area is comparatively new
and the search to produce higher molecular weight Q/M polymers
is continuing. The direct production of molecular weight
fragments in the range $3 \times 10^3 - 8 \times 10^3$ has been shown by the
trimethylsilylation of xonotlite [88] , a potentially useful
route by silylating water glass has also been reported [89] .

THE FUTURE

The technique of trimethylsilylation has been shown to be a useful tool in the study of silicate structures provided, of course, that the possibility of side-reactions is remembered in the interpretation of these results. Improvements in technique and a better understanding of the reactions involved can be expected to give more accurate information on the structure of the silicates and enable the expansion of the technique to the examination of more complex silicate structures.

Applications have not yet been firmly established for the polyorganosiloxanes prepared by trimethylsilylation and they are not, as yet, prepared commercially. However, it should not be very long before this useful chemistry leads to materials with, for specific applications, properties superior to the 'silicones' and therefore worthy of commercial preparation.

REFERENCES

[1]   N.H. Ray, Inorganic Polymers, Academic Press, London, (1978)

[2]   A.D. Wilson and S. Crisp, Organolithic Materials, Applied Science Publishers, London, (1977)

[3]   L. Holliday Ed., Ionic Polymers, Applied Science Publishers, London, (1975).

[4]   C.E. Carraher, J.E. Sheats and C.U. Pittman Jr., Organometallic Polymers, Academic Press, London, (1978)

[5]   C.W. Lentz, Inorganic Chemistry, 3, 574 (1964)

[6]   J. Gotz and C.R. Masson, J.Chem.Soc., (A), 2683 (1970)

[7]   F. Wu, J. Gotz, W.D. Jamieson and C.R. Masson, J. Chromatog., 48, 515 (1970)

[8]   J. Gotz and C.R. Masson, J.Chem.Soc., (A), 686 (1971)

[9]   S.K. Sharma, L.S. Dent Glasser and C.R. Masson, J.C.S. (Dalton), 1324 (1973)

[10] L.S. Dent Glasser and S.K. Sharma, Brit.Polymer. J., 6, 283 (1974)

[11] G. Garzo and D. Hoebbel, Z.J. Ecsery and K.J. Ujszaski, J. Chromatog., 167, 321 (1978)

[12] S.K. Sharma and T.C. Hoering, Yearbook, Carnegie Institute, Washington, 662 (1978)

[13] H.P. Calhoun and C.R. Masson, J.C.S.(Dalton), 1342 (1978)

[14] H.P. Calhoun and C.R. Masson, Chem. Comm., 560 (1978)

[15] G. Eglinton, J.N.M. Firth and B.L. Welters, Chem. Geol., 13, 125 (1974)

[16] G. Garzo and D. Hoebbel, J. Chromatog., 119, 173 (1976)

[17] L.S. Dent Glasser, E.E. Lachowski and G.G. Cameron, J. App. Chem. Biotechnol., 27, 39 (1977)

[18] C.B. Cook, Ph.D.Thesis (CNAA), Thames Polytechnic (1978)

[19] T. Shimono, T. Isobe and T.Tanitani, J.Chromatog., 179, 323 (1979)

[20] R.K. Harris and R.H. Newman, Org. Mag. Res., 9(7), 426 (1977)

[21] L.S. Dent Glasser, E.E. Lachowski, R.K. Harris and J. Jones, J. Mol. Struc., 51, 239 (1979)

[22] L.S.Dent Glasser and E.E. Lachowski, J.C.S.(Dalton), 393 (1980)

[23] L.S.Dent Glasser and E.E. Lachowski, J.C.S.(Dalton), 399 (1980)

[24] H. Yokoi and S. Enomoto, Chem. Pharm Bull., 26(6), 1846 (1978)

[25] K.E.Kolb and K.W. Hansen, J. Amer. Ceram. Soc., 48, 439 (1965)

[26] K.E.Kolb and K.W. Hansen, J.Amer.Ceram.Soc., 49, 105 (1966)

[27] J. Gotz, D.Hoebbel and W. Wieker, J. Non.Cryst.Solids, 20, 413 (1976)

[28] R. Nakamura, Kyushu. Daigaku Kogaku Shuho, 50(5), 635 (1977)

[29] E.A.Batyaev and N.P. Kharitonov, Zhur Prikhadnoi Khimi, 48(4), 906(1975)

[30] P.Balta and E. Balta, Materiale de Constructii,5(4), 182(1975)

[31] J.Gotz and D.Hoebbel, in Proc 8th Int.Conf.on Reactions in Solids Ed. J.Wood and V. Lindquist. 525 (1977)

[32] J.Gotz, C.R.Masson and L.M. Catelliz in Amorphous Materials Ed. D.Ellis, J. Wiley, London, (1972)

[33] C.R.Masson, J. Gotz, W.D. Jamieson, J.L. McLachlan and A. Volborth, Proc. 2nd Lunar Sci.Conf., 1, 957 (1971)

[34] C.R.Masson, I.B.Smith, W.D. Jamieson and J. McLachlan,
      Rev. 3rd Lunar Sci. Conf., 1029 (1972)

[35] R. Atwal, B.R. Currell, C.B. Cook, H.G. Midgley and
      J.R. Parsonage, A.C.S. Coatings and Platics Div. Preprints,
      37(1), 67 (1977)

[36] J. Gotz and C.R. Masson, Comptes Rend IX Congr. Int. de
      Verre Versailles, 261(1971)

[37] H.P.Calhoun, W.David Jamieson and C.R.Masson,
      J.C.S.(Dalton), 454 (1979)

[38] J.Gotz and C.R.Masson, J.C.S.(Dalton), 1134 (1978)

[39] C.W.Lentz, Nat.Acad. Sci. Nat. Res. Council Pub. No 1389,
      269 (1966)

[40] C.W.Lentz, Mag. Concr. Res., 18(58), 231 (1966)

[41] C.W.Lentz, Ind.Chem. Belge, 32, 487 (1967)

[42] W.Weiker, Z.Anorg.Allg.Chem., 360, 307 (1978)

[43] D.Hoebbel and W. Wieker, Z.Anorg.Allg.Chem.,405, 163 (1974)

[44] W.Wieker and H.Stade, Autoclaved Calcium Silicate Bldg.
      Prod. Pap. Symp. London, 125 (1968)

[45] F.Tamas, Cemento, 75(3), 357 (1978)

[46] B.R.Currell, H.G.Midgley and M.A.Seaborne, J.C.S.(Dalton,
      490 (1972)

[47] J.J.Fripiat and E. Mendelovici, Bull.Soc.Chim.Fr.,
      483 (1968)

[48] K.Komatsu, I.Toyama, A.Kawahara and T. Nakamura,
      Kogyo Kagaku Zasshi, 74(2), 160 (1971)

[49] L.S.Dent Glasser and E.E.Lachowski, Cement and Concr.Res.,
      8(6), 733 (1978)

[50] F.D.Tamas, A.K.Sarkar and D.M.Roy, Hung.J.Ind. Chem.,5,
      115 (1977)

[51] E.E.Lachowski, Cement and Conc. Res., 9, 111 (1979)

[52] N.B.Milestone, Cement and Concr. Res.,7, 345 (1977)

[53] L.S.Dent Glasser and E.E. Lachowski, Cement and Concr. Res.,
      6, 811 (1976)

[54] A.Bentur, N.B.Milestone and J.F.Young,
      Cement and Concr.Res.,8, 721 (1978)

[55] S.E.Memaly, K.Mohan and H.F.W.Taylor, Cement and Concr. Res.,
      8, 671 (1978)

[56] H.Mase, Bull.Chem. Soc.Japan, 34,214 (1948)

[57] K.Murata, J.Am.Mineralog., 28 545 (1943)

[58] A.Petzold, Wiss.Zeit.Hoch.Arch.Bauwesen Weimer,
      13(3), 343 (1966)

[59] B.R.Currell, H.G.Midgley, M.A.Seaborne and C.P.Thakur,
      Brit.Polym.J., 6, 229 (1974)

[60] J.A.Saunders, B.S.Cox and S.L.Slade, U.K.Patent 40474/73
      (1973)

[61] J.A.Saunders and B.S.Cox U.S.Patent 3,904,583 (1975)

[62] W.S.Bicknell, MSc Thesis (CNAA) Thames Polytechnic
      (1977)

[63] R.S.Atwal, S.Bicknell, C.B.Cook, B.R.Currell, H.G.Midgley,
      A.J.Mooney and J.R.Parsonage, Brit. Polym.J., 11(4),
      213 (1979)

[64] E.E.Lachowski and F.P.Glasser, Mineralog.Mag., 39,
      412 (1973)

[65] H.Yokoi, H. Kashiwagi, S.Enomoto and H.Takahashi,
      Nippon Kagaku Kaishi, 6, 768 (1978)

[66] S.E.Frazier, J.A.Bedford, J.Hower and M.E.Kenney,
      Inorg. Chem., 6, 1693 (1967)

[67] J.P.Linsky, T.R.Paul and M.E. Kenney, J. Polym.Sci.,
      Polym. Phys., 9(1), 143 (1971)

[68] M.E.Kenney, U.S.Patent 3,661,846 (1972)

[69] J.J.Fripiat, Ger.Offen 2,051,292 (1971)

[70] J.P.Linsky, Diss Abs. Int.B, 31(6), 3242 (1970)

[71] L. Zapata, J.Castelein, J.P.Mercier and J.J. Fripiat,
      Bull.Soc.Chem.Fr., 54 (1972)

[72] L.Zapata, J.J.Fripiat and J.P.Mercier, J.Polym.Sci.,
      Polymer Letter Edn., 11, 689 (1973)

[73] E.Mendelovici, Proc. 3rd European Clay Conf., 118 (1977)

[74] K.Kuroda and C.Kato, Polymer, 19(11), 1300 (1978)

[75] W.G.Davies and H.V.A.Beedle, Ger.Offen 2,542,425 (1976)

[76] B.R.Currell, H.G.Midgley and M.A.Seaborne, Nature,
      236(68), 108 (1972)

[77] K.Kuroda and C.Kato, Clays and Clay Minerals, 25, 407
      (1977)

[78] K.Kuroda and C. Kato, Mackromol.Chem., 179, 2793 (1978)

[79] K.Kuroda and C.Kato, Clays and Clay Minerals, 27,
      53 (1979)

[80] K.Kuroda and C.Kato, J.Inorg.Nuc.Chem., 41, 947 (1979)

[81] K.Kuroda and C.Kato, J.C.S.(Dalton), 6, 1036 (1979)

[82] K.Kuroda and C.Kato, Memoirs of the School of Sci.& Eng.
      Waseda Univ., 27 (1978)

[83] B.R.Currell, H.G.Midgley and J.R.Parsonage, Final
      Technical Report DAERO-74-9019, (1977)

[84] C.B.Cook, B.R.Currell, H.G.Midgley, J.R.Parsonage and
      C.P.Thakur, Brit. Polym.J., 11(4), 178 (1978)

[85] T.S.Bola, C.B.Cook, B.R.Currell, H.G. Midgley and
      J.R.Parsonage, Organic Coatings and Plastics Preprints,
      41, 183 (1979)

[86] R.S.Atwal, T.S.Bola, C.B.Cook, B.R.Currell, H.G.Midgley
      and J.R.Parsonage, Brit.Polym.J., 11(4), 182 (1979)

[87] C.B.Cook, B.R.Currell, H.G.Midgley and J.R.Parsonage
      Brit.Polm.J., 11(4), 186 (1979)

[88] A.Takahaski, Shikensko Kimoi, 22(4), 288 (1971)

[89] S.Kohama, Kagaku to Kogyo-Osaka, 51(11), 435 (1977)

Metal Vapor Synthesis of Organometal Polymers and Polymer-Supported Metal Clusters

Colin G. Francis[+] and Geoffrey A. Ozin[++]

[++]Lash Miller Chemical Laboratories and Erindale College
University of Toronto
Toronto, Ontario, Canada

[+]Department of Chemistry
University of Southern California
Los Angeles, California, USA

ABSTRACT

Metal vapor chemistry has matured from being merely a curiosity
and is now established as a viable synthetic technique. Recently
the method has been applied to the formation of organometallic
polymers and polymer-supported metal clusters. The purpose of
this chapter is to provide a short review of the initial studies
in the field. The synthesis of organometallic polymers is described via either formation of the organometallic polymer precursor or the reaction of metal vapor with a preformed polymer and
the two techniques are compared with respect to the chemical and
physical properties of the products. The generation of polymer-
stabilized small metal clusters at ambient temperatures, by the
deposition of a metal vapor into fluid polymer, thin films, is
then discussed with particular reference to the combination of
matrix-scale and macrosynthetic metal vapor synthesis. Some
thoughts regarding the mode of stabilization of metal clusters are
also presented. Finally, a brief description of the formation, by
evaporation techniques, of metal particle-polymer interactions is
included to indicate other types of experiment which may be used.

INTRODUCTION

Although the two fields of cryochemistry and organometallic
polymer chemistry have been studied widely [1,2], only recently
has some attempt been made to apply metal vapor synthesis to the
formation of organometallic polymers and polymer-supported
cluster species.

Two basic approaches to the functionalization of polymers by
means of metal-containing species have been used:

1) formation via interaction of a monomer with metal vapor,
   followed by polymerization of the organometallic polymer
   precursor, and

2) formation by the interaction of a polymer with metal
   vapor.

In addition, several workers have employed polymers as supports
upon which to deposit metal vapors in order to study metal clusters and clustering processes.

We have divided this chapter into three sections corresponding to these methodologies, but we have also subdivided the second
topic in order to distinguish between organometallic polymers and
polymer-supported metal clusters.

FORMATION OF METAL-CONTAINING POLYMERS
VIA INTERACTION OF A MONOMER WITH METAL VAPOR

The use of metal vapor synthesis to produce organometallic
polymers can be considered really to have begun with the work of
Middleton [3]. He observed an involatile, cream-colored material
as a product in the cocondensation of chromium vapor with styrene
and trifluorophosphine. I.r. spectra, supported by ^{19}F n.m.r.
spectra, showed that the product consists of a mixture of [1] and
[2]. A direct synthesis of bis(styrene)chromium was found to be
unsuccessful, the only product being a black, waxy material, the

$(\bigcirc : PF_3 = 1:6):$

$$Cr_{(g)} + \bigcirc + PF_3 \xrightarrow{\;77K\;} \bigcirc + \bigcirc$$

$$Cr(PF_3)_3 \qquad\qquad Cr(PF_3)_3$$

$$[\underline{1}] \qquad\qquad\qquad [\underline{2}]$$

i.r. spectrum of which resembles that of polystyrene.

Subsequently, Blackborow reported the cocondensation of iron with styrene at 77K [4]. The only product is polystyrene. However, on warming the condensate from 77K in an atmosphere of CO, compounds [$\underline{3}$] and [$\underline{4}$], as well as iron pentacarbonyl, are formed:

$$Fe(CO)_4 \qquad\qquad\qquad Fe(CO)_3$$

$$[\underline{3}] \qquad\qquad\qquad\qquad [\underline{4}]$$

Similarly, cocondensation of chromium with styrene produces polystyrene [5], but there is intractable material present in which chromium is bound to polystyrene. This was identified by its mass spectrum which contained peaks due to $(C_8H_8)_n H_m Cr$, where n = 2, m = 0 and n = 3-5, m = 4. Warming the cocondensate from 77K in an atmosphere of CO or PF_3 leads to the formation of small amounts of $(\eta^6 - C_6H_5CH=CH_2)CrL_3$ (where L = CO or PF_3), plus a noncrystallizable oil, identified by its ^1H n.m.r., i.r. and mass spectra as an analog of [$\underline{2}$]. From its oily nature and low molecular weight (M.Wt. = 860 in benzene) it is obvious that the polymerization has occurred only to a limited extent.

Cocondensation of nickel with styrene in a toluene matrix at 77K [5] yields red-brown crystals which appear to be tris(styrene)-

nickel, in which nickel is presumably coordinated to the side
chain of each styrene molecule. The compound is thermally unstable, decomposing above 253K. The product of decomposition
must be the nickel-polystyrene adduct obtained by Klabunde and
coworkers [6] from the nickel-styrene cocondensation. This was
reported to be an air-stable, black solid which is soluble in
toluene and acts as a selective, homogeneous hydrogenation catalyst. Analogous materials were obtained from the reaction of
nickel vapor with tetrafluoroethylene.

Lagowski [7] has shown that nickel atoms react with
propyne and pent-2-yne at 77K to yield various organic oligomers as well as a black, nickel-containing species which independently oligomerizes alkynes to form arene and cyclooctatetraene
derivatives. Interestingly, iron atoms were found to react with
pent-2-yne to yield a symmetric tetramer as the sole hydrocarbon
product.

In the previous reactions, the aim of producing an organometallic polymer precursor was defeated by the facile polymerization of the alkene or alkyne under the conditions of the metal
vapor process. Some other approaches have been taken in an effort to produce a more controlled polymerization process.

Hydrolysis of dichlorosilanes constitutes a major route to
organopolysiloxanes, via an initial reaction of the type:

$$(n+4)R_1R_2SiCl_2 + (n+4)H_2O \longrightarrow [R_1R_2SiO]_4 + [R_1R_2SiO]_n + 2(n+4)HCl$$

An alternative mode of formation is via rearrangement of the Si-O
bonds in some alkoxy-silanes [8]. With this in mind, reaction of
chromium vapor with PhMeSiCl$_2$ or PhSi(OC$_2$H$_5$)$_3$ at 140K was performed, using methylcyclohexane as a diluent in a rotary solution
reactor [10] to give the corresponding bis(η^6-arene)chromium complexes. However, in neither case does the complexed silane subsequently give the desired polymer.

A different route to organometallic polymers may occur where
the formation of the organometallic unit R_1-M-R_2, during the metal

vapor reaction, constitutes a chain propagation step. Examples of
this process are the cocondensation of chromium vapor with 1,4-
diphenylbutane [11] and the solution reaction of chromium or vanadium vapor with the fluid 1,1,3,3-tetraphenyl-1,2,2,3-tetramethyl-
trisiloxane [10]. As well as the corresponding cyclophanes [5]

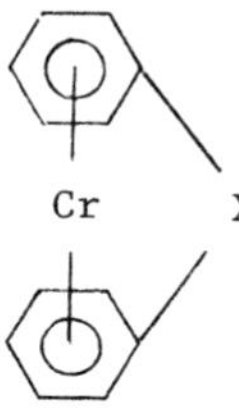

[5]

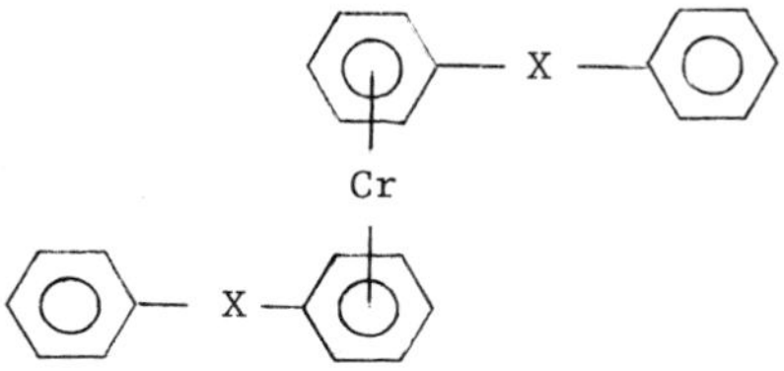

[6]

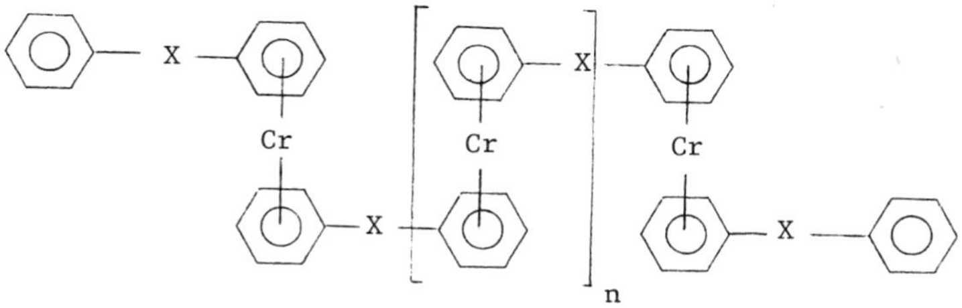

[7]

$$\text{(where} \quad X = -(CH_2)_4- \text{ or } -\underset{\underset{Me}{|}}{\overset{\overset{Ph}{|}}{Si}}-O-\underset{\underset{Me}{|}}{\overset{\overset{Me}{|}}{Si}}-O-\underset{\underset{Me}{|}}{\overset{\overset{Ph}{|}}{Si}}-)$$

and simple bis(arene)metal complexes [6], ladder polymers [7] may
be obtained, although the value of n is unknown.

McGlinchey [12] has also shown that $(\eta^6-C_6H_6)Cr(\eta^6-C_6F_6)$,
prepared by the cocondensation of chromium vapor with a mixture of
C_6H_6 and C_6F_6, reacts with Me_3SnLi to give oligomers [8] in which
a link of the type $C_6F_5-C_6H_5$ is formed.

The ability of sodium or copper atoms to abstract halogen

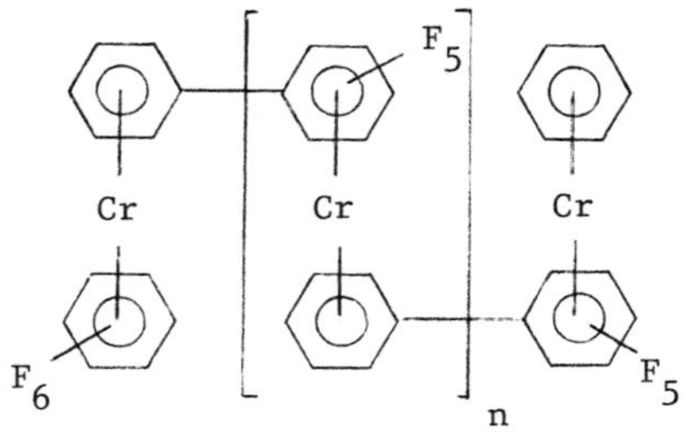

[8]

from coordinated aryl halides enables organometallic units to be coupled, as in the following reaction reported by Timms [13]:

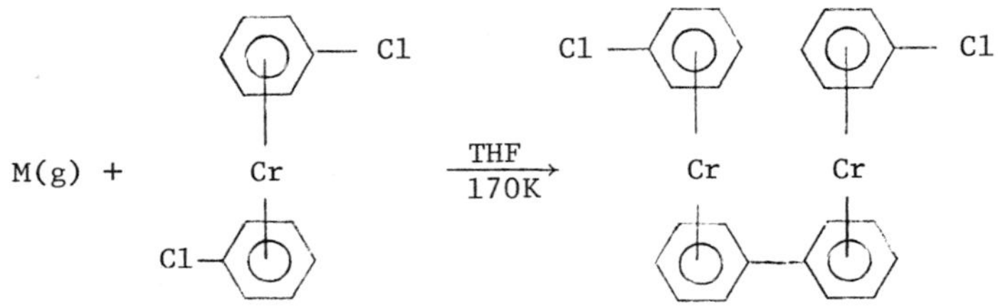

[9]

M = Na, Cu

The interest in this latter type of system presumably stems from the electro-conducting properties of some organometallic polymers such as poly(ferrocenylene) [14].

REACTIONS OF METAL VAPORS WITH POLYMERS

The development of the rotary solution reactor by Timms [15] played an important part in facilitating the use of polymers as ligands in metal vapor synthesis (Figure 1A). The fact that a polymeric ligand cannot be volatilized into the reactor under co-condensation conditions means that it is difficult to obtain an

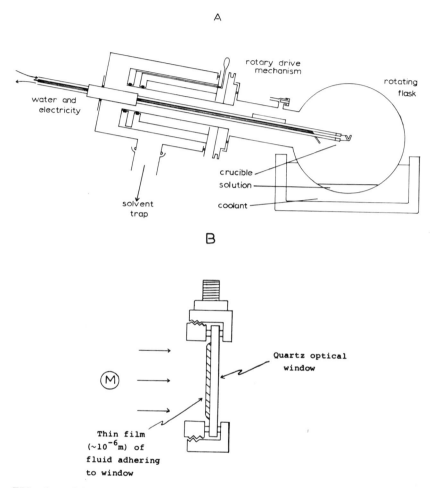

FIG. 1.   Schematics of (A) the metal vapor rotary reactor [15] and (B) the metal vapor fluid matrix method [16].

efficient reaction between metal vapor and a polymer. However, using the solution technique, organometallic polymers can be pre-pared directly by reaction with the metal vapor.

The extension of this principle to polymer-supported metal cluster species was achieved as a result of the development of the

Fluid Matrix Technique [16] (a modification of the spectroscopic
method of Blyholder [17]). In this procedure a liquid polymer
film is formed on a spectroscopic window and metal vapor is deposited into the film (Figure 1B). Spectroscopic observation of
the matrix at various stages during the reaction provides a means
for monitoring the course of the reaction.

(a) Formation of Organometallic Polymers

The polymers reported as ligands in metal vapor synthesis
[9,10] may be divided into three types: polysiloxanes, poly-
(phenylethers) and polystyrenes.

The polysiloxanes investigated are commercially available
fluids made by Dow Corning, designated DC510, DC550 and DC556.
The poly(methylphenylsiloxane) DC510 possesses a number average
molecular weight of 3300, with a ratio of phenyl to methyl substituents of 1:17. The metal vapors of Ti, V, Cr, Mo and W react with DC510 at 270K to give colored, air-sensitive fluids
which were shown to contain η^6-arene-metal species [10]. However,
metal vapors of Mn, Fe, Co and Ni give only metal slurries.

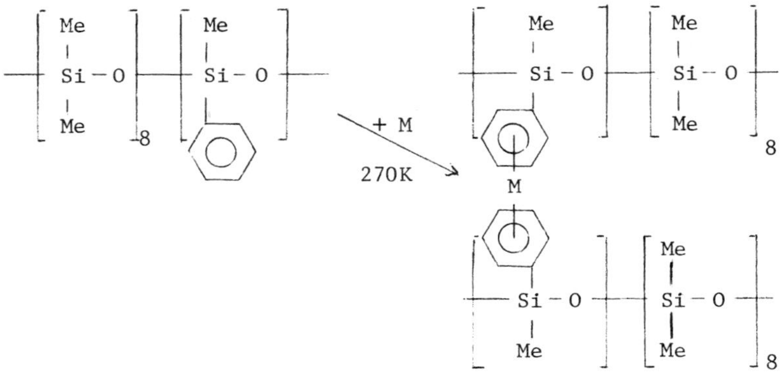

[10]

As the concentration of bis(arene)metal species is increased,
a corresponding increase in viscosity of the fluid occurs. Using
the chromium system as an example, the variation of viscosity
with arene-complexation was measured (Figure 2). The graph shows

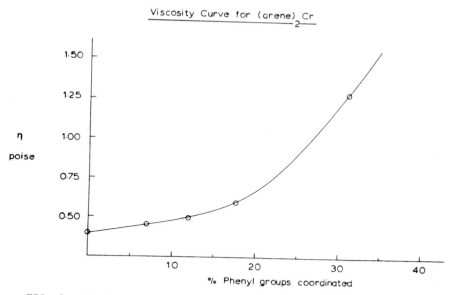

FIG. 2. Variation of viscosity with arene complexation in the Cr/DC510 system [10].

that at low metal loading (<20%) the viscosity increases slowly but at higher loadings a rapid increase in viscosity occurs.

It was observed that for low metal loadings, conversion of metal to complex is essentially quantitative but decreases until, at about 40% phenyl group coordination, it is no longer possible to coordinate more metal. Interestingly, this limit to the uptake of chromium coincides with a high and rapidly rising viscosity for the polymer. Similarly, reaction of chromium vapor with DC550 [a poly(methylphenylsiloxane) with a viscosity of 1.25 poise] leads to a poor yield of bis(arene)chromium and the presence of large metal aggregates.

As was mentioned previously, the products are very air-sensitive but, with the chromium-containing fluid, oxidation to $(arene)_2Cr(I)$ is followed by hydrolysis of the phenyl-silicon

bond to yield bis(benzene)chromium(I). Cleavage of the bond
occurs as well in the zerovalent metal state with methanol but
the rate of reaction is greatly reduced. The changes in reactivity are attributed to an increased polarization of the carbon-
silicon bond in the former case, favoring nucleophilic attack at
silicon.

Treating the Cr/DC510 system with 7,7,8,8-tetracyanoquinodi-
methane (TCNQ) yields a clear, green fluid containing a species
of formula $[(arene)_2 Cr]^+[(TCNQ)_2]^-$. Conductivity measurements
showed that the new fluid possesses a specific conductivity of
1.87×10^{-5} ohm^{-1} cm^{-1}, compared with a value of 2×10^{-16} ohm^{-1}
cm^{-1} for an uncomplexed polysiloxane [10]. It is difficult to
imagine any long-range order in this polymeric fluid so that an
increase of $\sim 10^{10}$ ohm^{-1} cm^{-1}, attributed to the $[(arene)_2 Cr]^+$-
$[(TCNQ)_2]^-$, is quite appreciable.

It was also demonstrated [10] that it is possible to form
various derivatives of DC510 by simple ligand displacement reactions from mixed naphthalene/DC510 sandwich complexes. These
were formed by reaction of chromium atoms with a mixture of an
alkylnaphthalene (R = methyl, hexyl) and DC510. The naphthalene
is then easily displaced by CO or Bu^{t}NC, affording the corresponding $[Cr(DC510)(CO)_3]$ and $[Cr(DC510)(Bu^tNC)_3]$ complexes as shown
below [11].

The DC556 fluid possesses -OH substituents in addition to
the phenyl and methyl substituents. However, Cr atoms appeared
only to react with the phenyl groups rather than forming a complex
of the type $Cr^{III}(OSiR_3)_3$.

The second class of polymers investigated was a poly(phenyl-
ether), Santovac 5, with structure [12]. Reaction at 300K with
chromium or vanadium vapors yields the corresponding bis(arene)-
metal complexes [10].

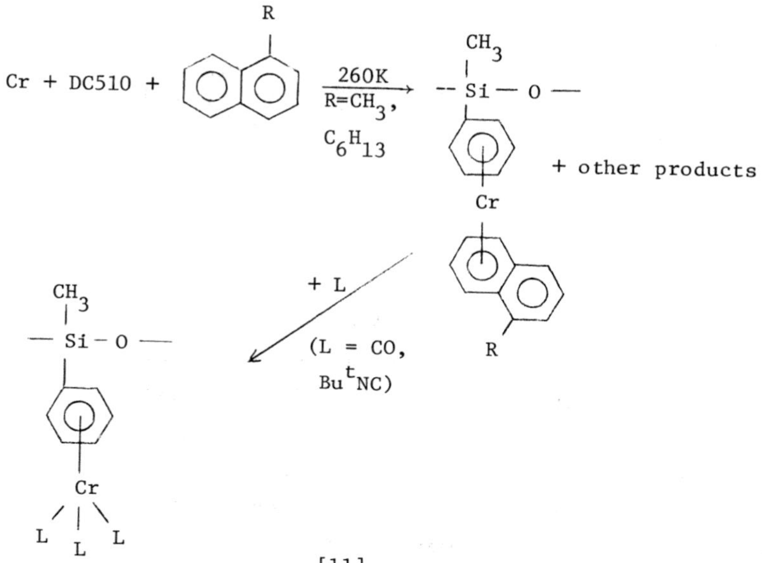

[11]

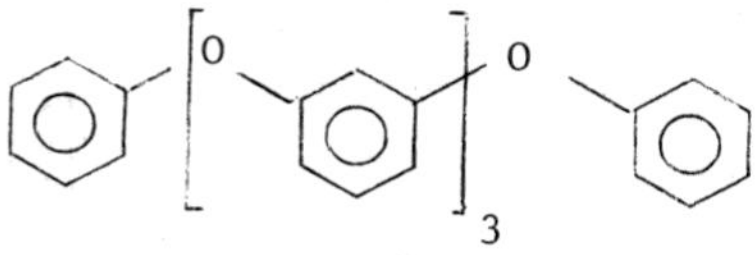

[12]

Finally, solid polystyrene (M.Wt. <u>ca</u>. 25,000) and poly(p-Bu[t]-styrene) were able to be used as ligands in the synthesis of bis-(arene)chromium complexes as shown below:

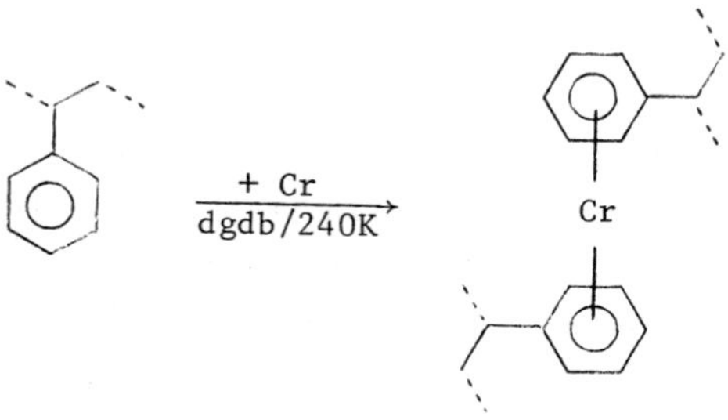

[13]

Diethylene glycol dibutylether was used as the diluent to obtain a
3% w/v solution which was reacted with metal at 230K [10]. The
use of polymeric solids as ligands does create some problems,
however. It is very difficult to obtain unreactive solvents which
are able to maintain the complexed polymer in solution while
possessing a low vapor pressure in order for the metal vapor reaction to be carried out.

(b) Formation of Polymer-Supported Metal Clusters

The manipulation of metal atom recombination reactions under
cryochemical conditions has been widely used to produce metal
clusters and cluster complexes [18,19,20]. Until recently [16],
however, there has been no procedure in the open literature for
stabilizing these cluster species for subsequent testing as real,
working catalysts. By extending the metal vapor-poly(methylphenylsiloxane) work, described in the previous section, to lower
temperatures and higher metal concentration, with in situ spectroscopic monitoring of product formation, it has been possible to
obtain some interesting results concerning metal cluster growth
and stabilization in liquid, polymeric media in the temperature
range 210-290K. In this section we will concentrate on the unimetallic vanadium-poly(methylphenylsiloxane) system [21] and its
extension to bimetal and trimetal vapor-polymer reactions [22].

(i) Unimetallic Systems. As mentioned earlier, the metal
vapors of Ti, V, Cr, Mo and W react with DC510 at 270K in a
metal vapor rotary reactor to give the respective polymer-anchored
bis(arene)metal complex. In the case of the V/DC510 system, the
strongest band observed in the optical spectrum occurs at 324 nm
and corresponds to a metal(d)-to-ring(π^*) charge transfer excitation of the polymer-supported bis(arene)vanadium complex.

The V vapor-fluid matrix experiment [21] involved low metal
deposition rate ($\leq 10^{-8}$ mol min^{-1}) into a thin film (roughly 10^{-6}
m) of DC510 adhering to a quartz window (diameter 1 cm) under
vacuum at 250K. As well as the intense MLCT band at 324 nm, the

growth of a new absorption at 453 nm (Figure 3a) was observed. The growth characteristics of these two bands as a function of vanadium loading are illustrated in Figure 3b. Significant points to note under conditions of comparatively low metal loading are (i) the absence of broad optical absorptions in the 200–300 nm region associated with bulk electronic excitations of colloidal vanadium (note that the reaction of V vapor with a polymethyl-siloxane DC200, or a paraffin oil, gives only a broad absorption in this region) and (ii) the absence of any absorptions in the region 200–700 nm associated with species other than those emerging at 324 nm and 453 nm, when conducted at temperatures $\geq$ 250K.

Under conditions of higher vanadium loading, the growth of a broad, underlying absorption in the region of 230–240 nm was observed (Figure 3a: E), close to that associated with colloidal vanadium generated in DC200 (Figure 3a: F). The product distribution for equivalent metal loadings, normalized with respect to polymer-anchored bis(arene)vanadium, in the V/DC510 reaction, as a function of the temperature of deposition in the range 210–270K, is shown in Figure 4. It can be seen that formation of the bis-(arene)vanadium species is favored in the range 250K, while formation of the other species becomes more important at both higher and lower temperatures in the region 210–270K. The origin of the $V_{colloid}$ absorption, however, appears to be different in the 210K and 270K extremes; namely, rapid metal aggregation in the surface layers of the film for the more rigid polymer at the lower temperature end and colloidal metal formation within the film for the more mobile polymer at the higher temperature end. In addition, when the reaction was performed at 210K (Figure 5), two new species were produced, absorbing at 550 nm and 635 nm.

Collecting together the data from the V/DC510 system (as well as a series of similar observations for Ti, Cr and Mo vapors condensed with DC510 and Santovac 5 [21]) and by analogy with the optical data for V_2 and V_3 trapped in solid argon at 10–12K [23], the most probable explanation for the bands in the visible region

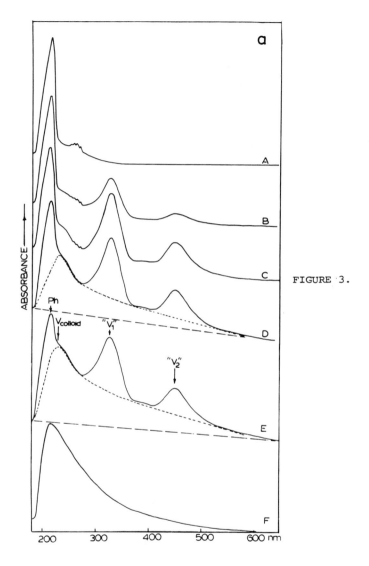

FIGURE 3.

of the V/DC510 spectrum is that they are primarily associated with
metal-localized excitations in very small, polymer-stabilized
vanadium clusters, with nuclearities in the range n = 2-4. A
graph of the energies for these transitions as a function of the

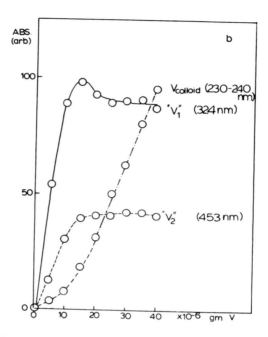

FIG. 3a.   (A-E) Ultraviolet-visible spectra of the products
formed with increasing loading of V vapor into DC510 deposited
at 250K and (F) the spectrum of colloidal vanadium generated in
DC200.   3b.   Graphical representation of the growth-decay behavior
of species labelled V_1, V_2 and $V_{colloid}$ in Figure 3a, as a function of vanadium loading into DC510.

reciprocal of the suggested metal nuclearity (Figure 6) displays a
monotonic correlation, as would be expected for the HOMO-LUMO band
gap with increasing cluster size. Similar effects have been observed for Mo_n (n = 2, 3, 4, 5) in DC510 [24].

    Current thoughts concerning probable modes of cluster growth
and stabilization in a polymer (or oligomer) are shown in Figure 7.
This scheme is intended to illustrate an initial chain-propagation
step and/or cross-linking process involving single metal atoms
complexed to two arene rings. These can then act as metal nucleation centers for subsequent cluster growth, either 'unsolvated'
(Figure 7, III) or stabilized by oxygen groups or arene rings on

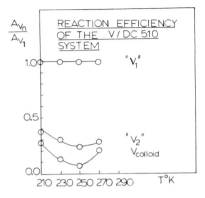

FIG. 4. Product distribution for equivalent V loadings, normalized with respect to polymer-anchored bis(arene)vanadium, in the DC510 reaction, as a function of the temperature of deposition in the range 210–270K [21].

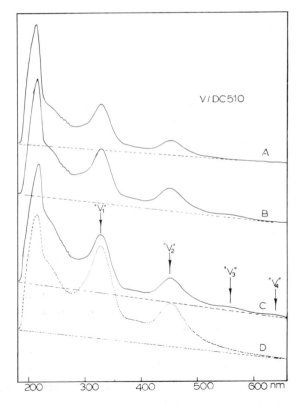

FIG. 5. Ultraviolet-visible spectra of the products formed with increasing loading of V vapor into DC510 at 210K.

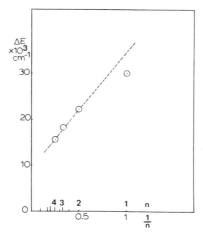

FIG. 6. Graphical representation of the transition energies observed for the DC510-stabilized V_n species as a function of n^{-1} (see text for details).

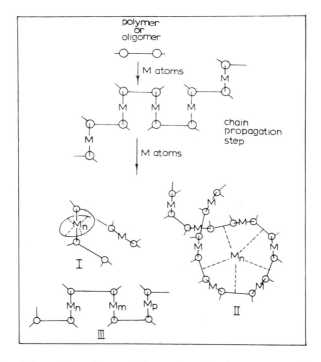

FIG. 7. Schematic of possible modes of cluster growth and stabilization in a functionalized polymer (or oligomer).

the polymer (Figure 7, I). An alternative mode of   cluster
stabilization involves physical entrapment within a high-viscosity
organometal polymer network (Figure 7, II).

From a preliminary kinetic analysis [25] of the growth decay
behavior of the metal cluster species in the DC510 system, a
scheme is currently envisaged in which M_n are formed in a series
of sequential nucleation events and stabilized by interaction of
each metal atom with at least one phenyl group on the polymer
chain (see Scheme I, below):

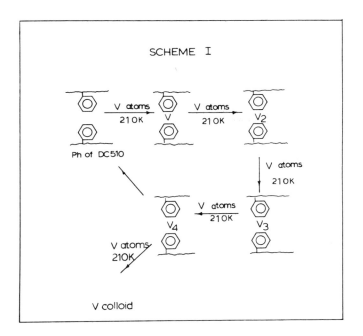

As the metal cluster develops, the metal-arene interaction dimin-
ishes in strength (cf. chemisorbed benzene on metal surfaces)
until at a critical cluster size (e.g., Ti: n = 2; V: n = 4; Cr:
n = 3; Mo: n = 5) the arene(s) is released from the cluster, allowing further aggregation of the metal into a size regime which is
capable of displaying broad optical absorptions characteristic of

colloidal metal. Expulsion of the metal clusters from their solvation shells can also be achieved by raising the temperature of
the polymer to around room temperature, the outcome usually being
rapid metal agglomeration to higher-nuclearity, bulk-like clusters. Additional experimental support for the existence of polymer-
stabilized metal clusters has very recently emerged from some preliminary laser Raman measurements on the vanadium and chromium
DC510 systems [25]. The spectra display a number of very intense
Raman lines in the region 260-100 cm^{-1} which are most probably
associated with metal-metal stretching modes of very low-nuclear-
ity vanadium clusters.

A point of special significance is illustrated in Figure 8
which demonstrates the feasibility of generating and studying

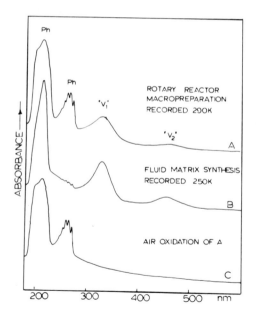

FIG. 8. Comparison of product formation in the V/DC510 system,
(A) using a macroscale rotary reactor, (B) using the fluid matrix
technique and (C) the air oxidation of the sample shown in (A)
[21].

polymer-supported, mononuclear and cluster systems by way of both
rotary reactor and matrix spectroscopic metal vapor-liquid polymer
techniques [21]. Similar results have recently been obtained for
the Mo/DC510 system [24].

(ii) Multimetallic Systems. It has been demonstrated that
organo multimetallic polymers containing mononuclear and/or cluster metal sites can be readily generated by the metal vapor-fluid
matrix technique [16,21,22]. In this brief discussion we will
focus attention on the V/Cr/DC510 and Ti/V/Cr/DC510 combinations,
using the MLCT transitions of the bisarene complexes (318, 324 and
355 nm for the Cr, V and Ti/DC510 systems respectively) and the
metal-localized excitations of polymer-stabilized dichromium (402
nm) and divanadium (453 nm) as spectroscopic probes of the extent
and site of metal attachment.

The initial deposition of Cr vapor into DC510 at 250K is
shown in Figure 9 and is seen to display the MLCT band of polymer-
supported bis(arene)chromium at 318 nm as well as polymer-

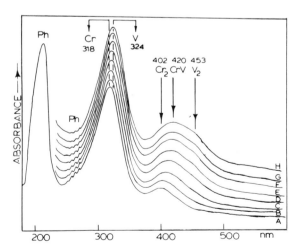

FIG. 9. The ultraviolet-visible spectra of the products of (A)
a Cr vapor deposition into a film of DC510 at 250K, followed by
(B-H) roughly equal increments of V vapor [22].

stabilized dichromium at 402 nm. Subsequent depositions of V
into the DC510 at 250K, close to the saturation loading, are depicted in Figure 9B-9H and show a small spectral shift of the
original 318 nm band to 324 nm, corresponding to the growth of
the polymer-anchored bis(arene)vanadium complex. Concomitant with
this growth in the MLCT region is the broadening and shifting of
the polymer-stabilized dichromium band at 402 nm to eventually
yield a structured band with maxima at 402, 420 and 453 nm, the
latter being associated with the previously identified divanadium
absorption. The new absorption, at energies intermediate between
those of Cr_2 and V_2 absorptions, can be ascribed to the polymer-
stabilized, bimetallic cluster CrV [22]. This is in keeping with
the observation of averaged metal-localized excitation energies
in heterobinuclear cluster carbonyl complexes [26] and hetero-
nuclear diatomic clusters [33].

A possible explanation of the way in which the chromium and
vanadium vapors interact with the DC510 under the conditions outlined above is illustrated in Scheme II.

Figures 10a and 10b illustrate the progress of sequentially
depositing Cr/V/Ti or Cr/Ti/V respectively into a thin film of
DC510 adhering to a quartz optical window held at 250K. The deposition of approximately 2×10^{-6} gm of chromium into the DC510
leads to the appearance of the characteristic 318 nm absorption
of polymer-anchored bis(arene)chromium (Figure 10a: A) together
with a small amount of the polymer-stabilized dichromium species.
Addition of a roughly equal increment of vanadium causes further
growth of the MLCT absorption assigned to bis(arene)chromium but
with a noticeable red shift of the peak maximum to 324 nm (Figure
10a: B), the wavelength of anchored bis(arene)vanadium [16,21].
Deposition of a roughly equal increment of titanium into the
polymer film (Figure 10a: C and D) causes additional growth in
the MLCT region but with a new peak at 355 nm, depicting the
generation of polymer-attached bis(arene)titanium [16]. In an
attempt to obtain clearer spectroscopic resolution of the bis-
(arene)chromium and vanadium sites, similar trimetal vapor

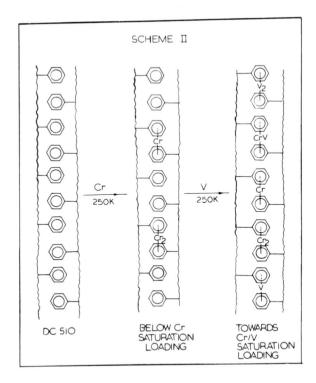

depositions were conducted using the deposition order Cr/Ti/V.   An
illustrative spectral trace is shown in Figure 10b. Following the
sequential Cr/Ti deposition (Figure 10b: B), the coexistence of
bis(arene)chromium and titanium on the polymer is easily discerned
by the MLCT bands at 318 and 355 nm respectively. Further deposition of V clearly defines the continued growth in the MLCT
region but with a new maximum at 324 nm (Figure 10b: C and D).

MISCELLANEOUS REACTIONS OF METAL VAPORS WITH POLYMERS

The previous section (b(II)) described how metal vapor-
polymer interactions, occurring within a fluid polymer, could lead
to a small, well-defined metal cluster species. Another group of

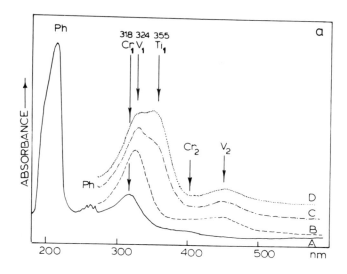

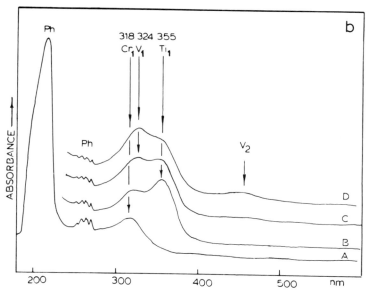

FIG. 10a. Ultraviolet-visible spectra of a thin film of DC510
at 250K following deposition of (A) Cr, (B) V, (C) Ti and (D) Ti
vapors in increments of approximately 2×10^{-6} gm [22].

    10b. Essentially the same conditions as Figure 10a ex-
cept that the metal vapor deposition sequence follows the order
(A) Cr, (B) Ti, (C) V and (D) V [22].

workers [27,28] has shown that small metal particles, with mean
sizes in the region below 10 nm, can be grown on the surface of
a polysiloxane oil. The apparatus consists of a rotating disc and
furnace assembly mounted within a vacuum chamber. Oil is supplied
at the center of the disc and flows outward due to the centrifugal
force, allowing metal vapor to interact with a continuously re-
plenished film of oil.

In the experiments reported, metal vapors of Al [27] or Fe
[28] were produced at a rate of $\sim$8 mmol min^{-1} and deposited on the
surface of the polysiloxane to give extremely fine particles. In
the Fe-polymer system, concentration of the metal-polymer suspen-
sion under air resulted in the isolation of mainly Fe_3O_4 particles
with average diameter of about 2.5 nm. Magnetic measurements
showed that these particles exhibit superparamagnetism, but that
the magnetic interaction between particles is quite small as a
result of their dispersion in the fluid.

In contrast, Burkstrand reported a series of experiments in
which copper vapor was deposited onto the surface of a solid polystyrene [29,30] or polyvinylalcohol [31] film. The chemical interactions taking place on the surface were monitored by x-ray
photoemission spectroscopy. With polystyrene, using low copper
coverages (<0.01 monolayer), the copper $2p_{3/2}$ core electron binding energy is 1.5 eV less than the value for bulk copper, indicative of a copper atom in a region of high electron density. This
could be a surface complex such as bis(arene)copper. At higher
copper coverages, the core lineshape showed the presence of a new
peak attributed to isolated small clusters of copper atoms. When
the polystyrene was pretreated with an oxygen plasma process [30],
generating carbon-oxygen bonds on the surface, subsequent deposition of copper vapor resulted in the formation of a copper-
oxygen-polymer complex. This showed increased adhesion of the
deposited metal films onto the polymer.

An alternative route for the production of thin films of
polymer-containing metal has been reported by Boonthanom and White

[32]. Evaporation of copper and cocondensation with organic fragments, such as those produced by heating polyethylene, gave films
(thickness $\sim$90 nm) consisting of polyethylene containing small
clusters of copper atoms. Annealing at 493K in air converted the
Cu to Cu_2O with accompanying loss of the polymer. This resulted
in the production of small "islands" of cuprous oxide, separated
by $\sim$10 nm, for which the conductivity characteristics were
treated as a model of the conductivity in bulk copper.

CONCLUSIONS

In summary, the formation of metal-containing polymers via
interaction of metal vapor with a monomer is complicated often by
the lack of control of polymerization of the polymer precursor
during the metal vapor process. It seems likely that the most
profitable studies in this area will come from an approach
similar to those of Timms [13] and McGlinchey [12].

Another way of alleviating this problem is the use of a preformed polymer as a ligand. Some of the advantages of this approach are shown from the early observations in the fields of
liquid-phase organometal polymers and polymer-supported metal
clusters, viz. (i) the experiments can be performed at either a
macrosynthetic or matrix spectroscopic scale of operation, (ii)
the experiments can be entirely conducted at or close to room
temperature, (iii) the resulting organometal polymers and polymer-
stabilized metal cluster compositions are homogeneous liquids (the
term "homogenized-heterogeneous catalysts" may turn out to be an
appropriate description for the latter) which have reasonable
stabilities at room temperature, and (iv) the methodology is
easily extended to bimetallic and trimetallic combinations.

Although solid polymers present some difficulties as ligands
in metal vapor synthesis, a combination of the standard solution
methods of metal vapor chemistry and the techniques of thin-film

technology should facilitate development of this area of research
in the future.

The use of metal vapor synthesis in collaboration with
polymer chemistry is still a novel field, but the tremendous possibilities which exist for such techniques suggest that the
future is extremely promising.

ACKNOWLEDGEMENTS

We would like to acknowledge the financial assistance of the
National Research Council of Canada, Imperial Oil, The Connaught
Foundation, Erindale College and the Lash Miller Chemical Labora-
tories.

REFERENCES

[1] M. Moskovits and G.A. Ozin, eds., Cryochemistry, Wiley, New
    York, 1976; J. R. Blackborow and D. Young, Metal Vapor Synthe-
    sis in Organometallic Chemistry, Springer-Verlag, Berlin,
    Heidelberg, New York, 1979.

[2] C.E. Carraher, Jr., J.E. Sheats, C.U. Pittman, Jr., eds.,
    Organometallic Polymers, Academic, 1978; P. Hodge, Chem. Brit.,
    1978, 14, 237.

[3] R. Middleton, Ph.D. Thesis, University of Bristol, 1974.

[4] J.R. Blackborow, C.R. Eady, E.A. Koerner Von Gustorf, A.
    Scrivanti and O. Wolfbeis, J. Organomet. Chem., 1976, 111, C3.

[5] J.R. Blackborow, R. Grubbs, A. Miyashita and A. Scrivanti, J.
    Organomet. Chem., 1976, 120, C49.

[6] K.J. Klabunde, T. Groshens, H.F. Efner and M. Kramer, J. Or-
    ganomet. Chem., 1978, 157, 91.

[7] L.H. Simons and J.J. Lagowski in Fundamental Research in
    Homogeneous Catalysis, Vol. 2, ed. Y. Ishii and M. Tsutsui,
    Plenum, New York, 1978.

[8] C. Eaborn, Organosilicon Compounds, Butterworth, 1960.

[9] C.G. Francis and P.L. Timms, J.C.S.--Chem. Comm., 1977, 466.

[10] C.G. Francis and P.L. Timms, J.C.S. Dalton, 1980, 1401.

[11] N.N. Zaitseva, A.N. Nesmeyanov, G.A. Domrachev, V.D. Zinov'ev, L.P. Yur'eva and I.I. Tverdokhlebova, J. Organomet. Chem., 1976, 121, C52.

[12] A. Agarwal, M.J. McGlinchey and T.-S. Tan, J. Organomet. Chem., 1977, 141, 85.

[13] P.L. Timms, paper delivered at the EUCHEM Conference, Venice, 1979

[14] C. U. Pittman, Y. Sasaki and T.K. Mukherjee, Chem. Letters, 1975, 383.

[15] R.E. MacKenzie and P.L. Timms, J.C.S.--Chem. Comm., 1974, 640.

[16] C.G. Francis, H. Huber and G.A. Ozin, J. Amer. Chem. Soc., 1979, 101, 6250.

[17] R. Sheets and G. Blyholder, Appl. Spectrosc., 1976, 30, 602, and references cited therein.

[18] G.A. Ozin, Acc. Chem. Res., 1977, 10, 21, and references cited therein.

[19] G.A. Ozin, Cat. Rev.--Sci. Eng., 1977, 16, 191, and references cited therein.

[20] G.A. Ozin. "Diatomic Metals and Metallic Clusters," Faraday Symposia of The Chemical Society, 1980, 14, 1, and references cited therein.

[21] C.G. Francis and G.A. Ozin, Proc. EUCHMOS Molecular Spectroscopy Conference, in Spectroscopy in Chemistry and Physics, Sept. 1979, Elsevier, 1979, and J. Mol. Struct., 1980, 59, 55.

[22] C.G. Francis, H. Huber and G.A. Ozin, Angew Chem. Int. Edn. 1980, 19, 402.

[23] H. Huber, T.A. Ford, W. Klotzbücher, M. Moskovits, E.P. Kündig and G.A. Ozin, J. Chem. Phys., 1977, 66, 524.

[24] C.G. Francis, H. Huber and G.A. Ozin, Inorg. Chem., 1980, 19, 219.

[25] M. Andrews, C.G. Francis, H. Huber and G.A. Ozin, unpublished work.

[26] H.B. Abrahamson, C.G. Frazier, D.S. Ginley, H.B. Gray, J. Lillienthal, D.R. Tyler and M.S. Wrighton, Inorg. Chem., 1977, 16, 1554.

[27] S. Yatsuya, K. Mihama and R. Uyeda, Japan J. Appl. Phys., 1974, 13, 749.

[28] S. Yatsuya, T. Hayashi, H. Akoh, E. Nakamura and A. Tasaki, Japan J. Appl. Phys., 1978, 17, 355.

[29] J.M. Burkstrand, Surf. Sci., 1978, 78, 513.

[30] J.M. Burkstrand, Appl. Phys. Lett., 1978, 33, 387.

[31] J.M. Burkstrand, J. Vac. Sci. Technol., 1979, 16, 363.

[32] N. Boonthanom and M. White, Thin Solid Films, 1974, 24, 295.

[33] W. Klotzbücher, G.A. Ozin, J.G. Norman, Jr., and H.J. Kolari, Inorg. Chem., 1977, 16, 2871.

Identification of Thermal Degradation Products of Titanium Polyethers Using Coupled Thermogravimetric Analysis-Mass Spectroscopy: Development and Evaluation of Instrumentation

Charles E. Carraher, Jr.
H. Michael Molloy
Thomas O. Tiernan
Michael L. Taylor and
Jack A. Schroeder

Department of Chemistry and
The Brehm Laboratory
Wright State University
Dayton, OH 45435

ABSTRACT

The construction and evaluation of a coupled Thermogravimetric Analyzer-Mass Spectrometer (TG-MS) instrument, capable of heating samples as small as one milligram to temperatures as high as 1200°C, while continuously monitoring the gaseous effluents using mass spectrometry, is described. Degradation of titanium polyethers was studied to assess the performance of the TG-MS.

INTRODUCTION

This report describes the construction and evaluation of a coupled Thermogravimetric Analyzer-Mass Spectrometer instrument as a tool for the detection and identification of thermal degradation products arising from the combustion of polymeric materials. In the studies described, infrared spectral analysis has

195

been employed to obtain supplemental information on combustion
residues.

The materials selected for study in evacuating the performance of the TG-MS are organometallic polymers. These polymers
were chosen for several reasons. First, since degradation of
these compounds typically occur by different routes in air and in
inert environments, the two environments can be clearly, easily
differentiated. Second, some of the organometallic polymers exhibit good high temperature stability, losing less than 20% of
their initial weight to the 800-1200°C range. Identification of
the degradation products from the polymer would be useful in
better understanding their good stability, and in designing and
synthesizing still more stable polymers. A third reason for
using organometallic polymers in assessing the TG-MS is that the
degradation of these materials occurs in a stepwise manner over a
wide temperature range (often over a range in excess of 800 C°)
with interspersed "stability plateaus" (temperature ranges where
little or no weight change occurs). Continuous monitoring of the
evolved chemical products, as is possible with the TG-MS described, is therefore necessary in order to understand these transitions in the degradation sequence. Finally, the organometallic
polymers are good candidates for study by TG-MS because they are
difficult to characterize by other techniques. Conventional C,
H, N elemental analysis often yields poor results, and since many
of these polymers are only sparingly soluble in most solvents,
characterization by NMR or ESR is not practical. Methods such as
those described herein are therefore needed for characterizing
the structures of such organometallic polymer materials.

A variety of TG-MS combinations have been used previously by
other investigators to evaluate the thermal properties of polymers and several publications have appeared which are concerned
with the application of these techniques for studying organo-

metallic products (1-7).   However, only Goldfarb, et al. (7)
have reported extensive investigations of the uses of TG-MS for
the study of organometallic polymers. The latter authors utilized this technique to identify thermally produced fragments of
polysiloxanes.

The organometallic polymers investigated here comprise a
family of titanium polyethers which are potentially useful in
exterior surface coatings, owing to their excellent resistance to
ultraviolet radiation (8-10).

INSTRUMENTATION

The TG-MS instrument constructed in the present study consists of a double-focusing duPont 21-491 Mass Spectrometer coupl-
ed through a single-stage glass jet separator to a duPont 951
Thermal Gravimetric Analyzer which is attached to a duPont 990
Thermal Analyzer Console.   The MS is equipped with a Hewlett-
Packard, HP-2216C computer having 24K core memory and a disc-
oriented data system specially developed for the duPont 21-491
Mass Spectrometer.   The MS system can be controlled by the computer system, which includes a dual 2.5M byte disc drive, a
Hewlett-Packard Cathode Ray Tube terminal, a Tektronix storage
scope (for display) driven by a dual 12 bit digital-to-analog
(D/A) converter, and a Versatec printer/plotter. Data is acquired using a 14 bit analog-to-digital (A/D) converter (13 plus
sign). This system can operate and process data at rates up to 8
KHZ.   Figure 1 is a block diagram of the TG-MS combination, and
the Hewlett-Packard 2116C computer and appropriate peripherals.
Also shown in Figure 1 (below the dotted line), as components of
this total system, are two other computer systems (HP-2100A and
Data General Nova 2) which are utilized for additional data processing and for mass spectral library searches and identifica-
tions. Data from the HP-2116C is manually transferred to other

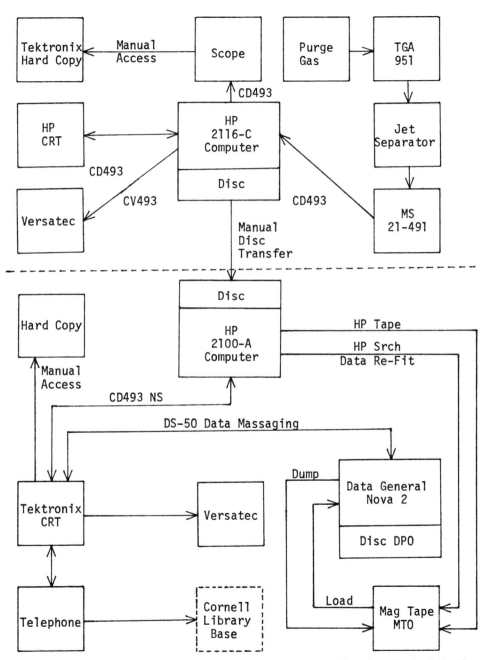

Figure 1. A Block Diagram of the TG-MS Combination Coupled with the
Hewlett-Packard 2100 Series Computer and Peripherals

computer banks by moving the disc storage cartridge from one system to the other.

The duPont 951 Thermal Gravimetric Analyzer, supported by four
flexible legs, was placed in a 22x16x3 inch deep sand bed to
reduce vibrations generated by other instruments in the laboratory and by personnel movements.

The line which transmits the purge gas to the TG is $\frac{1}{4}$-inch
o.d. stainless steel with a Swagelok quick disconnect which is
used during sample loading and preliminary adjustment of the system. Also incorporated in the purge gas line were a drying tube
and a shut-off valve to regulate the helium flow. The pressure of
the second-stage purge gas regulator was set at 40 P.S.I. to
reduce the possibility of a large volume of air entering and
subsequently damaging the TG-MS system in the event of a leak in
the gas line.

The TG quartz-furnace tube was attached to the MS as shown in
Figure 2. The quartz furnace tube was extended by adding $2\frac{1}{2}$
inches of $\frac{1}{4}$-inch o.d. quartz tubing in order to facilitate connection with the jet separator. Connection was made by means of a
Swagelok union with a $\frac{1}{4}$-inch ceramic-teflon front ferrule and a
stainless steel back ferrule to reduce the probability of breaking the tip of the quartz tube.

A 2-inch section of $\frac{1}{4}$-inch o.d. glass tubing was connected,
also using Swagelok unions, between the jet separator and quartz
tube and the entire interface assembly was placed in a specially
constructed heated oven. This arrangement permits the jet separator to be easily disconnected and cleaned by simply loosening
the connections on the Swagelok fitting which secure the 2-inch
glass tube (Figure 2). A glass-lined stainless steel transfer
line is used to connect the MS to the jet separator oven. This
line is thirty inches in length and 1/8-inch o.d. glass lined
stainless steel tubing is employed.

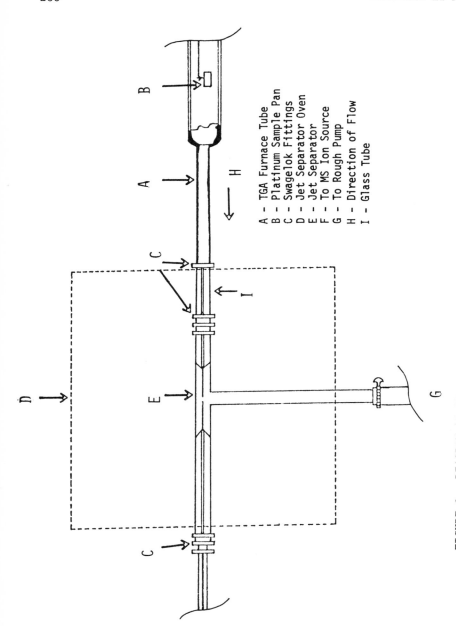

FIGURE 2. DIAGRAM OF THE TG-MS INTERFACE SHOWING THE COUPLING OF THE JET SEPARATOR TO THE TG QUARTZ FURNACE TUBE

A – TGA Furnace Tube
B – Platinum Sample Pan
C – Swagelok Fittings
D – Jet Separator Oven
E – Jet Separator
F – To MS Ion Source
G – To Rough Pump
H – Direction of Flow
I – Glass Tube

A thermocouple, interfaced with the Hewlett-Packard 2116C computer, was employed to sense the temperature during the course of the thermogravimetric analysis. This probe was attached to the TG in order to monitor the actual temperature of the sample during each run (Figure 1). During sample analysis, the TG-jet separator connection was wrapped with heating tape and maintained at a temperature of $125^{\circ}C$ while the jet separator oven was maintained at $135^{\circ}C$. The MS-jet separator transfer line was also wrapped with heat tape and maintained at $165^{\circ}C$ while the source temperature was held at $230^{\circ}C$.

The mass spectral data acquired using the TG-MS combination can also be compared with standard mass spectral files utilizing a low cost library search system which currently contains some 41,000 mass spectra. This time-sharing computer data system is located at Cornell University and is frequently capable of providing useful identification data even when the unknown spectra results from a mixture of sample components (11,12). The HP 2100A computer system is interfaced directly to this search system via an acoustic coupler, using normal telephone lines. (Figure 1). This mass spectral data system was used in the present study to facilitate analysis of the data obtained for several samples from the TG-MS system.

The infrared spectral data gathered on the original samples and degradation residues were obtained utilizing KBr pellets and Perkin-Elmer 237B, 457 and 621 Grating Infrared Spectraphotometers.

Instrument Calibration Procedures

The mass scale from m/e 12 to m/e 1000 of the dePont 21-491 Mass Spectrometer was calibrated over the range from m/e 12-1000 by introducing perfluorokerosene (PFK) into the MS system and then following the steps indicated in the block diagram given in Figure 3. PFK was used as a mass calibration standard since it is

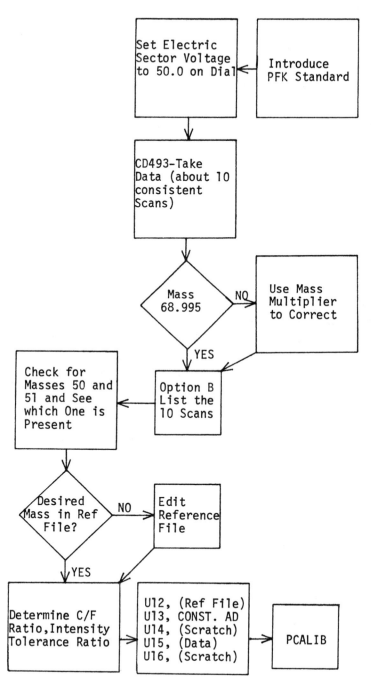

FIGURE 3. CALIBRATION FLOW CHART FOR THE DuPONT 21-491 MS

a widely utilized and well characterized reference standard which exhibits a large number of fragment ions (13). The MS is equipped with a Hall probe to measure the magnetic field strength. The computer monitors the Hall probe voltage (as an indicator of the magnetic field, and, upon calibration, the mass scale, the ion beam intensity and the thermocouple output simultaneously. The positions of ion peaks on the mass scale are determined by on-line centroiding of that portion of each peak with intensity greater than 50% of the maximum of that peak, which yields typical scan to scan reproducibility in mass assignments of 200 to 500 PPM. The Hall probe voltage centroids are converted to mass using a polynomial equation, typically eighth order. Repetitive scans of the PFK standard are acquired until ten consistent scans are obtained. The Hall probe voltages corresponding to the mass peaks detected in each scan are then averaged and displayed or printed out. A check of the print out is made to eliminate possible errors which may occur at this stage such as peaks of adjacent mass being averaged together, or one or more obviously erroneous values being included in the average. The peaks are then compared one by one with the reference spectrum stored on the disc. At this point, a nominal mass for each experimental peak is computed by using an approximate polynomial. The nominal mass is used only as an initial step in the identification of the masses and therefore does not need to be accurate. The poorer the approximate polynomial is, however, the greater will be the need to edit the calculated results. Next, each nominal mass in the unknown spectrum is compared (with specific tolerance limits) to a peak in the reference spectrum file and this list is displayed. At this point the operator can change or delete any of the matches made by the computer, as well as add any apparent matches which were missed. These data, consisting of a list of the Hall probe voltages, masses assigned, and weighting factors, determined from the mass of each peak, and the number of spectra in which it occurred, are recorded on the disc.

The data are then processed by a least squares curve fitting program which determines accurately the polynomial expression which best fits the data. A table is then displayed showing the coefficients of all the terms for polynomials ranging in order from two to eight, and indicating the accuracy of the fit for each. Finally, the desired order of smoothing is chosen, the computer automatically records the information on the disc and the mass multiplier is set to a predetermined value (for more information on this procedure see references 14-16). The calibration procedure typically requires thirty to sixty minutes, and yields accurate mass assignments. The mass spectrometer, when properly tuned, provides unit mass spectral resolution up to mass 1000. However, for the present study, only the mass range from m/e 12 to m/e 400 was utilized.

Calibration of the TG-MS instrument was accomplished using 2.5 to 3.5 milligrams of a standard calcium oxalate monohydrate, and entailed correlating the TG thermogram with the mass spectral data. Calcium oxalate monohydrate was used for calibration since it is the standard commonly used to calibrate the apparatuses. In addition, the degradation products are well known, with water, carbon monoxide and carbon dioxide being the three principle degradation products (17). Figure 4 illustrates the correlation between weight loss as indicated by the TG profile and the appearance of identifiable ions in the mass spectra. In Figure 4, the plot labelled T is a plot of the total ion current as a function of temperature, A is a plot of m/e 18 (indicative of water) as a function of temperature, B is a plot of m/e 28 (indicative of carbon monoxide) as a function of temperature, C is a plot of m/e 44 (indicative of carbon dioxide) as a function of temperature and G is a plot of per cent weight loss as a function of tempera-

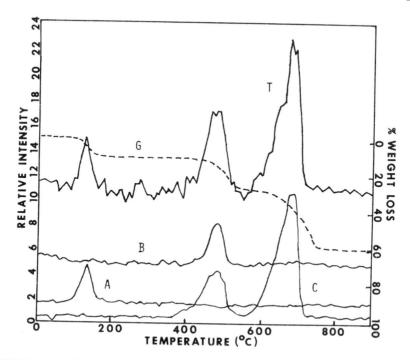

FIGURE 4. TG(---) AND MS(——) PLOTS FOR CALCIUM OXALATE MONA-
HYDRATE DECOMPOSITION. THE MS IONIZING VOLTAGE WAS 70eV, THE FLOW
RATE OF HELIUM WAS 55mL/MIN AND THE HEATING RATE WAS 20°C/MIN. THE
SAMPLING FREQUENCY OF THE DATA ACQUISITION SYSTEM WAS 41 KC.   T =
TOTAL ION CURRENT   A = PLOT OF m/e 18   B = PLOT OF m/e 28 C = PLOT
OF m/e 44   G = TG

ture. The TG-MS calibration is normally valid for several weeks,
if the system is used regularly and if no instrument malfunctions
occur.

During a typical TG-MS analysis requiring 45 minutes the TG
temperature was programmed to increase at a constant heating
rate of 20 C° per minute, over the range from ambient temperature
to 900°C, while volatile degradation products formed from degradation of the organometallic polymer were continuously swept from
the TG tube by the helium carrier gas (flowing through the furnace

tube at 55 milliliters per minute), then through the helium jet separator, and finally into the mass spectrometer ion source where ionization occurred and a characteristic mass spectrum was produced. The typical MS ion source temperature was 230°C and the ionizing voltage was 70 eV.

The TG-MS system was monitored periodically to ensure that sample "carry over" or "memory effects" did not occur. This was accomplished by conducting a run with no sample in the TG pan. The data gathered from such blank runs were checked and in all instances, indicated no sample residues from previous analyses.

RESULTS AND DISCUSSION

Titanium polyethers are of the general form I. As previously noted, these compounds have been suggested as ultraviolet stabilizers for exterior coatings (8-10). Also, titanium polyethers have been shown to be semiconductors, exhibiting specific resistivity in the range of 10^5 to 10^7 ohm-cm (18,19). It is particularly interesting that the titanium polyethers, as a family, exhibit poor low temperature stability (below 200°C) but moderate to good high temperature stability (retaining up to 80% weight at temperatures up to 900°C) (8,10). It was hoped that TG-MS analysis of these compounds would provide information on the decomposition mechanism, including identification of the major evolved degradation fragments and the sequence of bond breakage, to (potentially) identify the site of least stability, eventually enabling the design of more thermally stable polyethers and would reveal the weakest bond in the compounds. Results obtained from these degradation studies are summarized below.

Cp_2TiCl_2 and Hydroquinone

The polyether derived from the condensation of dicyclopentadienyl titanium dichloride, Cp_2TiCl_2, with hydroquinone exhibited an initial weight loss at a temperature of 89°C, with the most rapid weight loss (about 8%) occurring between 100 and 180°C, followed by an almost constant-slope weight loss of 20% over the next 600 C° range (Figure 5). The comparatively rapid degradation

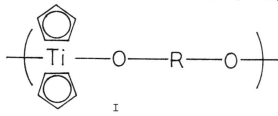

I

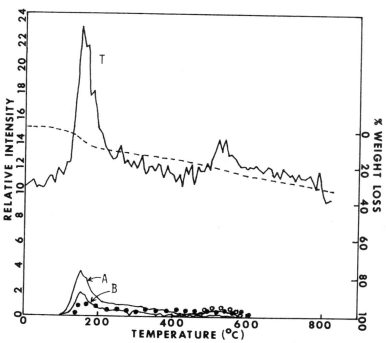

FIGURE 5. TG(---) AND MS (——) PLOTS FOR THE CONDENSATION PRODUCT FROM DICYCLOPENTADIENYL TITANIUM DICHLORIDE AND HYDROQUINONE DECOMPOSITION. THE MS IONIZING VOLTAGE WAS 70eV, THE FLOW RATE OF HELIUM WAS 55 mL/MIN AND THE HEATING RATE WAS 20 C°/MIN. THE SAMPLING FREQUENCY OF 41 KC WAS USED. T = TOTAL ION CURRENT A = PLOT OF m/e 66——,B = PLOT OF m/e 65——C = PLOT OF m/e 108.... D = PLOT OF m/e 78 ◯◯◯◯ E = PLOT OF m/e 94 •—•—•—•

occuring between 100 and 180°C gives rise to volatile degradation products which exhibit 15 major (normalized intensities greater than 1%) mass peaks (Table 1). Peaks are observed at m/e 66, 65, 39, 40, 26 63, 64, 31, 62, 67, 41 and 51 and these ions and their respective intensities correspond to those which are characteristic of the standard spectrum of cyclopentadiene as listed in the John Wiley compilation by McLafferty, et al. (Table 2). In addition, at temperatures between 120 and 180°C three additional major mass peaks appeared in the mass spectrum of the degradation products (m/e = 108, 54 and 82). Thus peaks at m/e 108, 54, 82, 26, 51, 41 and their respective intensities are consistent with the standard spectrum of p-benzoquinone (Table 2). Over this temperature range, however, the ions characteristic of cyclopentadiene are greater in intensity than those attributable to p-benzoquinone. Thus, degradation apparently occurs at moderate

Table 1

Major Ions in the Mass Spectra of the Degradation Products from the Pyrolysis of the Condensation Product of Dicyclopentadienyl Titanium Dichloride, Cp_2TiCl_2, and Hydroquinone, HQ, at 143°C

m/e	Apparent Degradation Product	Normalized Intensity	m/e	Apparent Degradation Product	Normalized Intensity
66	Cp	52.129	31	Cp	3.852
65	Cp	23.439	51[a]	HQ,Cp	1.257
39	Cp	17.985	26[a]	HQ,Cp	4.704
40	Cp	16.991	41[a]	HQ,cp	2.656
63	CP	4.217	108	HQ	7.117
64	Cp	3.427	54	HQ	4.542
67	Cp	3.021	82	HQ	2.595
62	Cp	3.183			

a. these fragment ions can be derived from Cp and HQ

Table 2

Comparison of Mass Spectral Data for Cyclopentadiene, p-Benzoquinone, Phenol and Benzene Compared with the TG-MS Spectra of
the Condensation Product of Dicyclopentadienyl Titanium Dichloride
and Hydroquinone

Cyclopentadiene

m/e	66	65	39	40	48	64	67	63	31
				Relative Intensity					
John Wiley Atlas Mass Spectrum	1000	473	316	273	82	82	63	58	49
TG-MS Degradation Product Spectrum	1000	450	345	326	–	81	58	61	74

p-Benzoquinone

m/e	108	54	82	26	50	51
		Relative Intensity				
John Wiley Atlas Mass Spectrum	1000	790	357	298	54	43
TG-MS Degradation Product Spectrum	1000	638	365	661	68	43

Phenol Mass

m/e	94	39	66	65	40	55
		Relative Intensity				
John Wiley Atlas Mass Spectrum	1000	531	467	395	219	151
TG-MS Degradation Product Spectrum	1000	430	–	319	612	665

Benzene Mass

m/e	78	51	52	50	39
		Relative Intensity			
John Wiley Atlas Mass Spectrum	1000	205	196	179	141
TG-MS Degradation Product Spectrum	1000	240	481	–	100

temperatures mainly through evolution of the cyclopentadiene moiety, and less so, but still to a significant extent, through the
evolution of p-benzoquinone. At the temperature of 227°C ions at
m/e of 141-142 appear, which are characteristic of the product
resulting from coupling of phenylene and cyclopentadiene was (Table
3).

At temperatures above 388°C, the mode of degradation has
changed and 13 additional major mass peaks appear in the mass
spectra of the degradation products (Table 3). Between 388°C and
588°C, peaks are observed at m/e 78, 51, 52, 50, 39 and 27 and the
respective intensities of these ions match those which are reported in standard spectrum of phenol (Table 2). Ions were observed at m/e 154, above 488°C, presumably indicative of the
phenylene dimer (Table 3). Therefore, higher temperatures apparently result in greater fragmentation of the oxyphenylene,-∅-
0, moiety.

Table 3

Major Ions in the Mass Spectra of the Degradation Products from
the Pyrolysis of the Condensation Product of Dicyclopentadienyl
Titanium Dichloride, Cp_2TiCl_2, and Hydroquinone, HQ, at 503°C

m/e	Apparent Degradation Product	Normalized Intensity	m/e	Apparent Degradation Product	Normalized Intensity
78	Benzene	8.816	55	Phenol	2.497
52	Benzene	4.238	40	Phenol	2.299
27	Benzene	0.860	43	Phenol	2.239
142	Benzene,Cp	1.857	41	Phenol	1.859
51	Benzene,Phenol	2.599	65	Phenol	1.200
39	Benzene,Phenol	2.479	154[a]	Benzene	2.039
94	Phenol	3.758			

a. mass 154 represents the coupling of benzene

The mass spectral ions characteristic of the cyclopentadiene and p-benzoquinone moieties reached maximum intensities at a degradation temperature of 151°C and then decreased steadily, becoming insignificant at temperatures above 480°C (Table 4). Ions indicative of benzene reached an intensity maximum at 511°C, decreased markedly at temperatures above 588°C, and fell off gradually in intensity to a temperature of 749°C (Table 4). The ion indicative of phenylene dimerization reached a maximum at 557°C, and was detectable to temperatures of 618°C (Table 4).

The order of degradation was deduced by noting the TG temperature at which selected ion fragments initially appeared (Table 4). The first significant ion detected in the mass spectrum of the degradation products was m/e 66, corresponding to cyclopentadiene (see A in Form II), followed shortly by m/e 108 corresponding to p-benzoquinone (see B in Form II). The ion next observed was m/e 141, derived from the coupling of the phenylene and cyclopentadiene moieties. At 388°C, benzene was detected as indicated by the appearance of m/e of 78 (see C in Form II), followed by observation of m/e of 94 attributable to phenol (see D in Form II). Finally, m/e 154, representing the product from coupling of phenylene, Ø-Ø, was detected. The fragmentations leading to the observed moieties which are described above are illustrated in form II, below, with their sequence of occurrence indicated by the alphabetical designations

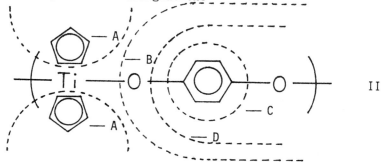

II

Table 4

Major Ions in the Mass Spectra of the Degradation Products from the Pyrolysis of the Condensation Product of Dicyclopentadienyl Titanium Dichloride, Cp_2TiCl_2, and Hydroquinone as a Function of Temperature.

m/e	Degra- dation Product	Initial Appearance		Maximum Intensity		Final Appearance	
		Temp. °C	Normalized Intensity	Temp. °C	Normalized Intensity	Temp. °C	Normalized Intensity
66	Cp^a	89	.523	151	55.122	480	2.298
108	HQ^b	120	1.127	151	7.988	726	1.055
142	$Cp-\emptyset^c$	227	.307	503	1.859	634	.574
78	$\emptyset$	388	.272	511	9.632	588	1.227
94	Phenol	457	1.831	503	3.758	749	.899
154	$\emptyset-\emptyset^d$	488	.424	557	2.323	618	.530

a. Cp represents cyclopentadiene
b. HQ represents hydroquinone
c. $\emptyset$ represents benzene
d. $\emptyset-\emptyset$ represents the product from the dimerization of phenylene

The absence of ions with significant intensities at m/e 47 (Ti), m/e 79 (TiO$_2$) and m/e 170-180 (Ti-Ø) indicate that titanium-containing, volatile species were not formed, which is consistent with the metal remaining within the TG residue.

In summary, the order of appearance of degradation products from the degradation of the condensation products of dicyclopentadiene titanium dichloride with p-benzoquinone are: A-cyclopentadiene, B-p-benzoquinone, C-benzene and D-phenol.

The infrared spectrum of the degradation residue of the product derived from the titanium polyether shows identifiable bands only at wave numbers below 800 cm^{-1}, and the observed spectrum is identical to that of titanium dioxide, TiO$_2$ (Figure 6). TiO$_2$ is white whereas the solid residue derived from the titanium polyether was light brown in color. If the resulting residue contained only Ti or TiO$_2$, then the weight of the residue would be 17 or 28% respectively of the initial sample. The observed residue weight from the TG was 56% of the original sample, however consistent with the inclusion of materials other than titanium (Figure 6). Polymeric "tars" typically yield spectra which exhibit broad bands having little if any definition. Therefore, such tars

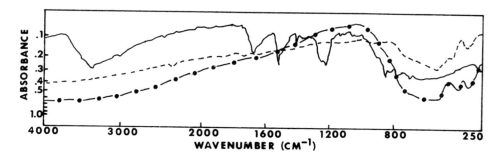

FIGURE 6. INFRARED SPECTRA OF THE CONDENSATION PRODUCT OF DICYCLO-PENTADIENYL TITANIUM DICHLORIDE AND HYDROQUINONE (——) OF THE RESIDUE AFTER HEATING TO 900°C IN HELIUM AT A RATE OF 20 C°/MIN (---), AND OF TITANIUM DIOXIDE (-•-•-).

yield essentially only background contributions to the infrared spectra of the residue. It is likely that the formation of TiO_2 occurs subsequent to the removal of the helium environment since oxygen is needed to form TiO_2 and several oxygen-containing moieties were detected in the evolved degradation products. Quantitative elemental analysis of the amount of titanium present in the residue as Ti or TiO_2 would have to be completed to satisfactorily resolve the question of whether or not the oxygen was derived from the sample itself, the surrounding atmosphere, or both.

Cp_2TiCl_2 and Bisphenol A

The polyether derived from the condensation of dicyclopentadienyl titanium dichloride, Cp_2TiCl_2, with bisphenol A, exhibited major weight loss in two temperature ranges (Figure 7). In the first range between 105 and 400°C the products evolved initially (at 105°C) yield ions characteristic of the cyclopentadiene moiety (m/e 66) (Figure 6). Ions characteristic of the propyl moiety (m/e 44) began to appear in significant (>1%) intensities at around 204°C and continue to be detected until the heating cycle is completed. Two maxima in the m/e 44 intensity are observed at 342 and 420°C. The second temperature range in which significant weight loss occurs is coincident with the observation of ions in the mass spectrum of the degradation product which are characteristic of the phenol moiety (m/e 94). There are initially observed at about 273°C, reach a maximum at around 419°C, and become insignificant at about 641°C. The probable origin of the moieties just mentioned are indicated in Form III where the sequence of elimination of these fragments is indicated by the alphabetical designations. Table 5 summarizes the order of appearance of the degradation products.

As in the case with the products from the condensation with hydroquinone, a significant quantity of nonmetallic residue was

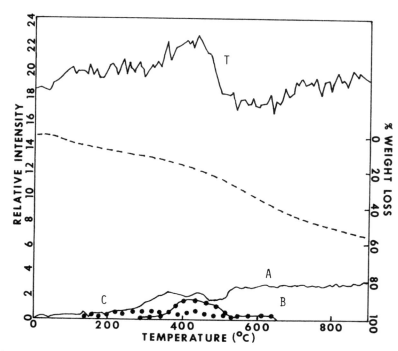

FIGURE 7. TG (---) AND MS(——) PLOTS FROM THERMAL DEGRADATION OF THE CONDENSATION PRODUCT FROM DICYCLOPENTADIENYL TITANIUM DICHLORIDE AND BISPHENOL A OBTAINED AT A MS IONIZING VOLTAGE OF 70 eV, TG FLOW RATE OF HELIUM AT 55 mL/MIN FOR A HEATING RATE OF 20 C^o/MIN AND A DATA SAMPLING FREQUENCY OF 41 KC.
T = TOTAL ION CURRENT   A = PLOT OF m/e 44 (——)
B = PLOT OF m/e 94 (-•-)   C = PLOT OF m/e 66 (...)

observed from the degradation of the Cp_2TiCl_2/bisphenol A condensation product at temperatures up to 900°C. The residue constituted 44% of the initial sample (see Figure 6) (Ti-12%; TiO_2= 20%).

In summary, the order of appearance of the degradation products from the pyrolysis of the condensation products of dicyclopentadienyl titanium dichloride with bisphenol A was: A-cyclopentadiene, B-propyl and C-phenol.

III

Degradation occurs through several kinetically controlled stability plateaus. Again, a stability plateau is a temperature range in which little or no weight loss occurs, and a kinetically controlled stability plateau is a region where weight loss occurs independently of heating rate, which is evident experimentally, because the plateau is not parallel to the temperature axis but exhibits a low pitch.

Cp_2TiCl_2 and Di-tert-butyl-p-hydrquinone

The order of appearance of degradation products from the pyrolysis of the condensation product of dicyclopentadienyl titanium dichloride and di-tert-butyl-p-hydroquinone is shown in Table 6 and illustrated schematically in Form IV, where the degradation sequence is indicated by the alphabetical designations.

Table 5

Major Ions in the Mass Spectra of Degradation Products from the Pyrolysis of the Condensation Product of Dicyclopentadienyl Titanium Dichloride, Cp_2TiCl_2, and Bisphenol A as a Function of Temperature

m/e	Degradation Product	Initial Appearance		Maximum Intensity		Final Appearance	
		Temp. °C	Normalized Intensity	Temp. °C	Normalized Intensity	Temp. °C	Normalized Intensity
66	Cp[a]	105	.668	335	5.364	488	.830
44	Propane	204	4.760	342	18.382	>900	-
94	Phenol	273	.387	419	20.453	641	.243

a. Cp represents cyclopentadiene

Table 6

Major Ions in the Mass Spectra of the Degradation Products from Pyrolysis of the Condensation Product of Dicyclopentadienyl Titanium Dichloride, Cp_2TiCl_2, and Di-tert-butyl-p-benzoquinone as a Function of Temperature

m/e	Degradation Product	Initial Appearance		Maximum Intensity		Final Appearance	
		Temp. °C	Normalized Intensity	Temp. °C	Normalized Intensity	Temp. °C	Normalized Intensity
66	Cp[a]	74	.947	434	3.404	848	.293
220	DTHQ[b]	74	1.375	258	27.479	> 900	–
205	THQ[c]	82	1.542	258	27.479	> 900	–
108	HQ[d]	89	.719	243	7.153	533	.435
94	Phenol	135	2.570	243	4.238	503	.771
44	Propane	404	12.473	572	21.595	787	5.212

a. Cp represents cyclopentaciene

b. DTHQ represents di-tert-p-benzoquinone moiety

c. THQ represents di-tert-p-benzoquinone moiety minus a methyl group

d. HQ represents p-hydroquinone minus two protons

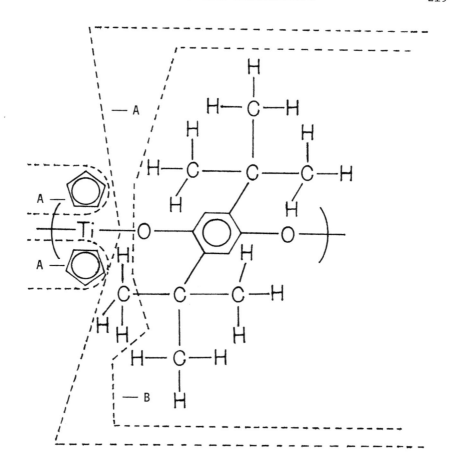

IV

Ions characteristic of the cyclopentadiene moiety were again the first to appear in the mass spectrum of the degradation products, initially at a temperature of 74°C. Unlike the two compounds discussed above, degradation in this case occurs rapidly largely over a narrow 50 C° temperature range (from 130 to 180°C), with about 80% weight loss from the sample (see Figure 8). Evolution of cyclopentadiene (m/e 66) and appearance of the m/e 220

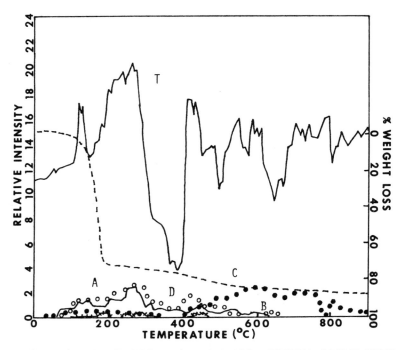

FIGURE 8. TG(---) AND MS(——) PLOTS FROM THERMAL DEGRADATION OF
THE CONDENSATION PRODUCT FROM DICYCLOPENTADIENYL TITANIUM
DICHLORIDE AND DI-TERT-BUTYL-P-HYDROQUINONE OBTAINED AT A MS
IONIZING VOLTAGE OF 70eV, TG FLOW RATE OF HELIUM AT 55 mL/MIN FOR
A HEATING RATE OF 20 C^{o}/MIN AND A DATA SAMPLING FREQUENCY OF 41
KC. T = TOTAL ION CURRENT   A = PLOT OF m/e 66 (-++-) B = PLOT OF
m/e 220 (——)   C = PLOT OF m/e 44 (...)   D = PLOT OF m/e 205 (OO)

ion, which corresponds to the entire di-tert-butyl-p-hydroquinone
moiety (less 2 protons), began at 74°C (see A in Form IV). The
mass spectral ion at m/e 205 corresponds to the di-tert-butyl-p-
hydroquinone moiety (less a methyl group) and appeared initially
at a temperature of 82°C. As expected, evolution of this product
closely parallels that which is indicated by the mass 220 moiety.
Other major ions observed in the mass spectra of the degradation
products of this compound were m/e 44 probably indicative of

propane, m/e 94 which apparently results from the phenol product, and m/e 108 which corresponds to the p-hydroquinone product.

At 97°C ion masses were detected in the degradation product spectra, 14 being greater than 1% in normalized intensity. The number of ion masses detected in the mass spectra maximized at a pyrolysis temperature of 235°C, and 50 of the 67 ions observed were greater than 1% in normalized intensity. The number and intensity of mass spectral peaks then decreased steadily, as the temperature increased to 400°C. It is clear, therefore, that degradation of this compound occurred abruptly with the generation of a large number of fragments. These fragments appear to be largely derived from the di-tert-butyl-p-hydroquinone moiety, and the mass spectra are in good agreement with tabulated mass spectral data for the 12 most abundant fragments.

The residue resulting from degradation of the Cp_2TiCl_2/di-tert-butyl-p-hydroquinone condensation products, in contrast to the previously discussed compounds, consists entirely of titanium (% residue = 12, % Ti = 13, % TiO_2 = 22) without organic content.

Cp_2TiCl_2 and Tetrachloro-p-hydroquinone

Degradation of the condensation product from dicyclopentadienyl titanium dichloride, Cp_2TiCl_2, with tetrachloro-p-hydroquinone began at a temperature of 105°C, as indicated by the observation of m/e 246-249 in the mass spectra of evolved products. These ions are indicative of the tetrachloro-p-hydroquinone moiety (m/e 246)(Table 7). The evolution of cyclopentadiene followed, initially at about 143°C, and continued to 404°C (Figure 9, Table 7). Degradation was rapid (similar to that of the condensation product of dicyclopentadienyl titanium dichloride with di-tert-butyl-p-hydroquinone) with 70% weight loss occurring in the range from 170 to 230°C. This was followed by a gradual weight loss of an additional 10% from 420 to 520°C. In

Table 7

Major Ions in the Mass Spectra of the Degradation Products from Pyrolysis of the Condensation Product of Dicyclopentadienyl Titanium Dichloride, Cp_2TiCl_2, and Tetrachloro-p-hydroquinone as a Function of Temperature

m/e	Degra-dation Product	Initial Appearance		Maximum Intensity		Final Appearance	
		Temp. °C	Normalized Intensity	Temp. °C	Normalized Intensity	Temp. °C	Normalized Intensity
246	TCHQ[a]	105	.549	741	1.331	>900	–
66	Cp[b]	143	4.616	166	6.064	404	1.032
36	Cl or HCl	618	.464	772	2.240	>900	–

a. TCHQ represents tetrachloro-p-hydroquinone moiety

b. Cp represents cyclopentadiene

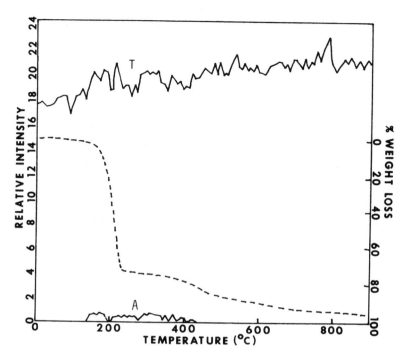

FIGURE 9. TG(---) AND MS(——) PLOTS FROM THERMAL DEGRADATION OF
THE CONDENSATION PRODUCT FROM DICYCLOPENTADIENYL TITANIUM DI-
CHLORIDE AND TETRACHLORO-P-HYDROQUINONE OBTAINED AT A IONIZING
VOLTAGE OF 70eV, TG FLOW RATE OF HELIUM AT 55 mL/MIN FOR A HEATING
RATE OF 20 C^o/MIN AND A DATA SAMPLING FREQUENCY OF 41 IC.
T = TOTAL ION CURRENT   A = PLOT OF m/e 66

contrast to the previously discussed compound (Cp_2TiCl_2 with di-
tert-butyl-p-hydroquinone) excessive fragmentation did not occur
either during or after the rapid weight loss (only 29 ion masses
were detected in the degradation product spectra and only 7 exceeded 1% in normalized intensity. Furthermore there did not
appear to be any significant change in the observed total ion
current over the entire temperature range. Trace quantities of
ions indicative of tetrachloro-p-hydroquinone (less one or two
chlorine atoms)(m/e 175 and 210) were detected at temperatures

near $300^{\circ}C$ and $700^{\circ}C$. Traces of hydrogen chloride and chlorine
m/e 36 were detected from 600 to $900^{\circ}C$.

Results obtained for this sample preclude the postulation of
a well-defined degradation sequence. During the heating cycle,
(at about $100^{\circ}C$) a cloud of gaseous material was visually evident
and a yellow deposit appeared on the inside tip of the TG quartz
tube. Thus, mass spectral ions detected at higher temperatures
may be due in part to the evolution and/or degradation of this
yellow solid, rather than arising solely directly from the condensation product. A tentative sequence of degradation is indicated by the date in Table 7 and illustrated schematically in
Form V, where the order of fragmentation is again designated by
the alphabetical notation.

The final weight of the residue at $900^{\circ}C$ was approximately
12% of the initial sample which corresponds to the value expected
if the residue consists only of titanium without any organic
residue (% Ti = 14, % TiO_2 = 24).

Cp_2TiCl_2 and 2-methyl-p-hydroquinone

The last titanium polyether studied was that derived from
the condensation of dicyclopentadienyl titanium dichloride,
Cp_2TiCl_2, and 2-methyl-p-hydroquinone. The postulated degrada-

tion sequence for this compound is indicated by the data shown in Table 8 and is illustrated schematically in Form VI.

VI

The mass spectral ions observed initially from the degradation products were m/e 18 and m/e 17 which were detectable in the temperature range from 51 to 160°C, reaching a maximum at about 97°C. This particular sample exhibited a strong infrared band at about 3300 to 3000 cm^{-1}, which is probably indicative of the presence of water. Therefore, the initial weight loss apparently resulted from the elimination of water from the hydrated sample. The initial evolved ions and associated fragments from the polymer chain itself which corresponded to the cyclopentadiene moiety began at about 105°C and continue to around 450°C. There were no other major ion masses detected in the degradation product mass spectra although trace intensities of m/e 78 (benzene), m/e 109 (p-hydroquinone), in the degradation product mass spectra, and m/e 123 (methyl-p-hydroquinone) were observed. No major change in the total ion current was evident and the total weight loss amounted to only 30%, which occurred gradually over an 800 C° temperature range (Figure 10). The major weight loss was largely due to cyclopentadiene elimination (see A in Form VI) (one cyclo-

Table 8

Major Ions in the Mass Spectra of the Degradation Products from Pyrolysis of the Condensation Product of Dicyclopentadienyl Titanium Dichloride, Cp_2, and 2-Methyl-p-hydroquinone as a Function of Temperature

m/e	Degra-dation Product	Initial Appearance		Maximum Intensity		Final Appearance	
		Temp. °C	Normalized Intensity	Temp. °C	Normalized Intensity	Temp. °C	Normalized Intensity
18	H_2O	51	14.475	97	26.638	212	16.188
66	Cp^a	105	.467	289	4.220	457	.310

a. Cp represents cyclopentadiene

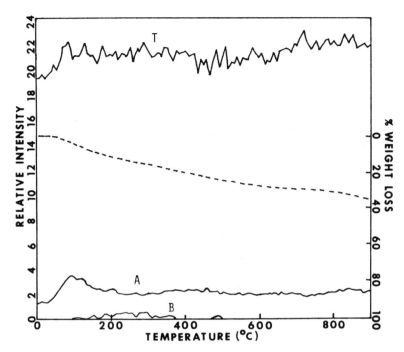

FIGURE 10. TG(---) and MS(——) PLOTS FROM THERMAL DEGRADATION OF THE CONDENSATION PRODUCT FROM DICYCLOPENTADIENYL TITANIUM DICHLORIDE AND 2-METHYL-P-HYDROQUINONE OBTAINED AT A IONIZING VOLTAGE OF 70eV, TG FLOW RATE OF HELIUM AT 55 mL/MIN FOR A HEATING RATE OF 20 C°/MIN AND A DATA SAMPLING FREQUENCY OF 41 KC.
T = TOTAL ION CURRENT   A = PLOT OF m/e 18
B = PLOT OF m/e 66

pentadiene = 22% and two cyclopentadienes = 44% of the residue). Therefore, the residue contained a substantial quantity of organic material (% Ti = 16, % TiO_2 = 26).

CONCLUSIONS

The results of the present study show that the coupled TG-MS instrument can provide much useful data relevant to determining the composition of polymeric materials of the type studied here.

Table 9

Temperature Range in which Cyclopentadiene was Observed for the Titanium Polyethers

Associated Diol	Initial Temperature of Appearance	Temperature at which Product Observation Maximized	Final Temperature at which Product was Observed
p-hydroquinone	89°	151°C	480°C
bisphenol A	105°C	335°C	488°C
di-tert-butyl-p-hydroquinone	74°C	258°C	848°C
tetrachloro-p-hydroquinone	143°C	166°C	404°C
2-methyl-p-hydroquinone	105°C	289°C	457°C

The data obtained here suggest structures for the titanium polyether which are consistent with those previously propsed (8-10).
Furthermore, this technique often permits detection of major
trends in the sample degradation sequence. Two general phenomena
could be tentatively identified for the titanium polyethers investigated in the present study. First, the initially evolved
degradation product has cyclopentadione, which typically appeared
at a temperature in the neighborhood of 100°C (Table 9). Secondly, this was generally followed by the evolution of the p-hydro-
quinone-containing moiety, along with the associated oxygen(s).
A great deal of variation was observed, however, with respect to
the extent and rate of degradation of the hydroquinone-containing
moiety from the several polymers studied, and the quantities of
organic components in the final residues from degradation also
varied markedly.

REFERENCES

[1]   R.G. Beimer, Organic Coatings and Plastics Chemistry, 35,
       428, (1975).

[2]   P. Gambert, C.E. James, C. Tallemant, B. Manne and P.
       Padieu, Recent Der. Mass Spectrum, Bio Chem. Med., 4 (1),
       361, (1978).

[3]   J. Chin and A. Beattie, Thermochem. Acta., 21, 263,
       (1977).

[4]   S. Morisaki, Therm. Anal. (Proc. Int. Conf.), 5, 297,
       (1977).

[5]   J.D. Smith, D. Phillips and T. Kaczmarek, Microchem. J.,
       21, 424, (1976).

[6]   C. Sceney, J.F. Smith, J.D. Hill and R. Magel, J. Thermal
       Anal., 9, 415, (1976).

[7]   T. Goldfarb, E. Choe and H. Rosenberg, "Organometallic
       Polymers", (Edited by C. Carraher, J. Sheats and C.
       Pittman), Academic Press, N.Y., (1978), Chapter 25.

[8]   C. Carraher, "Interfacial Synthesis", Vol. II (Edited by
       P. Millich and C. Carraher), Marcel Dekker, N.Y., 1977;
       Chpt. 20.

[9]   C. Carraher and S. Bajah, Polymer (Br), 14, 42 (1973) and
       15, 9 (2974).

[10] C. Carraher and S. Bajah, Br. Polymer J., 7, 155 (1975)

[11] F.W. McLafferty, R.H. Hertel and R.D. Villnock, Organic
       Mass Spectrum, 9, 690 (1974).

[12] K.S. Kwok, R. Venkataraghaven and F.W. McLafferty, J.
       Amer. Soc., 95, 4185 (1973).

[13] F.W. Lafferty, Anal. Chem., 28, 306 (1956).

[14] D.T. Terwilliger, D.C. Walters, M.L. Taylor and T.O.
       Tiernan, Proceedings of the 25th Annual Conference on Mass
       Spectrometry and Allied Topics, American Society for Mass
       Spectrometry, (1977), p. 652-4.

[15] B.M. Hughes, C.A. Davis, D.C. Fee, M.L. Taylor and T.O.
       Tiernan, Proceedings of the 22nd Annual Conference on Mass
       Spectrometry and Allied Topics, American Society for Mass
       Spectrometry, (1974) p. 119-22.

[16] D.T. Terwilliger, Computer System for the Hewlett-Packard
       2100 Series.

[17] DuPont, Instruction Manual for the 990 Thermal Analyzer
       and Modules, Chapter 9, page 8.

[18] C. Carraher, D. Leahy and S. Ailts, Organic Coatings and
       Plastics Chemistry, 37, (2), 201, (1977).

[19] C. Carraher, J. Schroeder, W. Venable and C. MacNeely,
       Organic Coatings and Plastic Chemistry, 38, (1), 544,
       (1978).

Structural Characterization of the Condensation Polymer of Dipyridine Manganese II Dichloride with 1,3-Di-4-piperidylpropane

Charles E. Carraher, Jr.[a]
Van R. Foster[a]
H. Michael Molloy[a]
Michael L. Taylor[a,b]
Thomas O. Tiernan[a,b]
Jack A. Schroeder[a]

Department of Chemistry[a]
and The Brehm Laboratory[b]
Wright State University
Dayton, Ohio 45435

ABSTRACT

The synthesis of polymers derived from the condensation of dipyridine manganese II dichloride with 1,3-di-4-piperidylpropane is described. Structural analysis of the polyamines is detailed utilizing results based on infrared, thermal, elemental and solution analyses.

INTRODUCTION

The previous paper describes the construction and operation of a new Thermogravimetric Analyzer-Mass Spectrometer, TG-MS, assembly suitable for identification of the sequence of products resulting from the thermal degradation of polymers. This paper

describes the initial synthesis of a manganese polyamine and describes the use of the TG-MS assembly for degradation product
identification.

Synthesis of dipyridine manganese II dichloride, Py_2MCl_2,
DMC, has been known for a long time (1). Its structure in the
solid state has been well studied using x-ray diffraction (2,3),
magnetic techniques (2-4) and spectral techniques (mainly infrared, ultraviolet and Raman spectroscopy; 5-12). DMC is octrahedral and is a linear polymer in its solid state (form I) with
bridging occurring through the chlorides. Gill, et al. (9) note
that while a tetrahedral arrangement is feasible for DMC (Mn II is
a d^5 atom) the observed octrahedral structure in the solid state
is presumably due to the low electronegativity of the Mn II atom
which favors six-coordination.

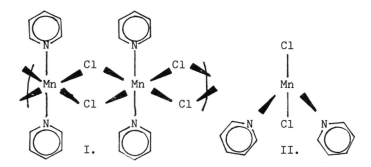

Even though much is reported on the physical properties of
the DMC in the solid state little is reported on any reactions or
solution properties of DMC. Dash and Ramana Rao (13) concluded
from magnetic susceptibility measurements in nitrobenzene that
DMC exists in a monomeric tetrahedral configuration of form II.

Here we report the initial synthesis of a manganese II polyamine, for which the degradation data and other analyses are
consistent with a structure of form IV.

IV.

EXPERIMENTAL

Dipyridine manganese II dichloride (ROC/RIC, Belleville, N.J.) and 1,3-di-4-piperidylpropane (a gift from Reilly Coal Tar Products; Di-Pip) were used as received without further purification. Reactions were carried out utilizing a one pint Kimex Emulsifying jar fitted onto a Waring Blendor (Model 700). Solutions of triethylphosphate, TEP, containing DMC were added to TEP solutions containing Di-Pip and triethylamine as the added base at stirring speeds around 20,000 rpm. Stirring is stopped and the reaction mixture poured into a beaker and left for about 24 hours. A brown precipitate begins to form after several minutes and continues to form for several hours. The reaction mixture is filtered using centrifugation, and the brown solid product is collected and washed repeatedly with TEP and then with diethylether. The solid is transferred to a preweighed petri dish and dried.

General elemental analyses were carried out utilizing typical wet analysis techniques. The products showed the presence of

Mn and the absence of chloride. Quantitative elemental analyses
were carried out by Alfred Bernhardt, Mikroanalytisches Labs and
Galbraith Labs, Inc. The results of these analyses for the condensation product between DMC and Di-Pip are consistent with a
product of form IV. (Thus, %Mn calc. = 12.1, found 13.5; %H calc.
= 8.11, found 8.13.)

Infrared spectra of the product were also obtained utilizing
KBr pellets and a Perkin-Elmer 457 Grating Infrared Spectrophotometer. The product shows bands characteristic of the presence of
both reactants (Di-Pip and DMC) and of water. The presence of
water is indicated by the presence of a strong band in the 3400-
3600 cm^{-1} region (reported to be ca 3450 cm^{-1} for water in organic
compounds and ca 3300 to 3400 cm^{-1} for uranyl-water compounds).
The presence of DMC is indicated by the presence of such bands as
those at 1580 and 1460 cm^{-1} which are characteristic of the aromatic ring stretching vibrations present in pyridinyl containing
compounds, and the presence at bands of ca 1040 and 1010 cm^{-1}
which are characteristic of C-H in-plane deformation. The presence of the Di-Pip moiety is indicated by the presence of such
bands as those at 2850 to 2975 cm^{-1} which are characteristic of
aliphatic C-H stretching, and the bands about 1390 cm^{-1}, a series
of bands in the 1100 to 1030 cm^{-1}, a series of bands in the 1100
to 1030 cm^{-1} region (in addition to those noted above), and bands
at 825 and 810 cm^{-1}, all present in both the monomer and polymer.
The N-H stretching band at 3225 cm^{-1}, present in the Di-Pip monomer, is not present in the polymer. Bands which are characteristic of the Mn-N and Mn-Cl moieties are currently in dispute, but
are below 250 cm^{-1}, the limit of the employed infrared spectrophotometer.

Light scattering measurements were also accomplished for the
polymer, utilizing serial dilutions and employing a Brice-Phoenix
2000 Universal Light Scattering Photometer. The dissymmetric

technique was used to calculate the data. Refractive index increments were determined utilizing a Bausch and Lomb Abbe Refractometer Model 3-L. The product chosen for study had a weight average molecular weight of 2.1×10^6.

Thermal stability of the condensation product was determined using Differential Scanning Calorimetry (DSC), Thermal Gravimetric Analysis (TG) and the combination TG-MS instrument mentioned above. The equipment employed included a duPont 900 Differential Scanning Calorimeter cell, attached onto a duPont 950 Thermal Analyzer Console, a duPont 900 Thermal Gravimetric Analyzer, and a duPont 21-491 Double-Focusing Mass Spectrometer. Details of the TG-MS assembly are presented in the previous paper.

RESULTS AND DISCUSSION

We have been incorporating metals into polymers for a variety of reasons, one being that such compounds are useful as delivery agents in biological systems (14-16). The portion delivered may be either the metal containing moiety, or the organic comonomer, or both. We wished to synthesize manganese-containing polyamines because manganese is an essential metal in many biological systems and there are a number of biologically active diamines which could be incorporated to give polymers of form IV. Further, there are a number of naturally occurring amides, which enhances the probability for biological acceptance and subsequent hydrolysis of manganese polyamines. Here, we will restrict our discussion to the condensation product between DMC and Di-Pip and its physical characterization.

As noted in the previous section, infrared spectroscopy and elemental analysis data for the polymer are consistent with a structure of form IV. In addition, Clark and Williams (8) noted that the frequencies of pyridine ring vibrations in the metal-pyridine complexes, including DMC, are, in general, indicative of the stereochemistry (octrahedral, square planer or tetrahedral) of the metal complex. DMC itself shows a pyridine ring vibration at 627 cm^{-1}. Product IV shows a band at about 620 to 630 cm^{-1}, consistent with an octrahedral structure. Similar aqueous compounds are known to be octrahedral in the solid state. For instance Richards, Quinn and Moros (3) utilized x-ray diffraction and ESR to prove the octrahedral structure of solid PyHMnCl$_3$H$_2$O.

The TG-MS data is also consistent with a product of form IV. Figure 1 shows plots of weight loss and total ion current as a

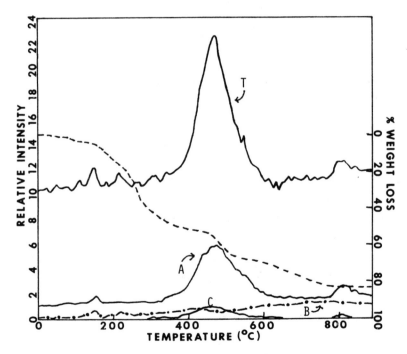

FIGURE 1.   TG(---) AND MS(——) PLOTS FOR THE CONDENSATION PRODUCT FROM DIPYRIDINE MANGANESE II DICHLORIDE AND DIPYRIDINE AT AN ACCELERATING VOLTAGE OF 70 eV, FLOW RATE OF HELIUM AT 55 ML/MIN FOR A HEATING RATE OF 20 C$^{\circ}$/MIN AND A DATA SAMPLING FREQUENCY OF 41 KC.   T = TOTAL ION CURRENT.   A = PLOT OF M/E 18.
B = PLOT OF M/E 44.   C = PLOT OF M/E 79.

function of temperature for the thermal degradation of the polymer. The plot of total ion current shows four maximums; the first occurs at a temperature of about 160°C, the second at about 250°C, the third and largest at about 460°C and the fourth at about 750°C. The TG thermogram shows five clear breaks (temperature regions where accelerated weight loss occurrs) at temperatures of approximately 130, 150, 250, 470 and 700°C, respectively. The breaks near 150, 250, 470 and 700°C also correspond to the maximums in the total ion current plot.

Initial weight loss upon degradation of the polymer occurs at about 120°C, and at the same time mass spectral peaks indicative of propylene are observed (Figures 1) followed shortly by the appearance of mass spectral peaks characteristic of water. Therefore, initial degradation apparently occurs through breakage of the propylene chain present in the Di-Pip moiety and the evolution of water.

The accelerated weight loss of 65% which occurs in the 200-250 °C range, corresponds with the appearance of a number of mass spectral peaks with relative ion intensities greater than 1%. The greatest accelerated weight loss occurs at about 453°C, concurrent with the appearance of 38 mass spectral ions with normalized intensities greater than 1% (Table 1). No other significant mass peaks are observed from degradation products over the entire 200 to 500 °C temperature region. The relative intensities of the mass spectral peaks detected at m/e 79, 52, 51, 50, 78, 53 and 39 are consistent with the fragmentation pattern of pyridine (17). The Di-Pip moiety apparently produces the fragment ions detected at m/e 119-115, 108-105, 94-91, 82-77, 69-64, 56-54, 43, 41 and 14. Many of these ions could result from several fragmentations of different portions of the Di-Pip moiety, and Table 1 lists the structure of only one set of these possible fragments.

Table 1

Major Mass Spectral Peaks Observed from the Degradation Products Joined at 453°C from the Condensation Product of Di-Pip with DMC.

m/e	Normalized Intensity	Assignment	m/e	Normalized Intensity	Assignment
14	3.6	CH_2	78	3.3	py
17	20.5	HO	79	15.1	py
18	95.6	H_2O	80	2.2	(N-ring) py
26	4.4	HC≡CH (py)	81	1.9	py
27	4.5	HC≡N (py)	82	2.8	py
39	6.4	HC≡CH-CH (py)	91	1.5	(N-ring)—CH_2
41	4.2	H_2C-CH-CH	92	3.1	py
43	2.2	H_3C-CH_2-CH_2	93	5.3	py
50	3.0	C≡C-CH≡CH (py)	94	1.1	py
51	5.8	C≡CH-CH≡CH (py)	105	2.5	(N-ring)—CH_2-CH_2
52	12.5	HC≡CH-CH≡CH (py)	106	4.3	"
53	2.8	HC≡CH-CH≡N (py)	107	2.1	"
54	2.3	HC-CH_2-CH_2-CH	108	1.5	"
55	5.3	HC-CH_2-CH_2-CH_2	115	1.2	(N-ring)CH_2-CH_2-CH_2
56	1.7	H_2C-CH_2-CH_2-CH_2	117	1.1	"
64	1.2	HC≡CH-C≡CH-CH	118	2.3	"
65	2.5	HC≡CH-CH≡CH-CH	119	1.5	"
66	2.2	HC-CH-CH_2-CH-CH			
67	2.3	HC-CH_2-CH_2-CH-CH			
68	1.4	HC-CH_2-CH_2-CH-CH			
69	2.2	H_2C-CH_2-CH_2-CH-CH_2			
77	2.2	py			

The differentiation between pyridine and the Di-Pip moiety is accomplished by comparing the appearance of m/e 79, 52, 51, 50, 78, 53 and 39 (all mass sequences are given in order of decreasing intensities) associated with the pyridine moiety and m/e 119-117,

108-105 and 91-90 characteristic of the Di-Pip moiety. Masses associated with pyridine are initially detected around 150°C reaching two maximums; one at 450°C and the second about 780°C. Masses associated with the Di-Pip moiety are detected around 350°C reaching a maximum around 460°C. Thus, degradation in the 200 to 500°C range began with the evolution of pyridine followed by the evolution of masses associated with the degradation of the Di-Pip moiety.

If fragmentation occurs in a statistical manner, predictable relationships between certain species should be evident. For instance, breakage of the Di-Pip chain at A or B (IV) should generate equal quantities of fragments A' associated with the quartet of ions at m/e 91-94 and B', associated with the quartet at m/e 105-108. Summation of the intensities of the ions associated with fragments A' and B' (corrected for difference in mass between the two fragments) shows that these agree within 1.5%, which suggests a random bond scission, at least at A and B.

The overall degradation sequence is depicted in Form V where the order of fragmentation is designated by the alphabetical indicator.

V.

The infrared spectra of the residue of the product remaining from the TG-MS analysis shows bands at 1060-1030, 1010, 640-590, 560, 500, 410-411 and 380-270 cm^{-1}. The 1010 cm^{-1} band is characteristic of inplane C-H stretching of pyridine. Other bands present in both the TG-MS residue and the original polymer are associated with DMC (620-570 and 380-270 cm^{-1}). Manganese dioxide, MnO_2, exhibits bands in the neighborhood of 700-450, 410-350 and 340-311 cm^{-1}. Bands present in the residue also correspond to many of the bands characteristic of MnO_2, indicating the possible presence of this compound. The final residue weight was 16% (%MnO_2 theory = 19, %Mn theory = 12). Therefore, the TG-MS residue probably consists of manganese and manganese oxide, with some organic material derived from the pyridine moiety.

In summary, the appearance of mass spectral peaks characteristic of the Di-Pip moiety, water, and pyridine in the mass spectra of the degradation products of the polymer, is consistent with the structure depicted in IV, and is also consistent with results obtained from elemental analysis and infrared spectral studies. While the infrared, elemental analysis and TG-MS data are consistent with structure IV, uncertainties remain with respect to the exact locations of the various ligands about the manganese II atom. Form IV is likely a major structure, since it is reasonable from steric considerations, that the cis or neighboring pyridines remain in these positions, and that when concurrent addition/replacement by either water or the diamine occurs, that such addition occurs to the trans position. Steric requirements are minimized if the trans positions are occupied by the diamines.

REFERENCES

[1]   F. Reitzenstein, Z. Anorg. Chem., 18, 253(1899).

[2]   F. Klaaijsen, H. Blote and Z. Dokoupil. Physica, 81B, 1 (1976).

[3]   P. Richards, R. Quinn and B. Morosin, J. Chem. Physics, 59(8), 4474 (1973).

[4]   S. Suzuki and W. Orville-Thomas, J. Molecular Structure, 37, 321 (1977).

[5]   J. Ruede and D. Thornton, J. Molecular Structure, 34, 75 (1976).

[6]   J. Allan, D. Brown, R. Nuttall and D. Sharp, J. Inorg. Nucl. Chem., 27, 1305 (1965).

[7]   M. Goldstein and W. Unsworth, Inorg. Nucl. Chem. Letters, 6, 25 (1970).

[8]   R. Clark and C. Williams, Inorganic Chem., 4 (3), 350 (1965).

[9]   N. Gill, R. Nyholm, G. Barclay, T. Christie and P. Pauling, J. Inorg. Nucl. Chem., 18, 88 (1961).

[10] M. Goldstein and W. Unsworth, Spectrochimica Acta, 28A, 1107 (1972).

[11] S. Akyuz, A. Dempster, J. Davies and K. Holmes, J. Chem. Soc. Dalton, 1746 (1976).

[12] T. Yoshihasi and H. Sano, Chem. Phys. Letters, 34(2), 289 (1975).

[13] K. Dash and D. Ramana Rao, Indian J. Chem., 3(11), 514 (1965).

[14] C. Carraher, "Organometallic Polymers" (Edited by C. Carraher, J. Sheats and C. Pittman), Academic Press, N.Y., 1978, pages 79-86.

[15] C. Carraher, T. Langworthy and W. Moon, Polymer P., 17(1), 1 (1976).

[16] C. Carraher, J. Schroeder, W. Venable, C. McNeely, D. Giron, W. Woelk and M. Fedderson, "Additives for Plastics, Vol. 2," (Edited by R. Seymour) Academic Press, N.Y., 1978, pages 81-91.

The Formation and Chemistry of New Functionally Substituted η^5-Cyclopentadienyl-Metal Compounds: Routes to New Vinyl Monomers

M. D. Rausch,[*] W. P. Hart and D. W. Macomber

Department of Chemistry
University of Massachusetts
Amherst, Masssachusetts   01003

ABSTRACT

Reactions of cyclopentadienylsodium with ethyl formate, methyl or ethyl acetate, or dimethyl carbonate in THF solution have produced high yields of the functionally substituted organosodium compounds $[C_5H_4C(O)R]^- Na^+$, where R = H, CH_3 or OCH_3. These organosodium reagents have proved to be valuable intermediates in the formation of functionally substituted cobaltocene, nickelocene, etc., sandwich complexes as well as for other η^5-cyclopentadienyl-metal systems such as $(\eta^5\text{-}C_5H_4CHO)M(CO)_2$ (M = Co, Rh) and $(\eta^5\text{-}C_5H_4CHO)W(CO)_3CH_3$. The latter aldehydes can be readily converted into corresponding acrylate and vinyl derivatives. The polymerization chemistry of these new organometallic monomers has been initiated.

INTRODUCTION

Soon after the discovery of bis-(η^5-cyclopentadienyl)iron (ferrocene) in 1951, it was demonstrated that this unique organometallic compound could undergo typical aromatic-type ring-substitution reactions [1,2]. The ability to introduce functional groups onto the η^5-cyclopentadienyl rings of ferrocene made possible the synthesis of vinylferrocene (1) in 1955, and its

243

polymerization reactions were studied by Arimoto and Haven [3].
A limited number of other η^5-vinylcyclopentadienyl-metal compounds have been synthesized and their polymerization chemistry
investigated, including vinylruthenocene (2)[4], vinylcymantrene
(3)[5], and vinylcynichrodene (4)[6,7]. Monomers 2-4 can be

prepared by virtue of the fact that the parent η^5-cyclopentadienyl-
metal compounds, like ferrocene (1), can be acetylated under
Friedel-Crafts conditions and converted into vinyl analogs by
standard organic procedures. The topic of vinylic organotransition
metal monomers and polymers has recently been reviewed by Pittman [8].
        It would seem worthwhile to investigate the formation and
polymerization chemistry of still other η^5-vinylcyclopentadienyl-
metal compounds, since the products might possess unusual and
potentially useful properties. Unfortunately, however, the ability
to functionalize η^5-cyclopentadienyl-metal compounds by means of
electrophilic aromatic substitution reactions is severely limited,
including the iron, ruthenium, manganese and chromium systems
mentioned above plus only a few other analogs [9,11]. The failure
of most η^5-cyclopentadienyl-metal compounds to undergo ring substitution may be ascribed either to their inherent lack of aromatic
character or to more facile reaction pathways under the reaction
conditions involved. In any event, the inability of most η^5-cyclo-
pentadienyl-metal compounds to form functionally substituted derivatives by ring-substitution routes has greatly impeded the
development of η^5-cyclopentadienyl-metal chemistry of the type
shown by ferrocene. This result in turn has also greatly hampered
the growth of vinylic organotransition metal monomers and polymers.

RESULTS AND DISCUSSION

We have recently developed a new and apparently general route for the formation of a wide variety of functionally substituted η^5-cyclopentadienyl-metal compounds [11]. The method provides a convenient means of introducing functional substituents onto η^5-cyclopentadienyl rings in systems which are incapable of undergoing electrophilic aromatic substitution. In the present paper, we shall discuss the general procedures by which new functionally substituted η^5-cyclopentadienyl-metal compounds can be synthesized, and especially the application of this method for the formation of new organometallic monomers.

Reactions of cyclopentadienylsodium with either ethyl formate, methyl acetate or dimethyl carbonate in refluxing THF solution for 2 h produce 70-95% yields of the respective compounds formyl-(5), acetyl-(6) or methoxycarbonylcyclopentadienylsodium (7)[12]. These reactions likely proceed via the intermediate formation of the respective 1-substituted cyclopentadienes, which react readily with the sodium alkoxide concurrently produced to afford the functionally substituted cyclopentadienylsodium derivatives 5-7.

$$5 \quad R = H$$
$$6 \quad R = CH_3$$
$$7 \quad R = OCH_3$$

These organosodium reagents are moisture sensitive, but are generally stable in air for short periods of time, in contrast to cyclopentadienylsodium itself. Their proton nmr spectra recorded in D_2O solution consist of two pairs of unresolved multiplets representing the ring protons as well as singlet resonances for the substituent protons [11].

Reactions of 5-7 with various transition metal halides have proved useful in the synthesis of functionally substituted sandwich compounds which are generally unattainable by other means.

Thus, reactions of 6 or 7 with either CoCl$_2$ or NiBr$_2$·2DME in
THF solution afforded 1,1'-diacetylcobaltocene (8), 1,1'-dicarbo-
methoxycobaltocene (9), 1,1'-diacetylnickelocene (10), and 1,1'-

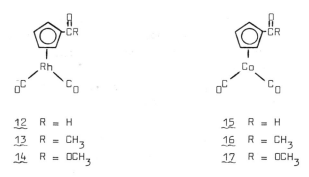

8   R = CH$_3$ 10   R = CH$_3$
9   R = OCH$_3$ 11   R = OCH$_3$

dicarbomethoxynickelocene (11), respectively, in yields of 30-
50% [13]. The ester derivatives 9 and 11 are currently being
evaluated as potential monomers for the production of organo-
cobalt- and organonickel-containing condensation polymers [14].
Similar routes to organometallic condensation polymers which
contain the ferrocene unit have been examined previously, although
only relatively low M$_n$ molecular weight products could be
obtained [15,16].

Organosodium reagents 5-7 are also convenient precursors
to functionally substituted η^5-cyclopentadienyldicarbonylmetal
derivatives, products which have likewise been unavailable
through electrophilic substitution reactions of the parent
compounds. Thus, reactions of 5-7 with [Rh(CO)$_2$Cl]$_2$ in THF
readily produce the corresponding formyl (12), acetyl (13) and

12   R = H   15   R = H
13   R = CH$_3$ 16   R = CH$_3$
14   R = OCH$_3$ 17   R = OCH$_3$

methoxycarbonyl (14) analogs, whereas the respective cobalt
counterparts (15-17) are available via reactions of 5-7 with
an equimolar mixture of $Co_2(CO)_8$ and I_2 in THF solution. These
products are obtained in yields of 50-98% as dark red liquids
which are distillable under reduced pressure.

     The availability of the functionally substituted organo-
rhodium and -cobalt derivatives 12-17 provides valuable new routes
to organometallic monomers of these metals. For example, 15 and
16 react with methylenetriphenylphosphorane in ethyl ether solution
to afford (η^5-vinylcyclopentadienyl)dicarbonylcobalt (17) and
(η^5-isopropenylcyclopentadienyl)dicarbonylcobalt (18) in moderate

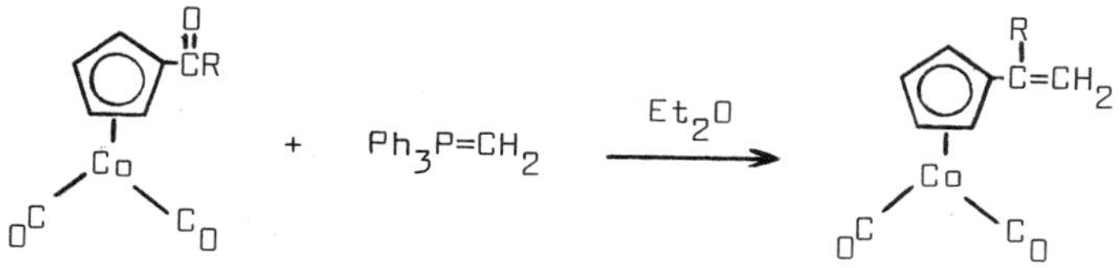

15 R = H
16 R = CH$_3$

17 R = H
18 R = CH$_3$

yields, respectively. Reduction of 15 or 16 followed by
reaction of the resulting alcohols with acryloyl chloride and
pyridine have likewise produced the corresponding acrylate
derivatives (19, 20) in good yield.

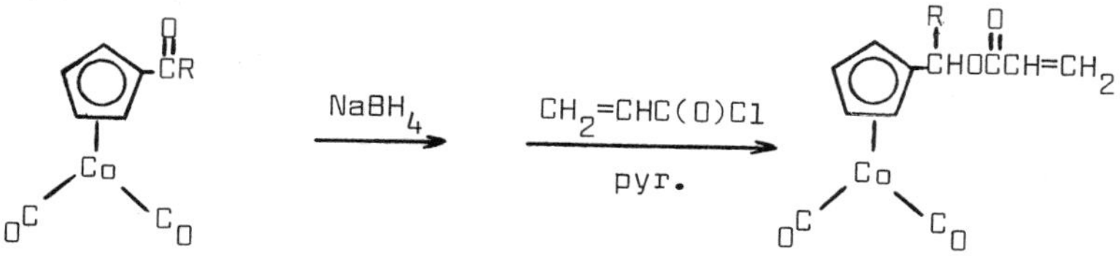

15 R = H
16 R = CH$_3$

19 R = H
20 R = CH$_3$

     Polymerization studies on these organocobalt vinyl monomers
under free radical and cationic conditions are currently in
progress, and will hopefully lead to new types of organometallic
polymers with potential catalytic properties [8,14,16].

RAUSCH, HART, AND MACOMBER

Previous studies in this laboratory have shown that the
novel monomer vinylcynichrodene (4) could undergo both homo- and
copolymerization. Monomer 4 was prepared from acetylcynichrodene
(21) by treatment with sodium borohydride followed by dehydration
of the resulting alcohol (22). The value of the Alfrey-Price

parameter e for 4 was evaluated by reactivity ratio studies
to be -1.98, indicating that 4 is an exceptionally electron
rich vinyl monomer, resembling vinylferrocene (1) and vinylcymantrene (3) [6,7].

    In order to further investigate the formation and react-
ivity of vinyl monomers of the Group IVB metals, formyl derivative 5 was allowed to react with tungsten hexacarbonyl in
refluxing DMF. Treatment of the resulting organometallic anion
with iodomethane in THF produced (η^5-formylcyclopentadienyl)-
tricarbonylmethyltungsten (23) in 82% yield. Reaction of aldehyde

23 in benzene solution with triphenylmethylphosphonium iodide
in the presence of 5N sodium hydroxide under phase transfer
conditions afforded (η^5-vinylcyclopentadienyl)tricarbonylmethyl-
tungsten (24) in 80% yield. Attempted formation of 24 under normal
Wittig conditions (e.g., $Ph_3P=CH_2$ generated in ethyl ether
solution from triphenylmethylphosphonium iodide and n-butyllithium) also produced 24, although in appreciably lower yield.

Reduction of aldehyde 23 by sodium borohydride in ethanol
proceeded smoothly to produce the corresponding alcohol in nearly
quantitative yield. Subsequent reaction of this alcohol with
sodium hydride followed by acryloyl chloride afforded the acrylate
derivative (25) in 65% yield. Homo- and copolymerization
reactions of 24 and 25, the first organotungsten vinyl monomers
to be synthesized and studied, are under present investigation [17].

It is therefore evident that the development of functionally
substituted cyclopentadienyl-metal compounds can lead to many new
organometallic monomers which would be unattainable by other means.
The polymerization chemistry of these new monomers and evaluation
of the properties of the resulting organometallic polymers should
be of interest to chemists for many years to come.

ACKNOWLEDGEMENT

Acknowledgement is made to the Donors of the Petroleum Research
Fund, administered by the American Chemical Society, and to a
National Science Foundation grant to the Materials Research Laboratory, University of Massachusetts, for support of this research
program. We are also grateful to Professor C. U. Pittman, Jr.,
for helpful suggestions and collaborative efforts.

REFERENCES

[1] R. B. Woodward, M. Rosenblum and M. C. Whiting,
     J. Am. Chem. Soc., 74, 3458 (1952).

[2] A. N. Nesmeyanov, E. G. Perevalova, R. U. Golovnya
     and O. A. Nesmeyanova, Dokl. Akad. Nauk SSSR, 97,
     459 (1954).

[3] F. S. Arimoto and A. C. Haven, Jr., J. Am. Chem. Soc.,
     77, 6295 (1955).

[4] T. C. Willis and J. E. Sheats, Org. Coatings & Plastics
     Chem., 41, 33 (1979).

[5] C. U. Pittman, Jr., G. V. Marlin and T. D. Rounsefell,
     Macromolecules, 6, 1 (1973).

[6] E. A. Mintz, M. D. Rausch, B. H. Edwards, J. E. Sheats,
     T. D. Rounsefell and C. U. Pittman, Jr., J. Organometal.
     Chem., 137, 199 (1977).

[7] C. U. Pittman, Jr., T. D. Rounsefell, E. A. Lewis, J. E.
     Sheats, B. H. Edwards, M. D. Rausch and E. A. Mintz,
     Macromolecules, 11, 560 (1978).

[8] C. U. Pittman, Jr., in "Organometallic Reactions And
     Synthesis," (E. I. Becker and M. Tsutsui, Eds.), Plenum
     Press, New York, 1977, Vol. 6, p. 1.

[9] M. D. Rausch, Can. J. Chem., 41, 1289 (1963).

[10] M. D. Rausch and R. A. Genetti, J. Org. Chem., 35,
     3888 (1970).

[11] W. P. Hart, D. W. Macomber and M. D. Rausch, J. Am. Chem.
     Soc., 102, 1196 (1980).

[12] Formyl derivative 5 was previously obtained by this
     general method, and was used in the formation and study
     of formylcyclopentadiene isomers: K. Hafner, G. Schultz,
     and K. Wagner, Justus Liebig's Ann. Chem., 678, 39 (1964).

[13] Compounds 9 and 17 were previously prepared by alternative,
     although less convenient procedures: M. Rosenblum, B.
     North, D. Wells, and W. P. Giering, J. Am. Chem. Soc.,
     94, 1239 (1972).

[14] W. P. Hart, C. U. Pittman, Jr., and M. D. Rausch,
     unpublished studies.

[15] E. W. Neuse and H. Rosenberg, "Metallocene Polymers,"
     Marcel Dekker, New York, 1970.

[16] C. E. Carraher, Jr., J. E. Sheats and C. U. Pittman, Jr.,
     "Organometallic Polymers," Academic Press, New York, 1978.

[17] D. W. Macomber, M. D. Rausch, T. V. Jayaraman, R. D.
     Priester and C. U. Pittman, Jr., unpublished studies.

Polymers as Matrices for Photochemical Reactions of Organometallic Compounds

by Marco-A. De Paoli

Instituto de Química,
Universidade Estadual de Campinas,
C.P. 1170, 13.100 Campinas, SP, BRASIL

ABSTRACT

A new matrix technique developed in our laboratory for the study of bimolecular photochemical processes and the isolation of unstable and air-sensitive organometallic compounds is described. This technique provides a simple and unexpensive method for studying these compounds at room temperature and ambient atmosphere. The matrices used are films of inert polymers, such as polytetrafluorethylene (PTFE) and polyethylene (PE). We have used this method to study, by means of infrared (i.r.) spectroscopy, the photochemical products and reactions of ironpentacarbonyl with olefins, such as: ethylene, acrylic acid, methylacrylate, norbornadiene, butadiene and isoprene; as well as to study the photofragmentation of ironpentacarbonyl. Evidence indicates that, in addition to permitting work with these compounds at ambient conditions, the method has other advantages over the usual low temperature frozen gas matrices.

INTRODUCTION

The first attempt to use polymers as matrices for photochemical reactions was made by Massey and Orgel, by dissolving $M(CO)_6$ (M = Cr, Mo or W) in a solution of polymethylmetacrylate and evaporating the solvent, in such a way that the polymer was trapped in the matrix (1). Upon irradiation with u.v. light the polymer assumed the color of the $M(CO)_5$ fragment. After a period

in the dark $M(CO)_6$ was regenerated. The same process was used to
study the flash-photolysis of $W(CO)_6$ in a polystyrene matrix (2).
The use of plastic films in photochemical experiments was also
reported by Kellogg and Bennett (3) to obtain low optical density
samples in such a way that the adsorbed acceptor did not absorb
the residual donor phosphorescence. Other attempts to study
photochemical processes in solid matrices at room temperature
were made by Pitts and co-workers (4) working with alkaline-
-halide pellets, in this case the effect of the matrix in the
photochemical process was the same as would be achieved by changing the solvent.
      The matrix isolation method using frozen gases has been ex-
tensively used to study photochemical processes of transition
metal carbonyls and related species (5), although there is no
report, to date, on the study of bimolecular processes in these
matrices, except for those cases where the second reagent is the
matrix itself (6). Unfortunately, low-temperature matrices such
as inert gases, methane or N_2 require the use of expensive low
temperature/high vacuum equipment. Thus a better matrix isolation method is needed.
      The sorption of ironpentacarbonyl by polytetrafluorethylene
(PTFE) was reported, and it was shown that the degree of sorption
depends inversely on the degree of cristallinity of the polymer
(7). It was also reported that this compound undergoes, when
sorbed, the same thermal reactions as in solution, without reacting with the polymer backbone (8). Later on, it was observed that
sorption in PTFE of a photopolymerizable monomer, followed by u.v.
irradiation, produces composite materials, such as polyvinylace-
tate in PTFE (PTFE-PVA) (9). Also the exposure to sun light of a
PTFE film treated with $Fe(CO)_5$ was shown to produce the dimer
$Fe_2(CO)_9$ (10).
      The photocnemical reaction of ironpentacarbonyl with dienes
has long been known to produce dieneirontricarbonyl compounds
(11). Also bis(diene)ironmonocarbonyl compounds are obtained by
means of irradiation of $Fe(CO)_5$ in the presence of an excess of
diene (12). The dieneirontricarbonyl compounds prepared with low
molecular weight dienes are rather air-sensitive and difficult to
handle. Known monoolefintetracarbonyliron(0) compounds are also
unstable, unless electron-withdrawing groups are attached to the
olefinic function. The reaction of eneacarbonyldiiron with
ethylene produces a very unstable and air-sensitive compound,
ethyleneirontetracarbonyl (13), which has also been identified in
an argon matrix (14). The reaction of ironpentacarbonyl with
substituted monoolefins has also been studied in solution (15).
      The study of carbonyl complexes is facilitated by their high
intensity absorption in the ν_{CO} frequency region of the infrared
spectra. Thus, for a matrix to be useful it must be transparent
in the region needed for spectral examination. Most polymers do

not absorb in the region of frequencies between 2200 and 1600 cm^{-1}. On the other hand, the polymer should not absorb the radiation used to induce the photochemical process, not only because it would hinder the reaction, but also because such absorption would lend to degrade the polymer. Saturated hydro- or fluorocarbon polymers fulfill these requirements and can also be selected to have different degrees of crystallinity. The use of polymers at room temperature also permits a higher mobility of the molecules within the bulk, reducing the need for the high reagent concentration which are required to study bimolecular reactions in frozen gases.

In this article we describe the use we have made of polymer matrices such as PTFE and PE to study photochemical substitution of carbon monoxide ligands of ironpentacarbonyl by olefinic ligands.

In PTFE matrices, we have observed the formation of diene-irontricarbonyl and bis(diene)ironmonocarbonyl compounds from the reaction of ironpentacarbonyl with butadiene and isoprene and the formation of ethyleneirontetracarbonyl from the reaction with gaseous ethylene (16). In PE matrices we have observed the formation of compounds of the type $Fe(CO)_4L$, where L is acrylic acid, methylacrylate (17) or norbornadiene (18). In these cases it is assumed that the primary step in the photosubstitution of CO is excitation of $Fe(CO)_5$ followed by the loss of a CO ligand. Following this, we studied the photofragmentation of $Fe(CO)_5$ in a PE matrix (17), confirming observations obtained using frozen gas matrices (19).

In all cases the polymer films containing the compounds were handled at room temperature and under the ambient atmosphere.

MATERIALS AND METHODS

The polymer films were used as furnished. The PTFE films were 0.2 mm thick, fabricated from DuPont Teflon by Incoflon. The PE films were 1.0 mm thick fabricated from Poliolefinas. Iron-pentacarbonyl was used as furnished by BASF do Brasil.

The sorption of ironpentacarbonyl in the polymer is carried out by soaking the film in a 10% solution of it in degased hexane. When using PTFE, 24 hours are necessary to obtain the highest degree of sorption, although this level, cannot be detected by weighing and does not saturate the i.r. spectrum (fig. 1a). With PE, five minutes of soaking in the same solution are sufficient to saturate the i.r. spectrum (fig. 1b). After 24 hours of soaking a 1% increase in the weight of the film can be detected. Polymers from other sources will probably require different sorption times due to variations in the degree of crystallinity. The polymer film must be washed with ethanol immediately after removal from the soaking solution, to prevent oxidation of iron-pentacarbonyl on the surface, forming a layer of ironoxide. Only during soaking is an inert gas atmosphere necessary. Once iron-pentacarbonyl is trapped inside the matrix it will not oxidize during the time required for the experiment.

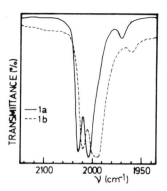

Fig. 1.   I.r. spectra of the polymer films treated with
ironpentacarbonyl, 1a) PTFE and 1b) PE.

Sorption of the second reagent, when used, depends on its
physical state.   For liquids, the film pretreated with $Fe(CO)_5$ is
soaked in the second reagent for the time required to saturate it.
For gases, the pretreated film is held in an immersion well irra-
diator filled with the gas under normal pressure.   For solids, the
pretreated film is soaked in a solution of the reagent and this
solvent is subsequently evaporated from the film.

        After this treatment the films are held in an immersion well
apparatus and irradiated.   A Philips HPK-125 W lamp was used for
the experiments requiring vycor filters and an adapted (20)
Philips HPLN- 125 W lamp was used when pyrex filters were required.

        The reactions were followed using a Perkin-Elmer model 399 B
double bean i.r. spectrophotometer using, for reference, a film
with similar thickness.   All spectra were measured with five
times expansion in the abcissa.   In our study of the photo-frag-
ments of $Fe(CO)_5$ we irradiated the films directly in the i.r.
spectrophotometer.

RESULTS AND DISCUSSION

        Figure 2 shows a typical sequence of spectra for an experi-
ment performed in PTFE.   Fig. 2a is the spectrum of a film
treated with ironpentacarbonyl and butadiene, Fig. 2b and 2c are
the spectra of the same film after one hour of visible light
irradiation and after two hours.   These spectra correspond to
those of butadieneirontricarbonyl (ν_{CO} = 2060, 1995 and 1984 cm^{-1})
and bis(butadiene)ironmonocarbonyl (ν_{CO} = 1985 cm^{-1}), respectively.
A similar experiment using isoprene gave similar results (ν_{CO} =
2055, 1991 and 1980 cm^{-1}; ν_{CO} = 1985 cm^{-1}).   The i.r. spectra of

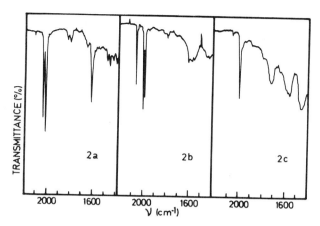

Fig. 2. I.r. spectra of a PTFE film: treated with Fe(CO)$_5$
and C$_4$H$_6$ (2a), irradiated during one hour (2b)
and irradiated during two hours (2c).

these films do not change after one week of exposure to ambient
conditions or even after pumping air through the films. Even
heating the films containing the dieneirontricarbonyl compounds
to 160°C produces no change in the spectra.

To use ethylene as the second reagent, the film pretreated
with ironpentacarbonyl was placed in the immersion well irradi-
ator, which was subsequently filled with ethylene under normal
pressure. After 45 minutes of u.v. photolysis, all absorptions
due to Fe(CO)$_5$ vanish and we observe four absorptions at: 2092,
2029, 2014 and 1993 cm^{-1} (fig. 3). These absorptions correspond
to the substitution of one carbon monoxide by an ethylene. The
substitution in the equatorial position of a trigonal bipyramid
leads to a C$_{2v}$ point group symmetry compound, which has four
active absorptions in the ν_{CO} frequency region of the infrared
spectrum. The product was previously reported by Murdoch and
Weiss (13), who report three peaks in the ν_{CO} region (2088, 2007
and 1986 cm^{-1}) and a shoulder at 2013 cm^{-1}. Other authors
working in an argon matrix observed only three bands (2090, 2009
and 1992 cm^{-1}) (14).

We also reacted acetylene with Fe(CO)$_5$ in the PTFE matrix,
using the same conditions as with ethylene. The modifications
observed in the spectrum can be assigned to a product, probably
acetylene-irontetracarbonyl. However, even after 24 hours of
photolysis the reaction is not complete and the strong absorp-

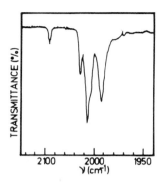

Fig. 3. I.r. spectrum of the ethyleneirontetracarbonyl
compound in the PTFE matrix.

tions of ironpentacarbonyl preclude a better interpretation of
the results.

As mentioned at the begining of this chapter, the degree of
sorption of molecules by polymers depends inversely on the degree
of cristallinity (8) and also on the square of the difference
(δ polymer $-$ δ sample), where δ is the Hildebrand's solubility
parameter (21). In order to increase the concentration of
reagents in the matrix we performed experiments with low-density
PE which is highly amorphous, has a Tg bellow room temperature,
and shows a high degree of sorption of liquids such as ironpentacarbonyl.

Irradiation of a PE film treated with $Fe(CO)_5$ and norbornadiene (NBD) produced interesting results. After three minutes
of photolysis, four new bands are observed in the ν_{CO} frequency
region of the i.r. spectrum (2098, 2032, 2020 and 2010 cm^{-1}) as
well as a band at 1730 cm^{-1}. After further irradiation the 1730
cm^{-1} band becomes stronger and the other ν_{CO} bands vanish. The
appearance of four ν_{CO} absorptions instead of three suggest the
formation of the compound $(\eta^2-NBD)Fe(CO)_4$ instead of the expected
$(\eta^4-NBD)Fe(CO)_3$ (23). A similar η^2-compound was reported (22) to
be obtained by irradiating a solution of 5,6-dimethylene-7-oxa-
bicyclo$|2,2,1|$hept-2-ene and $Fe(CO)_5$. This product shows four
ν_{CO} absorptions at 2097, 2032, 2017 and 1984 cm^{-1} (22). The band
at 1730 cm^{-1} corresponds to a carbonyl inserted into a dimer of
NBD (23). The formation of this dimer indicates that, in an
intermediate step, two NBD entities must be bound as η^2-ligands
to a single $Fe(CO)_3$, to subsequently dimerize and suffer carbonyl
insertion (24). This intermediate has not been identified.

Formation of compounds of the type $Fe(CO)_4L$ can be also observed in a PE matrix. After one minute of photolysis of a PE film treated with ironpentacarbonyl and methylacrylate we observed the complete disappearance of the ν_{CO} bands of $Fe(CO)_5$ and the formation of four new bands at 2098, 2032, 2020 and 1996 cm^{-1} (fig. 4). These bands correspond to the compound methylacrylate-irontetracarbonyl. Further photolysis does not produce any significant change in the i.r. spectrum. In the case of acrylic acid, exposure of the treated film to u.v. light for long periods does not lead to the total consumption of $Fe(CO)_5$. In this reaction we observed, after irradiation, four new absorptions: 2098, 2034, 1978 and 1940 cm^{-1}.

In solution the photochemical substitution of carbon monoxide in $Fe(CO)_5$ is assumed to follows a S_N1 type mechanism (25). The excited molecule decays by loosing a ligand, producing a coordenatively unsaturated sixteen electron species, $Fe(CO)_4$, which is very reactive. Since this species is probably also formed in the polymer matrices, we performed an experiment in order to observe it. The i.r. spectrum in the region between 2200-1800 cm^{-1} of a film of PE saturated with ironpentacarbonyl was recorded immediately after three minutes of broad-band u.v. light irradiation. Two new absorptions are observed at 2063 and 1973 cm^{-1} and the band at 1998 cm^{-1} (mainly due to $Fe(CO)_5$) is more intense, in relation to the 2020 cm^{-1} band, than before photolysis. Irradiation with a Nernst glower decomposes the new species formed, while the residual $Fe(CO)_5$ absorptions persist.

Kinetic measurements indicated that two different species are formed. One has an absorption at 1973 cm^{-1} and another covered

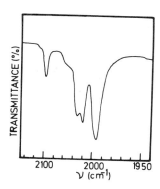

Fig. 4.   I.r. spectrum of the mehylacrylateirontetracarbonyl compound in the PE matrix.

by the strong 1998 cm^{-1} band of ironpentacarbonyl. The other
species absorbs at 2063 cm^{-1}. Kinetic measurements indicated a
first order rate constant of 6.5 and 2.5 x 10^{-3} sec^{-1} for the
disappearance of the bands at 2063 and 1973 cm^{-1}, respectively.
The bands at 1973 and 1998 cm^{-1} are assigned to the photofragment
$Fe(CO)_4$ and the other to a product of further fragmentation,
probably $Fe(CO)_3$ (17).

Evidence that the compounds are reacting in the bulk of the
polymer and not on the surface is given not only by the stability
of the compounds observed, but also by the detection of unstable
intermediates within the polymer matrix. The formation of composites (9) of PTFE also reinforces this conclusion. It is very
probable, although not measured, that, after sorption, the molecules of the reagents are lodged in the amorphous sites of the
polymers. This corroborates with the observation that sorption
increases as the degree of cristallinity of the polymer matrix
decreases. When lodged in the amorphous sites of the polymer,
the molecules have a greater stability than when exposed to air,
in the solid or in solution. This does not mean that they are
"frozen" in the polymer cage since they have sufficient mobility
to react. The substitution of four carbonyls of $Fe(CO)_5$ by two
butadiene or isoprene molecules provides evidence that the local
concentration of reagents is sufficiently high and that the carbon monoxide is lost at a sufficiently fast rate. The same reaction, in solution, requires 48 hours of irradiation with constant
bubling of argon to remove CO (12). It is probable that the rate
of diffusion of oxygen through the polymer is affected by this
"amorphous sites insertion", on the other hand, the carbon monoxide desorption is sufficiently fast to preclude the regeneration
of $Fe(CO)_5$ in the photofragmentation experiment (17, 26). Another
argument favoring the fast desorption of CO is that, in several
experiments, CO was not detected in the infrared spectra measured
immediately after photolysis.

Radiation polymer chemists have, for a long time, neglected
the use of light, using instead high energy radiation, based on
the argument that light does not penetrate polymers well anough
to produce significant bulk modification (27). In fact it is not
possible to measure the u.v.-visible transmission spectra of
thick polymer films (0.2-1.0 mm) by methods other than photo-
acustic spectroscopy (28). Although fluoro- or hydrocarbon chains,
such as PTFE and PE, scatter considerably light above 200 nm, the
occurrence of photochemical processes in the bulk of the polymer
films is evidence that there is a reasonable amount of light
penetration in the films. The relatively short photolysis time
required for these reactions, compared to solution experiments,
also corroborates this conclusion.

In the case of low-temperature matrices, the types of reactions that may occur are very limited by the so called "cage

effect" (5). It is generally difficult to produce a species by
an in situ photolysis in frozen gas matrices by photoejection of
a ligand, because the molecule is usually too large to squeeze
readily through the lattice intersticies and away from the newly
formed unstable species. Thus, in the low temperature matrices,
the two fragments are held together in the matrix cage, and
recombine to give back the parent molecule. This effect is seen
even in the case where the ligand is carbon monoxide, as is
evidenced by the ease of recombination of carbon monoxide with
$Fe(CO)_4$ in a frozen gas matrix (19). In the case of our room

temperature polymer matrix, the molecules held in the amorphous
sites of the polymer have, apparently, a much higher mobility.
Due to this mobility, the second ligand can approach the unstable coordenatively unsaturated species formed by the photolysis, probably, at a non-diffusion controled rate.

   These effects, higher mobility and high local concentration
of reagents, make possible bimolecular reactions in the polymer
matrices. These are among the advantages of this technique when
compared to the low temperature matrices where the molecules are
"frozen" in the lattice and the local concentration of light sensitive molecules is normally very low. We believe that our
method provides a valuable technique for the study of the infrared spectra of species formed by radiation processes since it
uses a highly inert matrix in a very convenient way.

Acknowledgements. The author acknowledges financial support from
the Conselho Nacional de Desenvolvimento Científico e Tecnologico
(grant number 30.0516/79-QU). He also thanks BASF do Brasil for
the donation of ironpentacarbonyl, the Alexander von Humboldt
Foundation for the donation of other reagents and Drs. F. Galembeck and C.H. Collins for profitable suggestions.

REFERENCES

(1) A.G. Massey and L.E. Orgel, Nature (1961), 1387.
(2) J.A. McIntyre, J. Phys. Chem., 74, 2403 (1970).
(3) R.E. Kellogg and R.G. Bennett, J. Chem. Phys., 41, 9042
     (1964).
(4) J.N. Pitts Jr., L.D. Hess, E.J. Baum, E.A. Schuck and J.K.S.
     Wan, Photochem. Photobiol., 4, 305 (1965).
(5) J.K. Burdett, Coord. Chem. Rev., 27, 1 (1978).
(6) M. Poliakoff and J.J. Turner, J.C.S. Dalton, 1974, 2276.
(7) F. Galembeck, J. Polym. Sci., Polym. Chem. Ed., 16, 3015
     (1978).
(8) F. Galembeck, J. Polym. Sci., Polym. Lett. Ed., 15, 107
     (1977).
(9) M.A. De Paoli, I.T. Tamashiro and F. Galembeck, J. Polym.
     Sci., Polym. Lett. Ed., 17, 391 (1979).

(10)   F. Galembeck, S.E. Galembeck, H. Vargas, L.A. Ribeiro,
       L.C.M. Miranda, and C.C. Glizoni, in "Surface Contamination",
       vol. 1, K.L. Mittal ed., Plenum Press, 1979, USA, pp 57-71.

(11)   R. Pettit and G.F. Emerson, Adv. Organometall. Chem., 1,
       1(1964).

(12)   E. Koerner von Gustorf, J. Buchkremer, Z. Pfeiffer and
       F.-W. Grevels, Angew. Chem., 83, 249 (1971).

(13)   H.D. Murdoch and E. Weiss, Helv. Chim. Acta, 46, 1588
       (1963).

(14)   M.J. Newlands and J.F. Olgivie, Can. J. Chem., 49, 343
       (1971).

(15)   W. Strohmeier, Angew. Chem., 76, 873 (1964).

(16)   M.A. De Paoli, S.M. Oliveira and F. Galembeck, J.
       Organometall. Chem. in press.

(17)   M.A. De Paoli and S.M. Oliveira, J.C.S., Chem. Comm.
       submited.

(18)   M.A. De Paoli and S.M. Oliveira, unpublished results.

(19)   M. Poliakoff and J.J. Turner, J.C.S. Dalton, 1973, 1351.

(20)   M.A. De Paoli and C.F. Rodrigues, Quimica Nova, 1, 1,
       16(1978).

(21)   J.H. Hildebrand and R.L. Scott, "The Solubility of Non-
       Electrolytes", 3rd ed., Reinhold, New York, 1950.

(22)   T. Boschi, P. Vogel and R. Roulet, J. Organometall. Chem.,
       133, C 36 (1977).

(23)   C.W. Bird, R.C. Cookson and J. Hudec, Chem. Ind., 1960, 20.

(24)   J. Mantzaris and E. Weissberger, J. Am. Chem. Soc., 96,
       1880 (1974).

(25)   E.A. Koerner von Gustorf, L.H.G. Leenders, I. Fischler and
       R. Perutz, Adv. Inorg. Chem. Radiochem., 19, 65 (1976).

(26)   We are actually studying this from the point of view of
       "how much the presence of molecules in the amorphous sites
       of the polymer afects the gas diffusion through it".

(27)   N.A.J. Platzer in "Irradiation of Polymers" R.F. Gould ed.,
       Amer. Chem. Soc. Publications, 1967, USA, pp vii.

(28)   F. Galembeck, C.A. Ribeiro, H. Vargas, L.C.M. Miranda and
       C.C. Glizoni, J. Appl. Polym. Sci., in press.

Linear Conjugated Coordination Polymers Containing Eight-Coordinate Metal Centers

Ronald D. Archer, William H. Batschelet, and Marvin L. Illingsworth
Dept. of Chemistry, University of Massachusetts, Amherst, MA 01003

ABSTRACT

Two linear coordination polymers are reported in which conjugated organic ligands and nonrigid eight-coordinate metal centers are linked to form macromolecules with molecular weights of greater than 10^4 Daltons. The tungsten(IV) chelate is an inert low-spin d^2 species with four oxygen and four nitrogen donors per metal with two bidentate blocking ligands and one bis-bidentate bridging ligand. The zirconium(IV) chelate is rendered inert through the use of a bis-quadridentate Schiff-base ligand. Future possibilities and potential uses are also discussed.

INTRODUCTION

Over the past quarter century a large number of four- and six-coordinate Werner coordination systems have been investigated for metal-coordination polymers (1-7). Fibers containing coordinated zinc (Enkatherm) (8-9) and greases containing metal phosphinate polymers (10) are examples of thermally stable coordination compounds. However, the vast majority of the systems which have been investigated have suffered from a retention of the brittleness associated with thermally-stable multielement inorganic condensed networks, only moderate oligomerization (due to stacking in planar systems in particular), intractability, or the lack of desired high temperature sta-

261

bility as a result of either organic linkages or weak coordinate bonds.
Because the four- and six-coordinate systems typically have rigid coordination spheres, organic single bonds are necessary for flexibility.
Unfortunately, the organic bonds which allow rotation are typically
less thermally stable than conjugated systems. The latter give thermal stability at the expense of flexibility.

Our approach is to use substitution-inert nonrigid metal centers
with thermally stable conjugated bridging ligands. Nonrigid coordination is the rule for coordination number 8 (11-12) because the D_{2d} do-
decahedron and the D_{4d} square antiprism polytopes (Figure 1) normally
possess very similar energies. Thermally stable and substitution-inert eight-coordinate tungsten(IV) 8-quinolinol chelates (13-16) and
zirconium(IV) quadridentate Schiff-base chelates (17) had been synthesized in our laboratory and seemed to be logical bases for thermally
stable coordination polymers. [Whereas most d^0 complexes are labile,
quadridentate ligands make zirconium(IV) an inert metal center. Low-
spin d^2 tungsten(IV) chelates are substitution-inert, even with bidentate ligands.] Ligands initially considered for conjugated bridges
were the anions of quinoxaline-5,8-diol and N, N', N'', N'''-tetrakis-
(salicylidene)-1, 2, 4, 5-tetraaminobenzene for tungsten(IV) and zir-
conium(IV), respectively. [Ligands such as phenazine-1,5-diol were
ruled out on steric grounds.] Whereas a double headed "bis-quadridentate" Schiff-base ligand completes the coordination sphere for the zirconium species, the "bis-bidentate" bridges would result in excessively

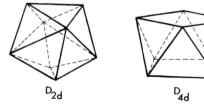

D_{2d} D_{4d}

Figure 1. The D_{2d} dodecahedral and D_{4d} square antiprismatic polytopes which are typically of similar energy for eight-coordination
complexes.

cross-linked brittle species unless two bidentate ligands are present to prevent the cross-linking. [The small amount of cross-linking desired in polymers will occur by a slight bit of ligand scrambling, which normally occurs in our inert monomer syntheses (18).] Therefore, we have developed specific methods for synthesizing $W(CO)_2$-$[P(C_6H_5)_3](\overparen{N\ O})_2$, where $\overparen{N\ O}$ is the anion of an 8-quinolinol derivative (19-20).[These tungsten(II) chelates can be oxidized with a bridging dione to form tungsten(IV)/bridging-diol species. (See below). Alternatively, $W(\overparen{N\ O})_2Cl_2$ chelates, which we have recently prepared (21),might be allowed to react directly with bridging ligands.

CURRENT STATUS

Tungsten(IV) Polymer

Bis(5,7-dichloro-8-quinolinolato)-5,8-quinoxalinediolatotungsten(IV), which we abbreviate as $[W(dcq)_2(qd)]_n$, has been synthesized by a unique two-electron redox reaction between the seven-coordinate (22) dicarbonylbis(5,7-dichloro-8-quinolinolato)(triphenylphosphine)-tungsten(II) (19) and quinoxaline-5,8-dione (23) in dichloromethane as follows:

$$W^{II}(CO)_2(PPh_3)(dcq)_2 \quad + \quad \longrightarrow \quad \left[W(dcq)_2(qd)\right]_n$$

Reaction at 0° for 40 hours followed by centrifugation and extraction with more dichloromethane yielded a dark blue polymeric material [which is analogous to the dark colored monomeric tungsten(IV) chelates] with elemental mole ratios within 3% of those anticipated for the infinite polymer (24). Solvation, hydrolysis, and inhomogeniety have precluded more perfect mole ratios. The anticipated coordina-

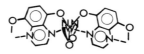

Figure 2. Schematic representation of the $[W(dcq)_2(qd)]_n$ polymer
with the 5,7-dichloro-8-quinolinol ligands represented as N⌢O. The
structure shown is for a mer isomer linkage, but the bifac isomer
is also possible (cf., reference 25).

tion sphere about the tungsten atom is shown schematically in Figure
2, with selection of positions (25) based on Orgel's rule (26) for
$W(N⌢O)_4$ type eight coordinate species.

Initial syntheses of the tungsten(IV) polymer at about 40°C led
to oligomers with inherent viscosities at about 0.1 dℓ/g. Synthesis
at 0° has led to a $[W(dcq)_2(dq)]_n$ polymer with an inherent viscosity
of 0.25 dℓ/g for the dimethyl sulfoxide (DMSO) soluble fraction. The
viscosity is similar to those observed for organometallic polymers a-
bove 50,000 Daltons (27). The analytical results can be fit to such
a molecular weight analysis if the terminal groups are hydroxide
groups. Metal:ligand ratios are also consistent with such molecular
weights. A slow hydrolytic decomposition of this polymer occurs because of the poor nucleophilicity of the qd^{2-} ligand and makes for
some variation in analyses.

The polymer appears black in the solid state and blue-black in
concentrated solutions. Absorption spectra are given elsewhere (24).
Whereas analogous monomeric tungsten(IV) eight-coordinate chelates
have strong metal to ligand charge-transfer absorptions near 700 nm
(16), the polymer exhibits absorption maxima well down into the near
infrared region with a tail on the absorption as low as 3500 cm^{-1}.
The possibility of a one-dimensional electron-transfer system with
this type of species is intriguing. Furthermore, the analogous
tungsten(V) monomers have low energy ligand to metal charge-transfer transitions (28); thus, electron conduction is a distinct possibility, especially for partially oxidized polymers (5).

Thermal analysis of the tungsten polymer in nitrogen indicates
an inherently stable polymer with a slow decomposition. Although

the decomposition begins near 250°C, the thermal profiles show incomplete decomposition to the oxide, even at 800°C. However, in air
decomposition to WO_3 occurs near 510°C, with a partial weight loss
even earlier near 230°C. Unfortunately, the nucleophility of the
bridging ligand is limited by the presence of two heterocyclic nitrogen atoms in the same aromatic ring. Other bridging ligands
which do not possess this weakness are currently under investigation.

Zirconium(IV) Polymer

Because of the low solubility of the Schiff base of 1,2,4,5-
tetraaminobenzene and salicylaldehyde, an indirect synthesis was used
in which the Schiff-base condensation reaction was made between
freshly prepared tetrakis(salicylaldehydato)zirconium(IV) (17) and
recrystallized 1,2,4,5-tetraaminobenzene in dry DMSO under nitrogen:

This condensation proceeds reasonably rapidly at room temperature in DMSO. The solvent interacts with the water produced in the reaction sufficiently for considerable polymerization to occur prior to hydrolysis at the zirconium centers. A red glossy material results upon removal of the DMSO.

The elemental analyses show that appreciable DMSO solvent is rigorously held after extensive drying and that excess oxygen is present relative to the infinite chain values, apparently from the water produced in the Schiff base reaction. This excess oxygen is assumed to be hydroxide end groups, which is consistent with the observed OH stretch (3500 cm^{-1}). The number of repeating units based on this assumption is approximately thirty, which corresponds to an average molecular weight near 2×10^4 Daltons. If some of the OH stretch and excess oxygen is due to water, then the molecular weight is higher. In fact, the zirconium to ligand ratio is consistent with approximately fifty-five repeating units per average polymeric molecule, or a molecular weight average closer to 4×10^4 Daltons (24).

An inherent viscosity of 0.15 (dℓ/g) for a 0.1% w/v DMSO solution of the red glossy material is consistent with the view that most of the OH groups are indeed present as end groups; i.e., with molecular weights of 20,000 to 40,000, although K and α haven't been determined for this system to date.

In addition to the OH stretch, the infrared spectrum exhibits a phenyl ring/oxygen stretch at 1310 cm^{-1}, which is shifted from 1280 cm^{-1} in the free bridging ligand. This shift is analogous to the shift observed in the model zirconium Schiff-base chelate, Zr(dsp)$_2$ (17), where H$_2$dsp = disalicylidene-o-phenylenediamine. Other infrared absorptions are also consistent with the solvated polymer.

Thin films and glassy layers of the zirconium polymer have been prepared. The films diffract light, but show no x-ray diffraction pattern, which is consistent with the amorphous nature of the polymer. At the short chain lengths obtained so far, the polymer appears somewhat brittle. A yellow-orange precipitate is obtained from the addition of acetone to DMSO solutions of the polymer.

Whereas H_4tsb and its derivatives have very limited solubility, the DMSO reaction between $Zr(sal)_4$ and TAB produces an oligomer which is orders of magnitude more soluble. Solvent removal in vacuo or the addition of a miscible nonsolvent such as acetone is necessary to isolate a solid product. Thus the stacking interactions of the aromatic systems have been overcome by the coordination to zirconium. This solubility difference appears to be related to the puckered, perpendicularly arranged, quadridentate ligands anticipated from the model compound x-ray structure (29), in which the ligands show no tendency to stack with either benzene solvate molecules or chelate ligands from adjacent molecules. Models of the polymer chains, assuming the described stereochemistry, also indicate significantly reduced lattice forces. Nonrigid molecular motion may further hinder packing of the polymer chains.

A condensation polymerization using the water scavenger, 2,2'-dimethoxypropane(DMP) was attempted, but no perceivable increase in the average polymer chain length was obtained. Thus, we conclude that DMP did not diminish the rate of chain termination, that imprecise stoichiometry limited the chain growth, or that DMP reacted with one of the other reactants. Further experimentation is necessary.

PROSPECTIVES

The isolation of polymeric species with molecular weights indicative of degrees of polymerization of from 20 to 50 units for step-growth polymerizations which involve either a two electron redox reaction or else a condensation reaction in the presence of a metal ion which can react with water means that the extent of reaction and stoichiometry are $\geq 95\%$. Reactions which are standard reactions for polymer formation could improve the degree of polymerization even further inasmuch as solubility of the polymers in DMSO is still reasonable at room temperature in both cases. Toward that end we have synthesized bis(N,N'-disalicylidene-3,4-phenylenediamine-1-ethylbenzoato)-zirconium(IV)(30).Preliminary reactions designed to form ladder polymers via condensation of aromatic amines to the free ester groups were unsuccessful. The reaction temperatures required $(280-360°)$ to produce aromatic amides and benzimidazoles resulted in side reactions and lower degrees of polymerization. The use of catalysts (31-32) should allow much lower temperatures. Conversely, more reactive functional groups could be used. Alternate polymerization modes for the tungsten·chelate polymer include the condensation of $W(q)_2(cq)_2$ [where q^- = the anion of 8-quinolinol and cq^- = the anion of 5-chloro-8-quinolinol] with Na_2S analogous to the synthesis of polyphenylsulfide from dichlorobenzene and sodium sulfide. Dilithio derivatives coupled with diiodo derivatives might also be possible, analogous to the polyferrocene preparations noted elsewhere in this volume.

Whereas these studies were initiated toward the end of producing thermally stable polymers, the potential usefulness of these species may be more toward the production of conducting polymers or a related photochemical electron-transfer use. Eight-coordinate tungsten(IV) chelates can be oxidized to analogous tungsten(V) species as noted above and partially oxidized polymers should function as electron carriers. The standard potential for the monomer chelates ranges from +0.2 to +0.4 volts vs. SCE depending on the ligands (33). The fluorescence of the zirconium Schiff-base polymer (34) is indicative of photochemical potential for these species as well. We have synthesized a tungsten(IV) Schiff-base polymer (16), but only in very low yields; therefore, tungsten(IV) Schiff-base polymers have not been investigated to date as no good high-yield pathway has been found. This challenging synthesis could couple the advantages of both systems we have been investigating. Further, the thermal weak points of the zirconium Schiff-base polymer appears to be a reverse reaction around 250°C, which could be avoided by reduction of the Schiff-base double bond with $NaBH_4$.

Overall, we feel that the potential for such chelate polymers has just barely been tapped. Bridging ligands such as 1,5-naphthyridine-4,8-diol and 1,5-diazaanthracene-9,10-diol offer marked nucleophilic advantages over the quinoxalinediol used herein. Synthetic complications have not yet allowed the synthesis of eight-coordinate polymers with these ligands, but the potential is there. New specific syntheses of mixed-ligand molybdenum chelates (35-36) should allow extension of our polymers to molybdenum. The advantage of molybdenum relative to tungsten is a reaction rate increase for molybdenum. Conversely, this implies less thermal stability in a hydrolytic sense. Extension to niobium and/or rheniun for which inert eight-coordinate cyano complexes are known should be obvious. The potential appears almost endless.

ACKNOWLEDGEMENTS

The support of the National Science Foundation through the University of Massachusetts Materials Research Laboratory and the support of the U.S. Army Research Office are gratefully acknowledged.

REFERENCES

[1] J.C. Bailar, Jr. in "Organometallic Polymers", C.E. Carraher, Jr., J.E. Sheats, and C.U. Pittman, Jr., Editors, Academic Press, N.Y., 1978, p. 313; and other examples elsewhere in the same volume.

[2] L. Holliday, Inorg. Macromol. Rev., 1, 3 (1970) and subsequent papers in the volume dated 1970-1972.

[3] E.M. Natanson and M.T. Bryk, Russ. Chem. Rev., 41, 671 (1972).

[4] S.B. Brown and M.J.S. Dewar, Inorg. Chim. Acta, 34, 221 (1979); J. Inorg. Nucl. Chem., 42, 140 (1980).

[5] J.T. Wrobleski and D.B. Brown, Inorg. Chem., 18, 2738 (1979), and references cited therein.

[6] J.B. Davison and K.J. Wynne, Macromolecules, 11, 186 (1978) and references cited therein.

[7] R.S. Bottei and C.P. McEachern, J. Thermal Anal., 6, 37 (1974) and references cited therein.

[8] D.W. Van Krevelen, Chem. and Ind., 49, 1396 (1971).

[9] D.W. Van Krevelen, Textile Progress, 8, 124-5, 156 (1976).

[10] B.P. Block, H.D. Gillman, and P. Nannelli, "The Synthesis and Characterization of the Poly(Metal Phosphinates)", ONR Contract N00014-69-C-0122, Final Report, June 1977, p. 2.

[11] M. G. B. Drew, Coord. Chem. Rev., 24, 179 (1977).

[12] J. K. Burdett, R. Hoffmann, and R. C. Fay, Inorg. Chem., 17, 2553 (1978).

[13] R. D. Archer and W. D. Bonds, Jr., J. Am. Chem. Soc., 89, 2236 (1967).

[14] W. D. Bonds, Jr., and R. D. Archer, Inorg. Chem., 10, 2057 (1971)

[15] W. D. Bonds, Jr., R. D. Archer, and W. C. Hamilton, Inorg. Chem., 10, 1764 (1971).

[16] R. D. Archer, C. J. Donahue, W. H. Batschelet, and D. R. Whitcomb, in "Inorganic Compounds with Unusual Properties, II", R. B. King, Editor, Am. Chem. Soc., Washington, D. C., 1979, p. 252.

[17] R. D. Archer and M. L. Illingsworth, Inorg. Nucl. Chem. Lttr., 13, 661 (1977).

[18] C. J. Weber, A. W. Kozlowski, and W. H. Batschelet, unpublished results.

[19] W. H. Batschelet, R. D. Archer, and D. R. Whitcomb, Inorg. Chem., 18, 48 (1979).

[20] W. H. Batschelet and R. D. Archer, unpublished results.

[21] W. H. Batschelet and A. W. Kozlowski, unpublished results.

[22] R. O. Day, W. H. Batschelet, and R. D. Archer, Inorg. Chem., 19, 0000 (1980) in press.

[23] K. H. Ford, M. S. Thesis, University of Pennsylvania, 1962.

[24] R. D. Archer, W. H. Batschelet, and M. L. Illingsworth, Org. Coatings and Plastics Chem., 41, 191 (1979); details to be published elsewhere.

[25] C. J. Donahue and R. D. Archer, J. Am. Chem. Soc., 99, 6613 (1977).

[26] L. E. Orgel, J. Inorg. Nucl. Chem., 14, 136 (1960).

[27] C. U. Pittman, Jr. and A. Hirao, J. Polym. Sci., Polym. Chem. Ed., 15, 1677 (1977).

[28] R. D. Archer, W. D. Bonds, Jr., and R. A. Pribush, Inorg. Chem., 11, 1550 (1972).

[29] R. D. Archer, R. O. Day, and M. L. Illingsworth, Inorg. Chem., 18, 2908 (1979).

[30] M. L. Illingsworth, Ph.D. Dissertation, University of Massachusetts, Amherst, 1980.

[31] E. W. Neuse in "Organometallic Polymers", C. E. Carraher, Jr., J. E. Sheats, and C. U. Pittman, Jr., Editors, Academic Press, New York, 1978, p. 95.

[32] M. Veda, A. Sato, and Y. Imai, J. Polym. Sci., Polym. Chem. Ed., 17, 783 (1979).

[33] R. Nowak, M. Stankovich, and R. D. Archer, unpublished results.

[34] J. C. Cooper, R. Nowak, and R. D. Archer, unpublished results.

[35] R. D. Archer, C. J. Weber, and C. J. Donahue in Proceedings of the Climax Third International Conference on the Chemistry and Uses of Molybdenum, H. F. Barry and P. C. H. Mitchell, Editors, Climax Molybdenum Co., Ann Arbor, Michigan, 1979, p. 64.

[36] C. J. Weber, unpublished results.

ELECTRICAL CONDUCTIVITY

Strategies for Control of Lattice Architecture in Low-Dimensional Molecular Metals: Assembly of Partially Oxidized Face-to-Face Linked Arrays of Metallomacrocycles

C. W. Dirk, E. A. Mintz, K. F. Schoch, Jr., and T. J. Marks[*]
Department of Chemistry and the Materials Research Center,
Northwestern University
Evanston, IL 60201

ABSTRACT

This paper discusses an approach to control molecular stacking interactions in low-dimensional mixed valence materials by locking partially oxidized metallomacrocycles together in a face-to-face orientation. Thus, doping of the cofacially linked oligomers $[M(Pc)O]_n$ (M = Si, Ge, Sn; Pc = phthalocyaninato) with halogen (I_2, Br_2) or quinone (e.g., TCNQ, DDQ) electron acceptors produces robust, electrically conductive polymers with a wide range of stoichiometries and properties. The new materials have been studied by a variety of physical methods including X-ray diffraction, resonance Raman and infrared spectroscopy, ESR, static magnetic susceptibility, and variable-temperature four-probe electrical conductivity. Evidence is presented that some of the polymers have "metal-like" conductivity in the stacking direction and that transport properties within the series can be readily manipulated by rational variation of lattice architecture (e.g., the identity of the metal, M) and acceptor characteristics. Additional information is presented on doping experiments with electron donors and on employing metallohemiporphyrazines as polymer building blocks.

INTRODUCTION

The past half-dozen years have witnessed intense scientific activity in the area of low-dimensional electrically conductive

materials [1-5]. Great excitement has been generated among chemists and physicists by the synthesis and properties of unusual
organic and metal-organic substances with metal-like properties.
These developments can be anticipated to lead ultimately to new
degrees of chemical control over collective solid state properties, to new methodology in chemical synthesis and in physical
measurements, and to better theoretical models for cooperative
phenomena in condensed matter. In the technological sphere, this
research may lead to a new generation of electronic materials with
applications as varied as sensors, rectifiers, fuel cell
components, solar energy conversion elements, and electrophoto-
graphic devices. The possibility of high temperature supercon-
ductors or at least highly conductive synthetic materials which
could replace metals in various applications, has contributed
additional impetus to the design and study of metal-like
materials.

Despite the advances that have been achieved, it is fair to
say that our understanding of those molecular and electronic
characteristics which control charge conduction is at a rather
primitive level. That is also true of the synthetic chemical
methodology required to tailor structures for testing current
theories about the molecular metallic state or for optimizing
materials performance and processing characteristics. A tremendous challenge exists in learning to control these factors, and
the purpose of this article is to report on new developments in
our Laboratory which address a number of chemical and physicochemical aspects of this problem.

STRATEGIES FOR THE DESIGN OF HIGHLY CONDUCTIVE LOW-DIMENSIONAL
MATERIALS

Two features now appear to be necessary for facile charge conduction in a molecular material [6,7]. First, the component molecules must be arrayed in close spatial proximity, with sufficient

intermolecular orbital overlap to provide a continuous electronic
pathway for carrier migration, and in crystallographically similar
environments, so that the pathway has a minimum of energetic hills
and valleys. Second, the arrayed molecules must exist in formal
fractional oxidation states ("mixed valence," "incomplete charge
transfer," "partial oxidation"). That is, the molecular entities
to be connected in series must have fractionally occupied electronic valence shells. Within the framework of a simple Hubbard
model [1-5], this requirement reflects the relatively narrow band-
widths (4t) and large on-site Coulomb repulsions (U) in such
systems. A simplified valence bond picture of this situation is
illustrated in Figure 1; partial oxidation facilitates charge
mobility by creating numerous electronic vacancies. A similar
picture could be constructed for partial reduction.

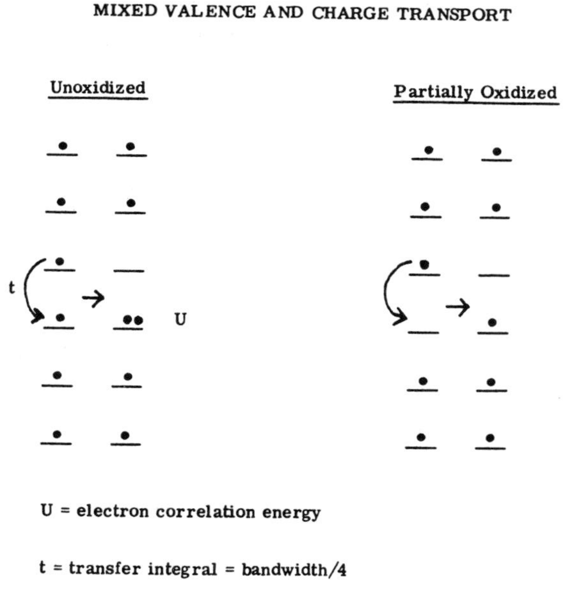

MIXED VALENCE AND CHARGE TRANSPORT

U = electron correlation energy

t = transfer integral = bandwidth/4

FIGURE 1. Schematic illustration of the effect of partial oxidation on charge mobility in a low-dimensional system composed of
molecular stacks.

A successful, first-generation strategy for the synthesis of mixed valent low-dimensional materials [6,7] has involved the co-crystallization of planar, conjugated metallomacrocyclic donor molecules (D) with halogen acceptors (A) as schematized below.

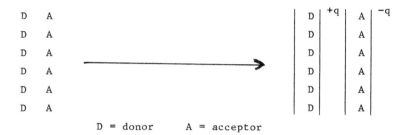

D = donor   A = acceptor

In optimum cases, the result has been lattices composed of segregated, partially oxidized metallomacrocyclic stacks and parallel arrays of halide or polyhalide counterions [6-8]. An important additional feature of this approach is that the form of the halogen (even if disordered) can be straightforwardly determined by resonance Raman/iodine Mossbauer spectroscopic techniques [6-11]. The degree of partial oxidation follows from this information and knowledge of the stoichiometry. This synthetic approach has enjoyed success for metal glyoximates [8,12,13], dibenzotetraazaannulenes [14], phthalocyanines [15-17], and porphyrins [18]. The crystal structure of an example, nickel phthalocyanine iodide ($[Ni(Pc)]I_{1.0}$) [15-17], is illustrated in Figure 2. The 300°K conductivity of this material in the molecular stacking direction (300-700 Ω^{-1} cm^{-1}) is high (carrier mean free paths are comparable to some of the most conductive "molecular metals") and the temperature dependence is "metal-like" ($\rho \sim T^{1.9}$) down to 60°K.

Although the above strategy is frequently effective, it suffers, as do all approaches based upon simple molecular stacking, from the weakness that the lattice architecture is

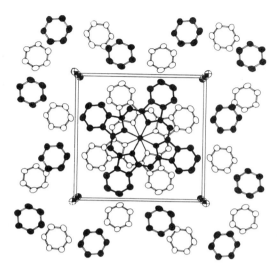

FIGURE 2. The crystal structure of nickel phthalocyanine iodide, $[Ni(Pc)]I_{1.0}$, viewed along the stacking direction. From ref. 7.

totally dependent upon the unpredictable and largely uncontrollable forces that dictate the stacking pattern, the donor-acceptor
orientations, and the stacking repeat distances. Figure 3
illustrates the complexity of the structural problem by
depicting some of the types of crystallization patterns which have
been identified. A common pitfall in the design of new materials
is the formation of integrated D-A stacks (3C,D), which invariably
leads to insulators [2,6,19]. As an example, attempts to substi-
tute oxidizing quinones for halogens in the aforementioned
phthalocyanine chemistry lead to integrated stack insulators [20].

A successful new approach to the control of molecular
stacking in low dimensional mixed valence materials is founded
upon the construction of macromolecules in which arrays of
metallomacrocycles are tightly locked in a "face-to-face" configuration by covalent bonds [21-23]. Followed by partial
oxidation, this approach capitalizes upon a great deal of accumu-
lated chemical and physical information about the subunits, and

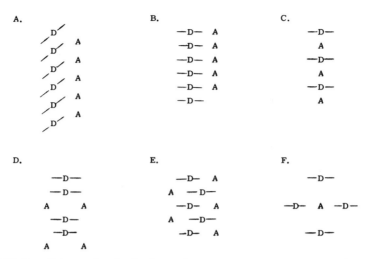

FIGURE 3. Schematic depiction of some common structures for donor-acceptor complexes. A. Segregated stacking, canted donors. B. Segregated stacking, D_{nh} donor stacking. C. Integrated stacking. D. Integrated stacking, donor dimers. E. Segregated stacking, zig-zag donor stacking. F. Ion clusters without stacking.

offers the real possibility of constructing robust new conductive polymers with well-defined and easily manipulated primary and secondary structures. The forces which hold the stacks together are now covalent linkages with bond energies on the order of 80-100 kcal/mol rather than weak packing, van der Waals, and bandwidth forces. As a result, it is now possible to both delve into those factors which stabilize the mixed valent state without fear of a breakdown in stacking, and to perturb systematically bandwidth and phonon dynamics. Although we focus our discussion primarily upon phthalocyanines (with additional remarks about hemiporphyrazines), it should be clear that the strategy has obvious generality.

SYNTHESIS AND PROPERTIES OF FACE-TO-FACE LINKED METALLO-
PHTHALOCYANINES

The Group IVA precursor phthalocyanines $Si(Pc)Cl_2$, $Ge(Pc)Cl_2$, and $Sn(Pc)Cl_2$ were synthesized as described elsewhere [22,24-26]. Hydrolysis in pyridine/NaOH solution converts the dichlorides to the corresponding dihydroxides, $Si(Pc)(OH)_2$, $Ge(Pc)(OH)_2$, $Sn(Pc)(OH)_2$ [22,24-26]. Condensation of these compounds to form phthalocyaninato polysiloxanes, polygermyloxanes, and polystannyloxanes (Figure 4) was carried out at $300-400°C/10^{-3}$ torr [22,24-26]. The resulting macromolecules have been characterized by a broad range of chemical and physicochemical methods. These $[M(Pc)O]_n$ materials have high chemical and thermal stability;

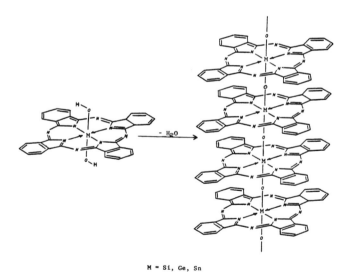

M = Si, Ge, Sn

FIGURE 4. Condensation reaction to produce cofacial arrays of Group IV metallophthalocyanines.

moreover, they are not significantly degraded by oxygen or
moisture. The polysiloxane polymer can be dissolved in concentrated sulfuric acid and recovered unchanged (typical of phthalocyanines containing non-electropositive metals). A rough estimate
of the minimum average chain length of $[Si(Pc)O]_n$ produced in the
condensation polymerization can be obtained by infrared spectro-
photometric analysis of the Si-O stretching region. For a typical
sample, the degree of polymerization is estimated to be on the
order of ca. 100 subunits or more. Structural information on the
face-to-face polymers can be derived from several lines of
evidence. X-ray powder diffraction patterns can be indexed in the
tetragonal crystal system [21,22,24] and are very similar to the
patterns exhibited by the columnar crystal structures of
$[Ni(Pc)]I_{1.0}$ [17] and $Ni(dpg)_2I_{1.0}$ (dpg = diphenylglyoximato)
[8]. The stacking intervals ($\underline{c}/2$) in these latter tetragonal
structures, determined in single crystal studies, are 3.244(2)
Å and 3.271(1) Å, respectively. The stacking intervals derived for
the $[M(Pc)O]_n$ materials from the powder diffraction data are found
to depend upon the ionic radius of the Group IV ion and vary from
3.33(2) Å (Si-O-Si) to 3.51(2) Å (Ge-O-Ge), to 3.95(2) Å (Sn-O-Sn)
[21,22,24]. These relationships are illustrated in Figure 5. The
reliability of these metrical parameters is further supported by
single crystal diffraction results on the model trimer
$[(CH_3)_3SiO]_2(CH_3)SiO[Si(Pc)O]_3Si(CH_3)[OSi(CH_3)_3]_2$ which contains
three cofacial Si(Pc)O units linked by linear Si-O-Si connections
at a distance of 3.324(2) Å [27]. In addition, the $[Ge(Pc)O]_n$ and
$[Sn(Pc)O]_n$ interplanar spacings obtained from diffraction data are
in good agreement with values estimated from ionic radii [28]
assuming linear Ge-O-Ge and Sn-O-Sn vectors, i.e., 3.58 Å for
$[Ge(Pc)O]_n$ and 3.90 Å for $[Sn(Pc)O]_n$ [21,22,24]. There is good
precedent for molecules with linear Si-O-Si, Ge-O-Ge, and Sn-O-Sn
linkages [29]. The $[M(Pc)O]_n$ polymers display vibrational and
optical spectra which are characteristic of metallophthalocyanines
[24].

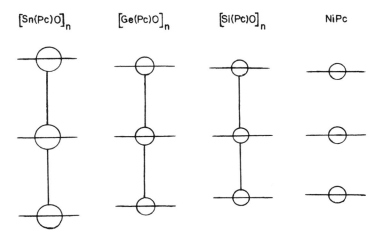

$[Sn(Pc)O]_n$ $[Ge(Pc)O]_n$ $[Si(Pc)O]_n$ NiPc

FIGURE 5. Scale drawing of the interplanar relationships
in the face-to-face phthalocyanine polymers and in $[Ni(Pc)]I_{1.0}$.

PARTIAL OXIDATION OF COFACIALLY LINKED METALLOPHTHALOCYANINES
WITH HALOGENS

Doping experiments on the $[M(Pc)O]_n$ polymers were first
carried out using iodination methodology developed in this
Laboratory for simple stacked systems [6,7]. Stirring the
powdered polymers with solutions of iodine in organic solvents or
exposing the powders to iodine vapor results in substantial iodine
uptake. Alternatively, $[Si(Pc)O]_n$ can be doped by dissolving in
sulfuric acid and precipitating with an aqueous I_3^- solution. The
stoichiometries which can be obtained depend upon the reaction
conditions; representative iodinated materials characterized by
elemental analysis are compiled in the left-hand column of
Table 1. A survey experiment also indicated that bromine-doped
material could be prepared. That partial oxidation of the cofa
cial array has indeed occurred is confirmed by resonance Raman
scattering spectroscopy in the polyiodide region (Figure 6). The

TABLE 1

Physical Data for Polycrystalline Samples of Halogen-Doped
$[M(Pc)O]_n$ Materials

Compound	$\sigma(\Omega^{-1}cm^{-1})300°K$	Activation Energy (eV)	Interplanar Spacing (Å)
$[Si(Pc)O]_n$	3×10^{-8}		3.33(2)
$\{[Si(Pc)O]I_{0.50}\}_n$	2×10^{-2}		
$\{[Si(Pc)O]I_{1.55}\}_n$	1.4	0.04 ± 0.001	3.33(2)
$\{[Si(Pc)O]I_{4.60}\}_n$	1×10^{-2}		
$\{[Si(Pc)O]Br_{1.00}\}_n$	6×10^{-2}		
$[Ge(Pc)O]_n$	$<10^{-8}$		3.51(2)
$\{[Ge(Pc)O]I_{1.80}\}_n$	3×10^{-2}	0.08 ± 0.006	3.51(2)
$\{[Ge(Pc)O]I_{1.90}\}_n$	5×10^{-2}	0.06 ± 0.003	
$\{[Ge(Pc)O]I_{1.94}\}_n$	6×10^{-2}	0.05 ± 0.007	
$\{[Ge(Pc)O]I_{2.0}\}_n$	1×10^{-1}		
$[Sn(Pc)O]_n$	$<10^{-8}$		3.95(2)
$\{[Sn(Pc)O]I_{1.2}\}_n$	1×10^{-6}		3.95(2)
$\{[Sn(Pc)O]I_{5.5}\}_n$	2×10^{-4}	0.68 ± 0.01	
$[Ni(Pc)]I_{1.0}$[a]	7×10^{-1}	0.036 ± 0.001	3.244(2)

[a] Reference 15.

characteristic totally symmetric stretching frequency of
I_3^- ($\nu =108$ cm^{-1}) [6,7,9] is observed along with an accompanying
overtone progression. For samples with I/M < 3, there are at most
only traces of I_5^- ($\nu \approx 160$ cm^{-1}) [8,10] and no indication of free
$I_2(\nu \approx 200$ $cm^{-1})$ [8,10]. The nature of the $[M(Pc)O^{\delta+}]_n$
electronic structure was also studied by electron spin resonance
(ESR). The symmetry of the lineshapes and the measured
g-values are consistent with π-radical cations, i.e., the unpaired
spin density is in molecular orbitals which are predominantly
ligand in character [30]. A similar conclusion was reached for

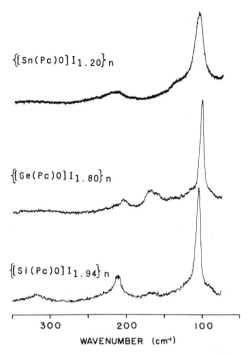

FIGURE 6.   Resonance Raman spectra (ν_0 = 5145 Å)of iodine-doped
phthalocyanine face-to-face polymers. From ref. 21.

[Ni(Pc)]$I_{1.0}$ [17]. ESR data for the {[M(Pc)O]I_x}$_n$ materials are
compiled in Table 2. In regard to structural changes which might
accompany doping, X-ray powder diffraction studies indicate that
iodination does not significantly alter the interplanar separations [21,22,24].

Four-probe electrical conductivity measurements on the
[M(Pc)]$_n$ powders using locally developed van der Pauw techniques
[31,32] show them to be insulators. However, iodine or bromine
doping results in large increases in electrical conductivity
(Table 1). The general trend in conductivity as a function of
metal is $\sigma_{Si} \gtrsim \sigma_{Ge} > \sigma_{Sn}$. Since the transport characteristics of
iodine-oxidized metallophthalocyanines are known to be largely

TABLE 2

Powder ESR Data for Iodinated Phthalocyanine Face-to-Face
Polymers

Compound	$g(300°K)$[a]	$\Gamma(300°K)(G)$[b]
$\{[Si(Pc)O]I_{1.40}\}_n$	2.003	5.1
$\{[Ge(Pc)O]I_{0.62}\}_n$	2.002	3.2
$\{[Sn(Pc)O]I_{1.20}\}_n$	2.002	6.0

a
 Average g-value; $g_{\parallel}$ and $g_{\perp}$ are not resolved.
b
 Observed linewidth.

ligand-dominated and relatively insensitive to the identity of the
metal [15-17], the metal dependence of the conductivity observed in
the face-to-face polymers is logically ascribed to structural differences such as the interplanar separation. Indeed, the
$\{[Si(Pc)O]I_x\}_n$ interplanar separation is within 0.1 Å of that in
$[Ni(Pc)]I_{1.0}$ and the room temperature powder conductivities of
the two materials are comparable (Table 1). The temperature
dependence of the $\{[M(Pc)O]I_x\}_n$ powder conductivities is thermally
activated (Figure 7), and least squares fits to Eq.(1) yield the

$$\sigma = \sigma_0 \, e^{-\Delta/kT} \tag{1}$$

activation parameters compiled in Table 1. Powder conductivities
are, of course, affected by interparticle contact resistance and
averaging over all crystallographic orientations. Thus, for low-
dimensional "molecular metals" such as $[Ni(Pc)]I_{1.0}$, powder conductivities are typically 10^2-10^3 less than single crystal conductivities in the stacking direction. Thus, "metal-like" temperature dependence is usually masked. From the powder data on
the $\{[M(Pc)O]I_x\}_n$ materials it can be anticipated that high,

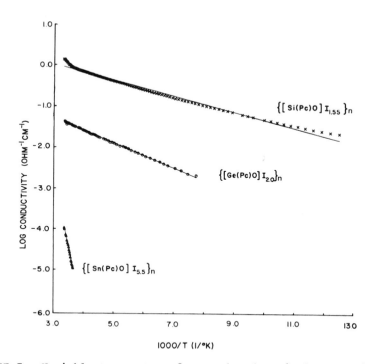

FIGURE 7.   Variable temperature four-probe electrical conductivity
data for powders of the iodine-doped, face-to-face phthalocyanine
polymers.

"metal-like" conductivities will be observed in the chain direction for the M = Si and possibly M = Ge materials. Voltage
shorted compaction (VSC) techniques [33] offer an effective qualitative means to sample stacking axis transport properties in
pressed powder samples by deliberately shorting out sources of
interparticle resistance.   In the present context it is important
to note that the VSC conductivity behavior of
$\{[Si(Pc)O]I_x\}_n$ samples is unequivocally "metal-like" [34].   The
results of the variable temperature conductivity studies also
underscore the robust thermal character of the cofacially arrayed
polymers. $\{[Si(Pc)O]I_x\}_n$ samples could be cycled to $300^\circ C$ with
only a minor decrease in room temperature conductivity (due primarily to vaporization of the iodine).

Static magnetic susceptibility by Faraday techniques reveals
another characteristic signature [1-5] of "molecular metals" in
the $\{[Si(Pc)O]I_x\}_n$ and $\{[Ge(Pc)O]I_x\}_n$ materials. Susceptibilities
are only weakly paramagnetic (χ_M = 300-500 x 10^{-6} emu after
diamagnetic corrections are made) and are only modestly dependent
on temperature down to $77°K$ [21,22]. Studies at lower temperatures are in progress.

RESPONSE OF THE COFACIALLY LINKED METALLOPHTHALOCYANINES TO OTHER DOPANTS

The strictly enforced molecular stacking in the $[M(Pc)O]_n$
materials offers an unprecedented opportunity to experiment with
some of the important forces [2,6,35-39] which stabilize or desta-
bilize the molecular metallic state. Thus, the electronic aspects
of the donor-acceptor interaction can be studied with far greater
control over the structural variables. It was first of interest
to learn whether non-halogen oxidants could produce a partially
oxidized, conductive metallophthalocyanine stack, or whether there
was something "magic" about iodine and bromine. High potential
quinones such as those shown below have electron affinities comparable to halogens, and produce conductive compounds with a

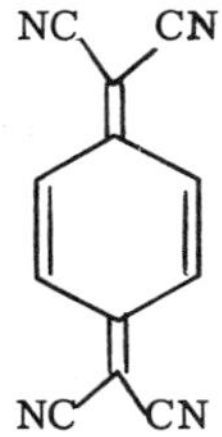

TCNQ 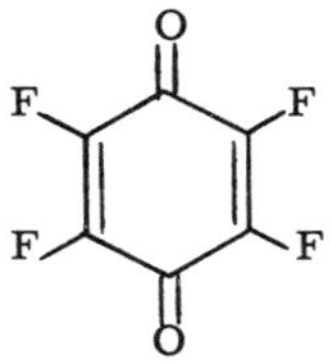   fluoranil 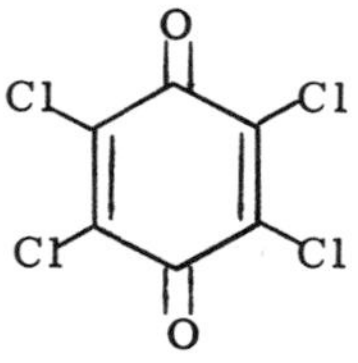   chloranil

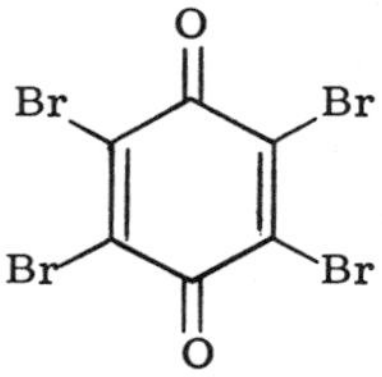

bromanil 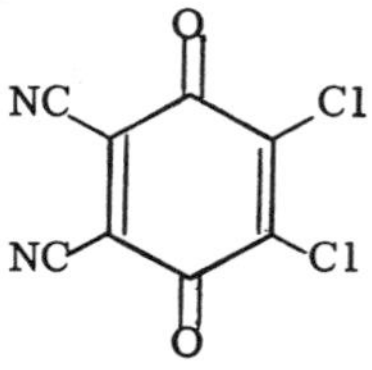   DDQ 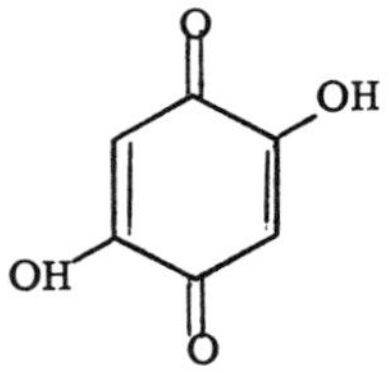   DHB

number of organic donors [1-5]. With simple metallophthalocyanines, however, only insulators are produced [20] and considerable structural evidence points to an integrated stack
crystal structure [40] as the reason for the low charge transport
capability. It was thus of interest to determine what donor-
acceptor interactions would occur in a metallophthalocyanine
system which was locked into stacks. Doping experiments were
conducted by stirring the $[Si(Pc)O]_n$ macromolecules with solutions of the above quinones. The products were characterized by
elemental analysis and vibrational spectroscopy. As can be seen
in Table 3, large increases in electrical conductivity accompany
the quinone doping. Indeed, the DDQ-doped materials are as conductive as many of the halogenated polymers [23,41]. The temperature dependence of the conductivity of some representative
samples is shown in Figure 8. The transport in these materials
is thermally activated and least-squares fits to Eq.(1) yield the
activation parameters compiled in Table 3. There are some notable
deviations from a linear ln σ vs. 1/T relationship (e.g.,
$\{[Si(Pc)O]TCNQ_{0.5}\}_n$) and further investigations of the reasons for
this behavior are in progress. Infrared spectral studies of the
TCNQ-doped materials reveal a displacement of ν_{CN} to lower frequencies [41], consistent with electron density uptake by the
quinone [42]. Clearly partial oxidation of the phthalocyanine
stack by high potential quinones occurs when a segregated stack
structure is enforced.

In principle, it should also be possible to partially reduce
phthalocyanines and to create conducting materials by injecting
nonintegral amounts of electron density per site [41]. A number
of attempts have been made to partially reduce metallophthalocyanines using alkali metals [20]. In all cases, the
resulting materials are insulators, and it was suspected that non-
stacked materials were being produced. A preliminary experiment
was conducted in which $[Si(Pc)O]_n$ was reacted with potassium vapor
in a sealed tube. The product was collected and handled at all
times in an inert atmosphere. As can be seen in Table 3, a signi-

TABLE 3

Electrical Conductivity Data for Polycrystalline Samples of
Cofacial Phthalocyanine Polymers with Various Dopants

Dopant[a]	Empirical Formula	$\sigma(\Omega^{-1}\ cm^{-1})300^{\circ}K$	Activation Energy(eV)
none	$[Si(Pc)O]_n$	3×10^{-8}	
I	$\{[Si(Pc)O]I_{1.55}\}_n$	1.4	$0.04\pm.001$
Br	$\{[Si(Pc)O]Br_{1.00}\}_n$	6×10^{-2}	
K	$\{[Si(Pc)O]K_{1.0}\}_n$	2×10^{-5}	
TCNQ	$\{[Si(Pc)O]TCNQ_{0.50}\}_n$	2.8×10^{-3}	$0.09\pm.002$
Flr	$\{[Si(Pc)O]Flr_{0.23}\}_n$	7.2×10^{-4}	$0.13\pm.001$
Chl	$\{[Si(Pc)O]Chl_{.037}\}_n$	6.9×10^{-4}	$0.13\pm.002$
Brl	$\{[Si(Pc)O]Brl_{0.84}\}_n$	5.8×10^{-4}	$0.15\pm.001$
DDQ	$\{[Si(Pc)O]DDQ_{1.00}\}_n$	2.1×10^{-2}	$0.08\pm.001$
DDQ	$\{[Si(Pc)O]DDQ_{0.35}\}_n$	6.2×10^{-2}	$0.05\pm.001$
DHB	$\{[Si(Pc)O]DHB_{0.13}\}_n$	3.8×10^{-5}	$0.19\pm.005$
ClA	$\{[Si(Pc)O]ClA_{0.14}\}_n$	1.8×10^{-3}	$0.11\pm.001$

[a] Flr = fluoranil; Chl = chloranil; Brl = bromanil; DDQ = dichlorodicyanoquinone; ClA = chloranilic acid.

ficant increase in electrical conductivity accompanies the potassium doping. Further efforts to refine the reductive doping procedure are now in progress.

SYNTHESIS AND HALOGEN DOPING OF COFACIALLY LINKED METALLO-HEMIPORPHYRAZINES

It has also been of interest to elaborate the face-to-face conducting polymer concept to include other types of

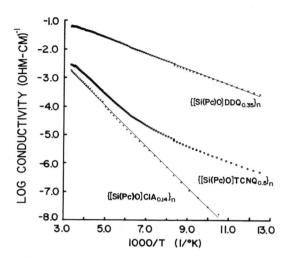

FIGURE 8. Variable temperature powder conductivities of the siloxane phthalocyaninato cofacial polymer doped with high potential quinones. ClA = chloranilic acid.

metallomacrocycles. A particularly intriguing question concerns whether such an approach could bring macrocycles which do not normally stack into a cofacial orientation and what the properties of the resulting materials would be. Previous results in this Laboratory indicated that metallohemiporphyrazines (M(hp), shown below) could be oxidized with iodine when M = Ni but the products

M(hp)

remained insulating [44]. Since the molecular orbital topology of
the hemiporphyrazine ligand, as probed by molecular orbital calcu-
lations at the Pariser-Parr-Pople SCF LCAO CI π-electron level
[45,46], is considerably different from phthalocyanines [45] (both
HOMO and LUMO are more localized in H_2hp [46]), it was of interest
to explore the effect of enforced molecular stacking on oxidation
state and charge transport. Face-to-face silicon and germanium
hemiporphyrazine macromolecules [46-48] were synthesized as
depicted in Eqs.(2)-(5). The macromolecules were characterized

$$2 \quad \bigodot \begin{smallmatrix} CN \\ \\ CN \end{smallmatrix} + 2 \quad H_2N \bigodot N \; NH_2 \longrightarrow H_2hp + 4NH_3 \qquad (2)$$

$$H_2hp + MCl_4 \longrightarrow M(hp)Cl_2 \qquad (3)$$

$$M(hp)Cl_2 + 2NH_4OH \longrightarrow M(hp)(OH)_2 + 2NH_4Cl \qquad (4)$$

$$nM(hp)(OH)_2 \longrightarrow [M(hp)O]_n + n/2 \; H_2 \qquad (5)$$

$$M = Si, \; Ge$$

by elemental analysis and vibrational spectroscopy. The available
structural data [47,49] indicate that $[M(hp)O]_n$ interplanar spacings should be comparable to those for the $[M(Pc)O]_n$ materials.
Iodine uptake was not as facile as for the $[M(Pc)O]_n$
macromolecules, and very large excesses of iodine as well as
heating of the sample were required to achieve significant dopant
levels $(\{[M(hp)O]I_x\}_n$ where $x > 1)$. Resonance Raman spectra
(Figure 9) do indicate that oxidation has occurred in the $x \approx 1.5$
materials, yielding formally mixed valent cofacial arrays, but the
degree of oxidation per added equivalent of iodine does not appear
to be as great as for the phthalocyanine analogues. ESR spectra
indicate the formation of π radical cations (Table 4), and the
measured g-values are not greatly different from the aforementioned phthalocyanine polymers. Thus, oxidation involves orbitals
which are predominantly ligand in composition. Room temperature

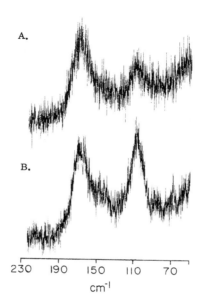

FIGURE 9. Resonance Raman spectra (ν_0 = 5145 Å) of A.
$\{[Si(hp)O]I_{1.51}\}_n$ B. $\{[Ge(hp)O]I_{1.48}\}_n$ as polycrystalline
samples.

TABLE 4

Powder ESR and Conductivity Data for Iodinated Hemiporphyrazine
Cofacial Polymers

Compound	$g(300°K)^a$	$\Gamma(300°K)(G)^b$	$\sigma(\Omega^{-1}\ cm^{-1})(300°K)$
$[Si(hp)O]_n$			$<10^{-12}$
$\{[Si(hp)O]I_{1.51}\}_n$	2.003	7.8	3.0×10^{-8}
$[Ge(hp)O]_n$			2.0×10^{-16} c
$\{[Ge(hp)O]I_{1.48}\}_n$	2.003	6.2	6.0×10^{-9}

a
Average g-value; $g_{\parallel}$ and $g_{\perp}$ are not resolved.

b
Observed linewidth

c
G. Meyer and D. Wöhrle, <u>Makromol. Chem.</u>, <u>175</u>, 714 (1974).

pressed powder conductivity data are also given in Table 4. A
significant increase in conductivity occurs upon iodination,
however charge transport values are still many orders of magnitude
below those for the analogous cofacial phthalocyanines. In some
respects the situation is similar to the halogenated glyoximates,
which form mixed valent stacks but are only semiconducting
[8,11,12]. Mixed valency and cofacial stacking are clearly a
necessary but not sufficient condition to form a highly conductive macromolecule.

CONCLUSIONS AND PROSPECTS

Cofacial metallomacrocycle assembly techniques represent what
is likely the most powerful approach yet developed for controlling
molecular stacking architecture in low dimensional materials. In
terms of fundamental understanding, we already have learned a
great deal about bandwidth-conductivity and donor-acceptor relationships in conductive materials composed of molecular stacks.
However, the surface has only been barely scratched in terms of the
exciting research opportunities which await exploitation in this
area. Further synthetic work offers the opportunity to make
drastic changes in macrocycle identity and electronic structure,
stacking distance and bandwidth, interplanar relationships and
phonon dynamics, and to correlate these chemical modifications
with physical observables. Already, efforts to introduce new
metal ions [50,51] and bridging functionalities [23,52] have
succeeded. Studies with new types of dopants should provide much
important information on donor-acceptor relationships and on the
cohesive forces which stabilize the mixed valent state. A wide
variety of magnetic, charge transport, and optical experiments
remain to be carried out which should ultimately provide invaluable information on how the chemistry and lattice architecture
are related to some of the fundamental characteristics of the
molecular metallic state. Finally, little is known about the pro-

cessing and fabrication properties of these new classes of
materials. Efforts in this direction are presently underway and
should provide interesting and useful information of technological
relevance.

ACKNOWLEDGMENTS

    This research was generously supported by the Office of Naval
Research and by the NSF-MRL program through the Materials Research
Center of Northwestern University (grants DMR76-80847A01 and
DMR79-23573). TJM is a Camille and Henry Dreyfus Teacher-Scholar.

REFERENCES

[1] J. T. Devreese, V. E. Evrard, and V. E. Van Doren, Eds.,
    Highly Conducting One-Dimensional Solids, Plenum Press, NY,
    1979.

[2] J. B. Torrance, Acct. Chem. Res., 12, 79 (1979).

[3] J. S. Miller, and A. J. Epstein, Eds., Synthesis and Proper-
    ties of Low-Dimensional Materials, Ann. NY Acad. Sci., 313
    (1978).

[4] H. J. Keller, Ed., Chemistry and Physics of One-Dimensional
    Metals, Plenum Press, New York, 1977.

[5] J. S. Miller and A. J. Epstein, Prog. Inorg. Chem., 20, 1,
    (1976).

[6] T. J. Marks and D. W. Kalina, in Extended Linear Chain Com-
    pounds, J. S. Miller, Ed., Plenum Publishing Corp., in
    press.

[7] T. J. Marks, Ann. NY Acad. Sci., 313, 594 (1978).

[8] M. A. Cowie, A. Gleizes, G. W. Grynkewich, D. W. Kalina,
    M. S. McClure, R. P. Scaringe, R. C. Teitelbaum, S. L. Ruby,
    J. A. Ibers, C. R. Kannewurf, and T. J. Marks, J. Am. Chem.
    Soc., 101, 2921 (1979) and references therein.

[9] R. C. Teitelbaum, S. L. Ruby, and T. J. Marks, J. Am. Chem.
    Soc., 102, 3322 (1980).

[10] R. C. Teitelbaum, S. L. Ruby, and T. J. Marks, J. Am. Chem. Soc., 101, 7568 (1979).

[11] D. W. Kalina, J. W. Lyding, M. S. McClure, C. R. Kannewurf, and T. J. Marks, J. Am. Chem. Soc., in press.

[12] L. D. Brown, D. W. Kalina, M. S. McClure, S. L. Ruby, S. Schultz, J. A. Ibers, C. R. Kannewurf, T. J. Marks, J. Am. Chem. Soc., 101, 2937 (1979).

[13] A. Gleizes, T. J. Marks, and J. A. Ibers, J. Am. Chem. Soc., 97, 3545 (1975).

[14] L.S. Lin, T. J. Marks, C. R. Kannewurf, J. W. Lyding, M. S. McClure, M. T. Ratajack, and T.-C. Whang, submitted for publication; Bull. Am. Phys. Soc., 25, 315 (1980).

[15] J. L. Petersen, C. S. Schramm, D. R. Stojakovic, B. M. Hoffman, and T. J. Marks, J. Am. Chem. Soc., 99, 286 (1977).

[16] C. S. Schramm, D. R. Stojakovic, B. M. Hoffman, and T. J. Marks, Science, 200, 47 (1978).

[17] R. P. Scaringe, C. J. Schramm, D. R. Stojakovic, B. M. Hoffman, J. A. Ibers, and T. J. Marks, J. Am. Chem. Soc., in press.

[18] S. K. Wright, C. J. Schramm, T. E. Phillips, D. M. Scholler, and B. M. Hoffman, Synth. Met., 1, 43 (1979).

[19] R. C. Teitelbaum, T. J. Marks, and C. K. Johnson, J. Am. Chem. Soc., 102, 2986 (1980), and references therein.

[20] K. F. Schoch, Jr. and T. J. Marks, unpublished results at Northwestern University.

[21] K. F. Schoch, Jr., B. R. Kundalkar, and T. J. Marks, J. Am. Chem. Soc., 101, 7071 (1979).

[22] T.J. Marks, K. F. Schoch, Jr., and B. R. Kundalkar, Synth. Met., 1, 337 (1980).

[23] C. W. Dirk, J. W. Lyding, K. F. Schoch, Jr., C. R. Kannewurf, and T. J. Marks, Polymer Preprints, in press.

[24] C. W. Dirk, K. F. Schoch, Jr., and T. J. Marks, manuscript in preparation.

[25] R. J. Joyner and M. E. Kenney, Inorg. Chem., 82, 5790 (1960).

[26] J. B. Davison and K. J. Wynne, Macromolecules, 11, 186 (1978), and references therein.

[27] D. R. Swift, Ph.D. Thesis, Case Western Reserve University, 1970.

[28] R. D. Shannon, Acta Cryst., A32, 751 (1976).

[29] C. Glidewell and D. C. Liles, J. Organometal. Chem., 174, 275 (1979), and references therein.

[30] E. A. Mintz and T. J. Marks, unpublished results at Northwestern University.

[31] R. C. Teitelbaum, Ph.D. Thesis, Northwestern University, August 1979.

[32] K. Seeger, Semiconductor Physics, Springer-Verlag, NY, 1973, pp. 483-487.

[33] L. B. Coleman, Rev. Sci. Instrum., 49, 48 (1978).

[34] K. F. Schoch, Jr., J. W. Lyding, C. R. Kannewurf, and T. J. Marks, manuscript in preparation.

[35] J. P. Lowe, J. Am. Chem. Soc., 102, 1262 (1980).

[36] H. A. J. Groves and C. G. DeKruf, Acta Crystal., A36, 428 (1980).

[37] A. J. Epstein, N. O. Lipari, D. J. Sandman, and P. Nielsen, Phys. Rev. B, 13, 1569 (1976).

[38] R. M. Metzger, Ann. NY Acad. Sci., 313, 145 (1978) and references therein.

[39] B. D. Silverman, Phys. Rev. B, 16, 5153 (1977).

[40] L. Pace, A. Ulman, and J. A. Ibers, private communication.

[41] K. F. Schoch, Jr. and T. J. Marks, unpublished results; Bull. Am. Phys. Soc., 25, 315 (1980).

[42] R. P. Van Duyne, M. R. Suchanski, J. M. Lakovits, A. R. Siedle, K. D. Parks, and T. M. Cotton, J. Am. Chem. Soc., 101, 2832 (1979), and references therein.

[43] K. F. Schoch, Jr. and T. J. Marks, unpublished results at Northwestern University.

[44] D. W. Kalina, D. R. Stojakovic, and T. J. Marks, unpublished results at Northwestern University.

[45] T. J. Marks and D. R. Stojakovic, J. Am. Chem. Soc., 100, 1695 (1978).

[46] C. W. Dirk and T. J. Marks, unpublished results at North-
     western University.

[47] J. N. Esposito, L. E. Sutton, and M. E. Kenney, Inorg. Chem.,
     6, 1116 (1967).

[48] R. D. Joyner and M. E. Kenney, Inorg. Chem., 1, 717 (1962).

[49] E. C. Bissel, Ph.D. Thesis, Case Western Reserve University,
     1970.

[50] P. M. Kuznesof, K. J. Wynn, R. S. Nohr, and M. E. Kenney,
     J. Chem. Soc., Chem. Comm., 121 (1980).

[51] R. S. Nohr, K. J. Wynne, and M. E. Kenney, Polymer Preprints,
     in press. We thank these authors for a preprint.

[52] C. W. Dirk and T. J. Marks, Bull. Am. Phys. Soc., 25, 315
     (1980).

Conducting Polymers: Partially Oxidized Bridge-Stacked Metallophthalocyanines

P. M. KUZNESOF

Department of Chemistry
Agnes Scott College
Decatur, Georgia 30030

R. S. NOHR

Department of Chemistry
George Mason University
Fairfax, Virginia 22030

K. J. WYNNE

Chemistry Program
Office of Naval Research
Arlington, Virginia 22117

M. E. Kenney

Department of Chemistry
Case Western Reserve University
Cleveland, Ohio 44106

ABSTRACT

The synthesis and characterization of a novel class of
polymeric phthalocyanines (Pc), $(PcMX)_n$ (M=Al,Ga,Cr; X=F
and M=Si,Ge,Sn; X=O) of exceptional thermal stability are
summarized. These materials possess a linear MX backbone
surrounded by a sheath of cofacial M-centered Pc rings.

299

$(PcAlF)_n$ and $(PcGaF)_n$ are sublimable (10^{-3}mmHg,540°C and 430°C, respectively) allowing for thin film formation. Iodine-doping leading to compositions $(PcMXI_y)_n$ with y ranging from 0.06 to 5.5 is reported. Thermogravimetric analysis has proven useful for iodine analyses and has revealed that the order of thermal stability with regard to loss of iodine is $(PcCrFI_y)_n$ < $(PcGaFI_y)_n$ ~ $(PcAlFI_y)_n$ < $(PcSiOI_y)_n$. Raman spectra point to I_3^- and I_5^- as the principle polyiodide species, though their relative proportions vary depending on M and doping level. Increases in the electrical conductivity by as much as 10^9 with maximum conductivities in the range of 0.01-1 $ohm^{-1}cm^{-1}$ result from iodine doping. Conduction appears to be thermally activated (77-300K) with an apparent activation energy of 0.04eV. It is likely that electron transport is primarily ligand based and is metal-like in character.

INTRODUCTION

Development of new semiconducting and conducting organic and organometallic polymers continues to be sparked by the multitude of potential applications in solid state electro-optical systems (1). These advanced technology materials may assume prime roles in devices such as sensors, detectors, and electro-photographic units where low density, combined with thermal, photo and chemical stability will be demanded. The use of chemically modified electrode surfaces for studies of electro-chemical electron transfer, electrocatalysis, and photoassisted electrode reactions is also undergoing intensive exploration (2). Investigation of these processes employing coatings of electroactive polymeric materials can be expected to be highly rewarding. Currently, widespread application of these new substances is limited by the necessity for specialized film deposition procedures and the lack of materials which demonstrate ease of fabrication and processibility. Evidence of progress in these areas is beginning to appear (3).

Phthalocyanines (4), because of their thermal, photo, and hydrolytic stabilities, intense colors, and ready availability

have received their share of attention in the quest for new
electroactive materials (5-14). The purpose of this account
is to summarize recent studies on a novel class of bridge-
stacked phthalocyanine polymers (Figure 1) which have been made
highly conducting ($\sigma_{RT} \sim 1$ ohm^{-1}cm^{-1}) by halogen (I_2,Br_2))
doping (5-7,10,11). These systems meet criteria established
for the preparation of new conductive materials in that (a) the
metal-macrocycle units are stacked cofacially and (b) these
stacks can be partially oxidized so that each unit is formally
in a non-integral oxidation state. Furthermore, the bridged
polymers are unique in that they are exceptionally robust both
chemically and thermally. This stands in contrast to simple
metallomacrocyles for which there is no guarantee of stacking

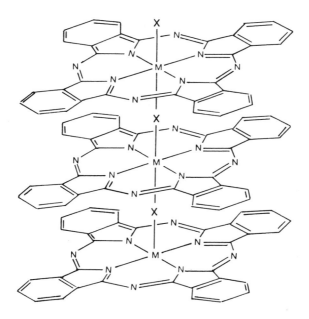

FIGURE 1. Proposed Structure for (PcMX)$_n$ (M=Al,Ga,Cr;X=F and
M=Si,Ge,Sn;X=O).

ab initio in linear arrays or of remaining stacked upon treatment
with typical oxidizing or reducing agents.

SYNTHESIS and CHARACTERIZATION

The phthalocyanine polymers $(PcMX)_n$ (M=Si,Ge,Sn, X=O;
M=Al,Ga, X=F) were originally prepared by Kenney and
co-workers (15-18). The oxo-bridged group IVA phthalocyanines
(18) are easily obtained in high yield and purity by vacuum
polymerization (300 - 400°C) of the dihydroxo precursor:

$$nPcM(OH)_2 \xrightarrow{-nH_2O} (PcMO)_n \qquad M=Si,Ge,Sn.$$

$(PcSiO)_n$ is an exceptionally stable material: it is unaffected
by aqueous HF at 100°C, aqueous 2M NaOH at reflux and H_2SO_4 at room
temperature (18). The fluorobridged aluminum and gallium
phthalocyanines are prepared (15,16) first by converting
PcAlCl and PcGaCl to their hydroxo derivatives followed by
repeated evaporation to dryness with concentrated (48%) aqueous
hydrofluoric acid:

$$PcMCl \xrightarrow[\text{reflux}]{NH_3(aq), \text{ py}} PcMOH \cdot xH_2O$$

$$nPcMOH \cdot xH_2O \xrightarrow{48\% \text{ HF}} (PcMF)_n \qquad M=Al,Ga.$$

Heating these substances several hours at 300°C in vacuo
results in good yields of satisfactory purity. High purity
$(PcMF)_n$ is obtainable by vacuum sublimation (540°C, M=Al;
430°C, M=Ga). $(PcCrF)_n$ is available commercially (Eastman)
or may by synthesized from $Cr(CO)_6$ and phthalonitrile (19)
followed by hydrofluoric acid treatment of the resulting
PcCrOH (7).

The extremely high thermal stabilities of the $(PcMX)_n$
compounds are apparent from thermogravimetric analysis (TGA)

investigations (5-7). The onset of initial weight loss (i.e. 5% loss in weight) was recorded at ca. 430°, 450°, 490°, and 540°C for $(PcSiO)_n$, $(PcCrF)_n$, $(PcGaF)_n$, and $(PcAlF)_n$, respectively. A second weight loss between 600-700°C occurs for all of these compounds after a 35-40% decrease in initial sample weight.

A variety of physical evidence points to the structure shown in Figure 1 for the oxo- and fluorometallophthalo-cyanines. Certain relatively broad bands in the infrared spectra have been attributed to M-X stretching along the $(MX)_n$ polymer backbone (16-18). Interplanar spacings which were derived from x-ray powder diffraction data are compared in Table 1. The expected increase in ring-ring distance upon descending a family is evident and there is excellent agreement with the distances calculated from ionic radii (11,16) assuming linear M-X-linkages.

From electron microscopy studies (15, 16) on sublimed crystals of $(PcAlF)_n$ and $(PcGaF)_n$, micrographs were obtained which exhibited lattice image lines. These lines, spaced by ca. 1300 to 1500 pm - which corresponds to the van der Waals

TABLE 1.

Ring-Ring (M-X-M) Spacing (pm) for $(PcMX)_n$

	$(PcSiO)_n$	$(PcGeO)_n$	$(PcSnO)_n$	$(PcAlF)_n$	$(PcGaF)_n$	$(PcCrF)_n$
obs:	332(2)[a]	350(2)[a]	383(3)[a]	366[b]	386[b]	387[c]
obs[d]:	333(2)	351(2)	395(2)			
calc[e]:	332[d]	358[d]	390[d]	364[b]	381[b]	380

(a) Ref. 18. (b) Ref. 16. (c) Ref. 7. (d) Ref. 11.
(e) Calculated using the pertinent ionic radii: R. D. Shannon, Acta. Crystallogr., A, **32**, 751(1976).

width of the phthalocyanine ring (16) - are oriented parallel
to the long axis of the needle-like crystals. This suggests
that linear chains of $(PcAlF)_n$ and $(PcGaF)_n$ are packed
parallel to the long crystal axis and indicates that the
$(PcAlF)_n$ structure may be regarded as pseudo-one dimensional.

Halogen doping (partial oxidation) of low dimensional
materials has proved to be an effective strategy for transform-
ing a normally insulating (or poorly semiconducting) substance
(σ_{RT} <10^{-6}ohm^{-1}cm^{-1}) to a conductive material (σ_{RT} >10^{-2}
ohm^{-1}cm^{-1}). Uptake of iodine by $(PcMX)_n$ to yield $(PcMXI_y)_n$
has been accomplished (5-7,10,11) either by exposure of the
powdered material to iodine vapor or by stirring slurries of
the phthalocyanine in iodine-saturated solutions of heptane,
carbon tetrachloride, ethanol, or chlorobenzene. Depending on
reaction conditions and $(PcMX)_n$ purity, a wide range of
dopant concentrations can be obtained. Thus, I/M ratios of
0.061 for one sample of $(PcCrF)_n$ (7) to 5.5 for a sample of
$(PcSnO)_n$ (11) have been reported. Conductivity data for
maximum halogen/M compositions are summarized in Table 2.
For the Al, Ga and Cr derivatives, maximum iodine uptake is
achieved within 24 hours if the reaction is carried out using
the slurry technique (5-7) whereas the solid/vapor route
requires one to three weeks (5,6).

Physical methods including TGA, IR, Raman, mass spectrometry,
and magnetic susceptibility measurements have been applied to
the characterization of these materials. TGA has been shown to
be particularly useful for determining the iodine content of
doped oxo- and fluorphthalocyanines (5-7). The thermograms
in Figure 2 for pure $(PcAlF)_n$ (curve D) and at several iodine
doping levels are typical. Thus, thermogram A exhibits two
weight losses between room temperature and the onset of the
plateau at 240°C which signals complete loss of iodine. The

TABLE 2.

Composition and Conductivity Data for Halogen-Doped $(PcMX)_n$[a]

Compound	$(ohm^{-1}cm^{-1})$[b]	$E_a(eV)$
$(PcAlFI_{3.4})_n^c$	0.59(0.51)	0.03
$(PcGaFI_{2.1})_n^c$	0.15(0.11)	0.04
$(PcCrFI_{3.2})_n^d$	0.62	f
$(PcSiOI_{4.60})_n^e$	$(0.1)^i$	0.04[g]
$(PcSiOBr_{1.00})_n^e$	(0.06)	f
$(PcGeOI_{2.0})_n^e$	(0.1)	0.05[h]
$(PcSnOI_{5.5})_n^e$	$(2x10^{-4})$	0.68

(a) Data are present only for maximally doped samples.
(b) Room temperature linear four point probe and van der Pauw,
( ), conductivities, pressed discs. (c) References 5,6.
(d) Reference 7. (e) Reference 11. (f) Temperature dependence of conductivity not reported. (g) for I/Si ratio of
1.40. (h) for I/Ge ratio of 1.94. (i) A conductivity of 0.2
$ohm^{-1}cm^{-1}$ is reported for $(PcSiOI_{1.4})_n$.

44% weight loss to this point translates to I/Al = 3.4. Above
240°C the thermogram is identical to that for pure $(PcAlF)_n$.
The iodine-doped gallium derivatives exhibit weight loss
processes in the same temperature regimes as the aluminum
species. Thermograms for $(PcCrFI_y)_n$ (7) also reveal
evidence for two weight loss processes at 130 and 170°C prior
to removal of all iodine. The weight loss-temperature profiles
for two moderately doped samples of $(PcSiO)_n$ (I/Si = 1.3 and
and 1.4) show a low temperature (ca. 80°) loss of iodine, but
the second stage loss is not completed until ca. 340°C (6,20).
In terms of thermal stability with regard to loss of iodine,
the order is $(PcCrFI_y)_n$ < $(PcGaFI_y)_n$ ~ $(PcAlFI_y)_n$

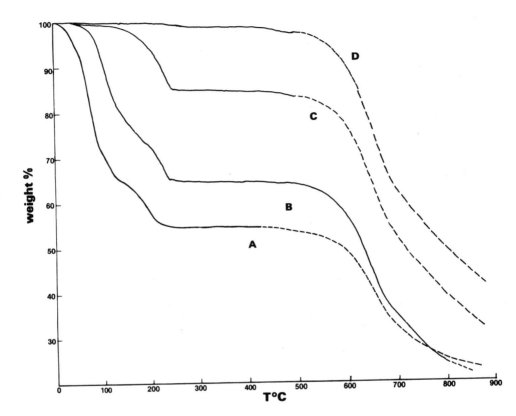

FIGURE 2. Thermograms for (A) $(PcAlFI_{3.4})_n$, (B) $(PcAlFI_{2.4})_n$, (C) $(PcAlFI_{0.75})_n$ and (D) $(PcAlF)_n$. Heating rate 5°/min (——) and 10°/min (---), N_2 flow rate 50 ml/min.

$< (PcSiOI_y)_n$. It is worth noting that all doped samples of the Al, Ga, and Cr fluorophthalocyanines including those brought to constant weight under dynamic vacuum, evolved iodine when stored at ambient temperature and pressure, as evidenced by the discoloration of the polyethylene caps of the storage vials (6). This stands in contrast to the report that the iodine-doped Si, Ge, and Sn oxophthalocyanines are

indefinitely stable in air, and that iodine can only be
driven off by prolonged heating above $100^{\circ}C$ (11).

Raman spectroscopy has proved indispensable for characterizing the polyiodide species formed upon oxidation of a wide
variety of low dimensional materials (9,14,20,21). Raman data
for the doped oxo-bridged group IVA phthalocyanines (11) show
I_3^- to be the principle polyiodide species; at moderate to
high dopant levels, scattering due to I_5^- can be observed.
On the other hand, for all $(PcCrFI_y)_n$ compositions scattering attributable to I_5^- dominates the spectrum although
a band due to I_3^- is clearly evident (7). The spectra for
highly doped $(PcAlF)_n$ and $(PcGaF)_n$ exhibit intense scattering for both I_3^- and I_5^- with the former dominant
(6). The presence of two polyiodide species shown by Raman
spectroscopy correlates with TGA data such as shown in Figure 2
for $(PcAlFI_y)_n$. Curve B shows the TGA of $(PcAlFI_{2.4})_n$.
Two weight loss processes at 90 and $200^{\circ}C$ are apparent. After
heating this sample to $90^{\circ}C$ under dynamic vacuum $(PcAlFI_{0.75})$
was obtained. In contrast to the parent composition, little
weight loss is observed below $150^{\circ}C$ for this composition
(Figure 2, Curve C) and the Raman spectrum shows a less prominent I_5^- peak relative to I_3^-. Thus, the combination of TGA
and Raman data indicate that the first weight loss process for
the heavily doped materials can be primarily associated with
I_5^- decomposition; the second stage is prinicpally due to
loss of I_3^-. One particular composition, $(PcGaFI_{0.62})_n$,
exhibited a thermogram with a single broad weight loss beginning
at ca. $60^{\circ}C$. The Raman spectrum for this sample showed the
I_5^- peak to be dominant, consistent with the recorded
thermogram. To summarize, all work reported to date on iodine-
doped oxo- and fluorophthalocyanines (5-12) shows I_3^- and
I_5^- to be the primary polyiodide species present, presumably

occupying channels between the $(PcMX)_n$ chains. No evidence
for I_2 (detectable by Raman spectroscopy) has been noted and,
although Mössbauer spectra have not been reported for
$(PcMXI_y)_n$, there does not seem to be any precedence for
significant concentrations of I^- in any iodine doped conducting phthalocyanine material.

For iodine-doped $(PcAlF)_n$ and $(PcGaF)_n$, variable
temperature (ambient to 300°C) mass spectral data were
collected (6) which confirmed iodine as the predominant
volatile species during thermolysis. Profiles of I_2^+ ion
intensity vs temperature also showed I_2 was lost in two
stages, in agreement with the Raman and TGA data cited above.

ELECTRICAL CONDUCTIVITY

$(PcMX)_n$ compounds are typical insulators, but partial
oxidation results in increases in conductivity by as much as 9
orders of magnitude compared to the undoped material (5-7,
10,11). The conductivity rises rapidly as the I/M ratio
approaches 0.2 with relatively slight increases thereafter up
to the maximum doping level. Room temperature conductivities
(compressed discs) for these compositions are collected in
Table 2 and, with the exception of the tin phthalocyanine,
fall between 0.01 and 1 $ohm^{-1}cm^{-1}$.
Trends within a family are apparent if conductivities for a
given I/M ratio are compared. Thus, σ (Al) $> \sigma$ (Ga); σ (Si)
$> \sigma$ (Ge) $> \sigma$ (Sn). These trends correlate with the inter-
phthalocyanine ring spacings given in Table 1 (smaller
spacing, higher conductivity) and support the idea that the
conductive pathway is primarily ligand based via a conduction
band generated by inter-ring pi-orbital overlap. These systems,
then, along with related highly conductive stacked metallomacrocyclic systems (22), stand in contrast to Krogmann salts

and related compounds which are low dimensional conductors in which charge transport occurs mainly along a transition metal backbone.

The variation of conductivity with temperature (77–300K) has been examined (6,11) for various $(PcMXI_y)_n$ compositions, ($M{\neq}Cr$). The four-probe studies indicated that conduction is thermally activated, and the linearity of $\log \sigma$ vs $1/T$ plots allowed for evaluation of the apparent activation energies listed in Table 2. The remarkably low activation energies and high conductivities are comparable to those observed for compressed discs of $PcNiI_{1.0}$, single crystals of which were found to behave as a "molecular metal" (8). Furthermore, Faraday-type magnetic susceptibility data collected for the iodine-doped Si, Ge, and Sn compounds (11) appear consistent with magnetic behavior typical of many low-dimensional mixed valence systems – weak paramagnetism at 300K and only weakly temperature dependent.

SUMMARY

Bridge-stacked $(PcMX)_n$ compounds constitute a new class of low-dimensional materials which can be made highly conducting by halogen doping. That this conductivity is probably metal-like along the chain direction can be inferred from the physical studies which have been performed and from studies of related systems for which metallic conduction has actually been observed. Although iodine-doped $(PcAlF)_n$ and $(PcGaF)_n$ appear to have lower thermal stability with regard to loss of iodine compared to iodine-doped $(PcSiO)_n$, the volatility of the former materials, due to polymer chain linkage via coordinate covalent bonds, allows for formation of thin films. Because of this processing advantage $(PcMF)_n$ in films may find application in solid state devices or as electro-optic materials.

310   KUZNESOF ET AL.

ACKNOWLEDGEMENT

We thank the Office of Naval Research for partial support of this work.

REFERENCES

(1) For recent reviews on low-dimensional materials and
    conducting polymers see: (a) "Molecular Metals," W. E.
    Hatfield, ed., NATO Conference Series VI, vol 1, Plenum
    Press, NY, 1979; (b) "Chemistry and Physics of One-
    Dimensional Metals," H. J. Keller, ed., NATO Advanced
    Study Institute Series, B-25, Plenum Press, NY, 1977;
    (c) J. S. Miller and A. J. Epstein, Prog. Inorg. Chem.,
    20, 1 (1976); (d) "Synthesis and Properties of Low-
    Dimensional Materials," J. S. Miller and A. J. Epstein,
    Eds., Ann. N.Y. Acad. Sci., 313 (1978); (e) J. B.
    Torrance, Accts Chem. Res., 12, 79 (1979); (f) A. F.
    Garito and A. J. Heeger, ibid., 7, 232 (1974); (g) E. P.
    Goodings, Chem. Soc. Revs., 5, 95 (1976); Endeavor, 34,
    123 (1975); (h) J. O. Williams, Adv. Phys. Chem., 16,
    159 (1978).

(2) R. W. Murray, Accts. Chem. Res., 13, 135 (1980); F. B.
    Kaufman, A. H. Schroeder, E. M. Engler, S. R. Kramer,
    and J. Q. Chambers, J. Amer. Chem. Soc., 102, 483 (1980).

(3) J. F. Rabolt, T. C. Clarke, K. K. Kanazawa, J. R. Reynolds,
    and G. B. Street, J. C. S. Chem. Comm., 347 (1980); R. R.
    Chance, L. W. Schacklette, G. G. Miller, D. M. Ivory, J.
    M. Sowa, R. L. Elsenbaumer, and R. H. Baughman, ibid.,
    348 (1980).

(4) A. B. P. Lever, Adv. Inorg. Chem. Radiochem., 7, 27
    (1965); G. Booth in "The Chemistry of Synthetic Dyes,"
    vol 5, K. Venkataraman, ed., Academic Press, NY, 1973,
    Chapt. IV; A. A. Berlin and A. J. Sherle, Inorg. Macromol.
    Rev., 1, 235 (1971).

(5) P. M. Kuznesof, K. J. Wynne, R. S. Nohr, and M. E.
    Kenney, J. C. S. Chem. Comm., 121 (1980).

(6) P. M. Kuznesof, K. J. Wynne, P. G. Siebenmann, R. S.
    Nohr, and M. E. Kenney, J. Amer. Chem. Soc., in press.

(7)   R. S. Nohr, K. J. Wynne, and M. E. Kenney, Second Chemical
      Congress of the North American Continent, San Francisco,
      CA, August, 1980.

(8)   C. J. Schramm, D. R. Stojakovic, B. M. Hoffman, and
      T. J. Marks, Science, 200, 47 (1978).

(9)   J. L. Peterson, C. J. Schramm, D. R. Stojakovic,
      B. M. Hoffman, and T. J. Marks, J. Amer. Chem. Soc., 99,
      286 (1977).

(10) K. J. Schoch, Jr., T. J. Marks, B. R. Kundalkar, L.-S.
      Lin, and R. C. Teitelbaum, Bull. Am. Phys. Soc., 24, 326
      (1979); T. J. Marks, B. R. Kundalkar, L.-S. Lin, and K.
      F. Schoch, Jr., "IBM Symposium on the Structure and
      Properties of Highly Conducting Polymers and Graphite,"
      IBM, San Jose, CA, March, 1979.

(11) K. F. Schoch, Jr., B. R. Kundalkar, and T. J. Marks,
      J. Amer Chem. Soc., 101, 7071 (1979).

(12) W. A. Orr and S. C. Dahlberg, ibid, 101, 2875 (1979).

(13) J. H. Brannon and D. Magde, ibid, 102, 62 (1980).

(14) B. M. Hoffman, T. E. Phillips, C. J. Schram and S. K.
      Wright, Reference 1a, p. 393.

(15) J. P. Linsky and T. R. Paul, Ph.D. Theses, Case Western
      Reserve University, 1970 and 1971, respectively.

(16) J. P. Linsky, T. R. Paul, R. S. Nohr, and M. E. Kenney,
      Inorg Chem., in press.

(17) R. D. Joyner and M. E. Kenney, J. Amer. Chem. Soc.,
      82, 5790 (1960); J. E. Owen and M. E. Kenney, Inorg.
      Chem., 1, 334 (1962); R. D. Joyner and M. E. Kenney,
      ibid., 1, 717 (1962).

(18) W. J. Kroenke, L. E. Sutton, R. D. Joyner, and M. E.
      Kenney, Inorg. Chem., 2, 1064 (1963).

(19) E. G. Maloni, L. R. Ocone, and B. P. Block, Inorg.
      Chem., 6, 424 (1967).

(20) W. Kiefer, Appl. Spectrosc., 28, 115 (1974) and reference therein.

(21) A. Gleizes, T. J. Marks, and J. A. Ibers, J. Amer. Chem. Soc., 97, 3545 (1975); M. A. Cowie, A. Gleizes, G. W. Grynkewich, D. W. Kalina, M. S. Mclure, R. P. Scaringe, R. C. Teitelbaum, S. L. Ruby, J. A. Ibers, C. R. Kannerwurf, and T. J. Marks, ibid., 101, 2921 (1979); T. J. Marks reference 1d, p. 594; R. C. Teitelbaum, S. L. Ruby, and T. J. Marks, J. Amer. Chem. Soc., 100, 3215 (1978).

(22) T. E. Phillips, R. P. Scaringe, B. M. Hoffman, and J. A. Ibers, J. Amer. Chem. Soc., 102, 3435 (1980) and references therein.

Synthesis, Structure, and Properties of One-Dimensional Partially Oxidized Tetracyanoplatinate Complexes

Arthur J. Schultz, Jack M. Williams and Richard K. Brown
Chemistry Division
Argonne National Laboratory
9700 South Cass Avenue
Argonne, Illinois 60439

ABSTRACT

The crystal structure of anion-deficient partially oxidized tetracyanoplatinate complexes generally can be classified into one of two categories. The electrical properties of compounds in the two categories are also observed to be different. We have prepared a new complex which does not fall into either of the previous categories. Its synthesis, structure and properties are described.

As recently as 1968, Krogmann [1] reported the preparation, x-ray crystal structure and chemical bonding analysis of the unusual anion-deficient complex $K_2[Pt(CN)_4]Br_{0.3} \cdot 3H_2O$, KCP(Br). Although partially oxidized tetracyanoplatinate (POTCP) complexes had been reported over 100 years ago [2], the "rediscovery" of these metallic appearing materials by Krogmann came at a time when interest in one-dimensional (1-D) conductivity was rapidly increasing. This was due, in part, to a theory of 1-D high-temperature superconductivity proposed by Little [3]

---

*Work performed under the auspices of the Office of Basic Energy Sciences of the U. S. Department of Energy

313

in 1964. In the intervening years, an expanding number of POTCP
complexes have been synthesized and studied [4]. In this article,
we describe some recent findings and new conclusions concerning
1-D metallic properties in Pt-chain forming complexes.

A common property of all POTCP materials studied to date is
that they contain linear, or nearly linear, Pt-atom chains, as
depicted schematically in Figure 1. Typical intrachain Pt-Pt
separations are in the range of 2.80-2.95 Å, which may be compared with 2.78 Å in Pt metal, with interchain separations of
~10 Å. The removal of a fractional portion of the electrons from
the 1-D band formed by the overlapping Pt d_{z^2} orbitals is
responsible for the unusual 1-D metallic or semiconducting properties exhibited by these compounds. The partial oxidation is
stabilized in the solid state by anion deficiency, as in KCP(Br),
or by cation deficiency, as in $K_{1.75}[Pt(CN)_4] \cdot 1.5H_2O$, K(def)TCP.

An important finding relevant to the structure-property
phenomena in these salts is that the crystal structures of the
anion deficient materials generally can now be classified into

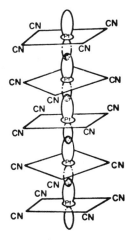

FIG. 1. Schematic diagram of overlapping Pt d_{z^2} orbitals which form
         the metallic "chain".

one of two categories. The hydrated compounds crystallize in the noncentrosymmetric primitive tetragonal space group P4mm, which contains cations between the $Pt(CN)_4$ groups in the "upper-half" of the unit cell while the water molecules reside in the "lower-half", as depicted in Figure 2 for KCP(Br) [5-7]. Thus the Pt-Pt intra-chain distance can be alternatively compressed or expanded in a pairwise fashion due to the asymmetric packing of cations and water molecules in the unit cell. This is clearly evident in $Rb_2[Pt(CN)_4]Cl_{0.3} \cdot 3H_2O$ [Pt-Pt = 2.924(8) and 2.877(8) Å] [8] and $(NH_4)_2[Pt(CN)_4]Cl_{0.3} \cdot H_2O$ [Pt-Pt = 2.910(5) and 2.930(5) Å] [9].

The second class of anion deficient POTCP complexes are usually anhydrous and crystallize in the centrosymmetric body-centered tetragonal space group 14/mcm. With the exception of $Cs_2[Pt(CN)_4]Cl_{0.3}$ [10] and the azide salt $Cs_2[Pt(CN)_4](N_3)_{0.25} \cdot 0.5H_2O$ [11], these materials contain the linear (assumed) triatomic bifluoride anion (FHF)⁻ [12]. An illustration of this structure type for $Rb_2[Pt(CN)_4](FHF)_{0.40}$ [13] is given in Figure 3. The

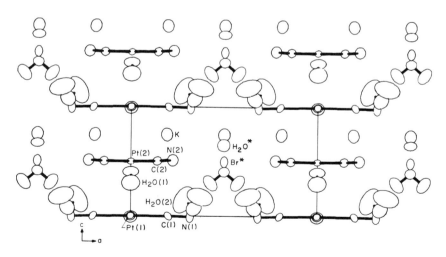

FIG. 2. A b-axis half-cell projection of the structure of $K_2[Pt-(CN)_4]Br_{0.3} \cdot 3H_2O$.

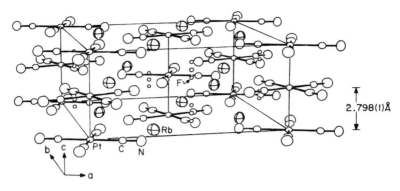

FIG. 3. Perspective view of the unit cell of $Rb_2[Pt(CN)_4](FHF)_{0.40}$.
The Pt-Pt spacing is the shortest of any known POTCP salt.

cations reside in the plane of the $Pt(CN)_4$ groups, which serves to
tie the $Pt(CN)_4$ groups together directly through $CN-M^+-NC$ interactions, rather than between planes (see Fig. 2). In addition, all
Pt-Pt distances are required to be crystallographically equivalent.

An important and common property displayed by all POTCP complexes is the existence of modulated lattice distortions arising
from electron-phonon coupling. At low temperature, a 3-D ordering
of the lattice instabilities occurs (at T_{3D}) and leads to an
electrically insulating state (Peierls insulator) [4]. Detailed
analysis by Wood and Underhill [14] of the conductivity data shown
in Figure 4 reveals that T_{3D} occurs at $\sim100°K$ for KCP(Br), and
$\sim73°K$ for $CsCP(FHF)_{0.4}$. Thus, we are able to conclude that in relations to the *hydrated* POTCP compounds, the anhydrous complexes
generally have higher degrees of partial oxidation (DPO), smaller
Pt-Pt repeat distances, higher conductivities at all temperatures,
and lower 3-D ordering temperatures. The last property may be
due to the absence of interchain hydrogen bonding interactions in
the anhydrous crystals, which results in less interchain coupling
and higher anisotropy.

Finally, we have recently prepared a new POTCP salt $Rb_3[Pt-$
$(CN)_4](O_3SO·H·OSO_3)_{0.49}·H_2O$, RbCP(DSH), which is highly unusual

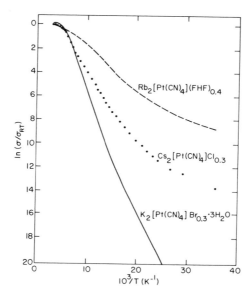

FIG. 4. The temperature dependence of ratio of the conductivity to the room temperature conductivity.

for many reasons. It contains a large polyatomic trianion and a ratio of three alkali metal ions per Pt atom, instead of the usual two. The crystals are triclinic ($\overline{P1}$), have a short Pt-Pt repeat distance (2.826(1) Å), and a relatively high room temperature conductivity parallel to the Pt-atom chain (2000 ohm^{-1} cm^{-1}). This conductivity is approximately triple that of KCP(Br). Although the material is hydrated, the water molecules do not couple the chains of Pt(CN)$_4$ stacks, but hydrogen bond to the disulfatohydrogen anions, linking them together parallel to the Pt chain direction (see Fig. 5). The physical properties of this new material are presently under study.

The DPO for Pt, derived from the least-squares refinement of the sulfate ion occupancy in the x-ray structure, is +0.49. From the stoichiometry derived from elemental analyses, the DPO is $\sim$0.47. A more reasonable value of $\sim$0.40, roughly predictable from the

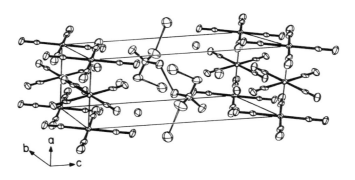

FIG. 5. Perspective view of the unit cell of $Rb_3[Pt(CN)_4](O_3SO\cdot H\cdot OSO_3)_{0.49}\cdot H_2O$. Hydrogen bonding interactions are indicated by unshaded bonds and the O-H-O separation in the disulfatohydrogen ion is 2.52(2) Å. The Rb ions are shown without any bonds drawn to them.

Pt-Pt distance, may be accommodated by the addition of $\sim$0.1 H^+ ions per Pt to form HSO_4^- or H_3O^+ ions. Because of the highly acidic medium used in the preparation, this is not entirely unexpected.

EXPERIMENTAL

Electrochemical [15] Synthesis of RbCP(DSH): A 0.8 M solution of Rb_2SO_4 is prepared by adding 1.34 g of Rb_2SO_4 to 6 ml of water. The solution is then saturated with 1.08 g of $Rb_2[Pt(CN)_4]\cdot1.5H_2O)$ and acidified to a pH of less than 1 with 0.25 ml of 9 M H_2SO_4. The clear solution is transferred to a 50 ml polyethylene beaker with Pt electrodes and electrolyzed at 0.8 volts for 72 hours. The reddish-bronze crystals were isolated by filtration and washed with cold water.

Crystal Structure Analysis of RbCP(DSH): The compound crystallizes in the triclinic space group $P\bar{1}$ with cell constants of a = 5.652(1) Å, b = 9.372(2) Å, c = 13.762(2) Å, α = 71.19(1)°, β = 81.94(2)°, γ = 73.44(1)°, V_c = 660.40 Å^3 and z = 2. The structure was solved with x-ray diffraction data obtained on a syntex $P2_1$ diffractometer using MoKα radiation.

REFERENCES

[1]   Krogmann and H. D. Hausen, Z. Anorg. Allg. Chem., 358, 67 (1968); K. Krogmann, Angew. Chem. Internat. Edit., 8, 35 (1969).

[2]   W. Knop, Justus Liebig's Ann. Chem., 43, 111 (1842).

[3]   W. A. Little, Phys. Rev. A 134, 1416 (1964).

[4]   For reviews, see J. S. Miller and A. J. Epstein, Prog. Inorg. Chem., 20, 1 (1976); G. D. Stucky, A. J. Schultz and J. M. Williams, Ann. Rev. Mater. Sci., 7, 301 (1977); J. S. Miller and A. J. Epstein, Eds., Synthesis and Properties of Low-Dimensional Materials, Annals N. Y. Acad. Sci., 313, (1978).

[5]   J. M. Williams, J. L. Petersen, H. M. Gerdes and S. W. Peterson, Phys. Rev. Lett., 33, 1079 (1974); J. M. Williams, M. Iwata, S. W. Peterson, K. A. Leslie and H. J. Guggenheim, Phys. Rev. Lett., 34, 1653 (1975).

[6]   C. Peters and C. F. Eagen, Inorg. Chem., 15, 782 (1976).

[7]   G. Heger, H. J. Deiseroth and H. Schultz, Acta Cryst. B34, 725 (1978).

[8]   J. M. Williams, P. L. Johnson, A. J. Schultz and C. C. Coffey, Inorg. Chem. 17, 834 (1978).

[9]   P. L. Johnson, A. J. Schultz, A. E. Underhill, D. M. Watkins, D. J. Wood and J. M. Williams, Inorg. Chem., 17, 839 (1978).

[10] R. K. Brown and J. M. Williams, Inorg. Chem., 17, 2607 (1978).

[11] R. K. Brown, D. A. Vidusek and J. M. Williams, Inorg. Chem., 18, 801 (1979). Although there is evidence of a small degree of hydration, and the actual space group is P4b2, the crystal structure of the azide salt is pseudo body centered and is nearly isomorphous with the I4/mcm compounds.

[12] J. M. Williams, A. J. Schultz, K. B. Cornett and R. E. Besinger, J. Am. Chem. Soc., 100, 5572 (1978).

[13] A. J. Schultz, C. C. Coffey, G. C. Lee and J. M. Williams, Inorg. Chem., 16, 2129 (1977).

[14] A. E. Underhill and D. J. Wood, private communication.

[15] R. E. Besinger, D. P. Gerrity and J. M. Williams, Inorg. Synth., in press.

CATALYSIS

Catalysis by Transition Metal Derivatives Bound to Structurally Ordered Polymers

Carlo CARLINI and Glauco SBRANA

Centro di Studio del C.N.R. per le Macromolecole Stereordinate ed

Otticamente Attive, Istituto di Chimica Organica Industriale,

Università di Pisa, Via Risorgimento 35, 56100 Pisa, Italy.

INTRODUCTION

It is well known that metalloenzymes are able to catalyze chemical reactions with a very high activity and selectivity[1-3].
Therefore the synthesis of complexes between transition metal derivatives and structurally ordered macromolecular ligands to give catalytic systems having high activity and stereoselectivity is of large interest from both applicative and speculative point of view. In this connection the main aim of this article is to emphasize the role of the constitutional and configurational order of the macromolecular ligand in determining the properties of the corresponding complexes with transition metal derivatives. This is of particular importance for designing catalytic systems displaying peculiar features as far as activity, selectivity and stereochemistry are concerned. Moreover a better insight into environment and coordination sphere of the polymer attached metal derivative

323

can be of great help for mechanistic studies of the reactions involved.

COMPLEXES OF TRANSITION METAL DERIVATIVES WITH POLY(α-AMINO ACID)S

A large amount of work has been carried out dealing with complexes of transition metal ions and synthetic low and high molec-
ular weight peptides|1,4|. However, few cases have been investigated
up to now concerning complexes of transition metal derivatives with
poly(α-amino acid)s, looking at possible relations between catalyt
ic, stereochemical properties and primary as well as secondary
structure of the polymer ligand.

Pecht et al. have reported|5,6| that poly(L-histidine)-Cu(II)
|PLHCu(II)| complexes are catalytically active in the homogeneous
oxidation of ascorbic acid(I), (2,5-dihydroxyphenyl)acetic acid(II)
and p.hydroquinone(III)(Scheme 1) in the range of pH 4.3-5.0, depending on the substrate used. PLHCu(II) complexes exhibited higher
catalytic activity than the aquo copper(II) complex, the activity
ratio being in the range 6.2-172 (Table 1).

The structure proposed by the authors|6,7|, on the basis of po-
tentiometric titrations and circular dichroism data, for the
PLHCu(II) complex formed at pH< 5 involves, in addition to three
imidazole nitrogens, at least one deprotonated peptide nitrogen at
the square planar coordination positions of Cu(II) ion. Recently|8|
Peggion et al. have better defined such structure showing that it
is not compatible with the β structure of poly(L-histidine) but
with a random coil polypeptide chain. According to these last results the higher activity of PLHCu(II) complexes seems to be attrib
uted not to a peculiar ordered conformational environment of the

TABLE 1

Catalytic Oxidation and Hydrolysis by Water Soluble Complexes of Copper(II) with Different Poly(α-amino acid)s

Catalyst	Reaction	pH	Substrate	Stereoselectivity index[a]	Activity ratio[b]	Ref.
PLHCu(II)	Oxidation	4.3	ascorbic acid	–	6.2	6
		5.0	(2,5-dihydroxyphenyl)-acetic acid	–	172	6
		4.8	p.hydroquinone	–	95	6
PLLCu(II)	Oxidation	10.5	DL- and L-3,4dihydroxy-phenylalanine	1.53	7[c]	9,10
		6.9	phenylalanine	1.00	2.9[d]	9
PLLCu(II)	Hydrolysis	7.0	D- and L-phenylalanine methyl and ethyl esters	2.41	552[e]	13

[a] Expressed as the ratio between initial oxidation rate or pseudo first order hydrolysis constants of D and L enantiomers.

[b] With respect to the aquo copper(II) complex.

[c] With respect to the Cu(ethylenediamine)$_2$ complex.

[d] With respect to the Cu(n.butylamine)$_4$ complex.

[e] With respect to the Cu(bipyridyl)$_2$ complex.

active sites but to the fact that the polyelectrolyte behaves as a
second phase in which the concentration of the reagents is higher
than that in the bulk of solution|6|.

Scheme 1

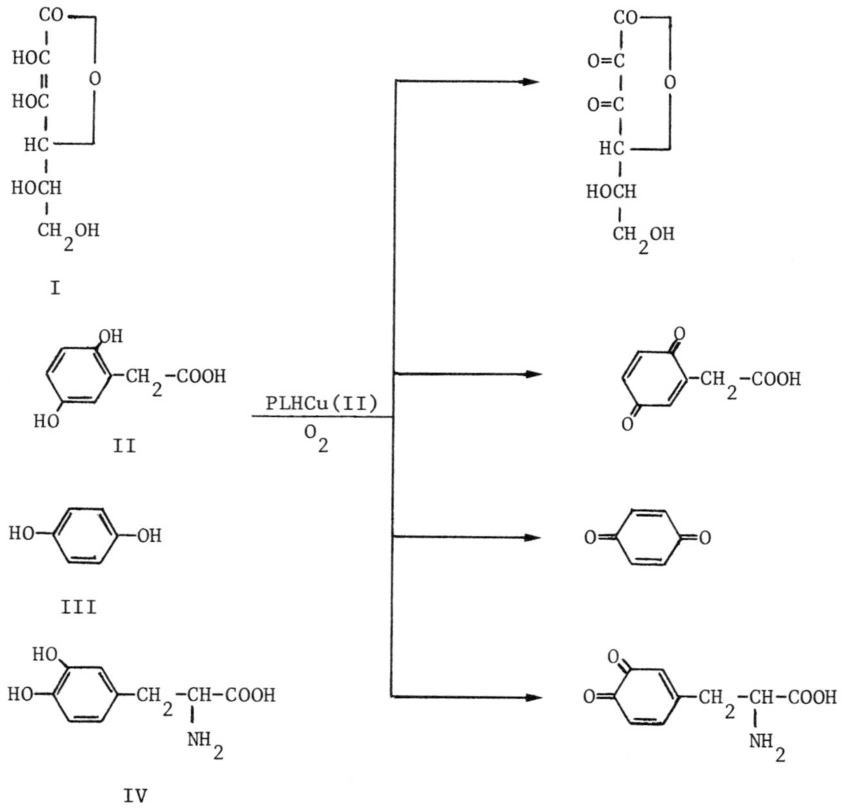

Hatano et al. showed that poly(L-lysine)-Cu(II)|PLLCu(II)| complex
exhibits catalytic activity|9,10| in the homogeneous oxidation of
3,4-dihydroxy-phenylalanine(IV) (Scheme 1).

The oxidation rate of PLLCu(II) complex was found to be larger than that of the corresponding complexes with ethylendiamine or butyl-amine at both pH of 6.9 and 10.5. However, only at pH 10.5 PLLCu(II) complex behaves as an asymmetric selective catalyst for the oxidation of D-IV (Table 1).

The higher catalytic activity displayed by the PLLCu(II) complex with respect to the corresponding derivatives with low molecular weight amines at pH 10.5 was related by the authors|9| to the fact that the former can maintain cupric ions in solution, whereas in the latter case insoluble $Cu(OH)_2$ is formed. In addition the polymer chain may affects the electrostatic field around Cu(II) ions with an enhancement of the oxidation rate; this last effect might play an important role when the reaction is performed at pH 6.9 .

The investigation of the structure of the PLLCu(II) complex shows that it is depending on the pH value|11,12|. In fact at pH < 8.5 the cupric ion is coordinated to four side chain amino groups, whereas at pH 10.5 the coordination of peptide nitrogens to the cupric ion occurs with displacement of protons.

The above authors reported also that at pH 10.5 the main chain of the polymeric ligand is arranged in helical conformation and the asymmetric selectivity in the oxidation of D-IV is proportional to the content of α-helix in the polymer. Accordingly the complex of Cu(II) ions with poly(L-α,γ-diaminobutyric acid), which does not as sume one screw sense helical conformations, does not display any stereoelectivity in the oxidation of D-IV|10|.

The α-helix structure of the polymer ligand is necessary, but not sufficient, to explain the stereoelectivity observed in the above catalytic process, the distance between coordinated Cu(II) ions along the main chain of PLL being also very important.

Therefore it was suggested |10| that two or more Cu(II) ions bound
to PLL in a relative fixed disposition could act cooperatively with
a bifunctional catalysis. According to this hypothesis IV may coordinate to PLLCu(II) through both amino acid and cathecol groups at
the same time. This bifunctional activation could allow to bind
preferentially D-IV with respect to its enantiomer.

Hatano et al. |13,14| have also reported that the catalytic activ
ity of the PLLCu(II) complex at pH 7 for the hydrolysis of phenyl-
alanine methyl and ethyl esters(V) (Scheme 2) is higher of two order of magnitude with respect to the corresponding $Cu(bipyridyl)_2$
complex and aquo copper(II) ion (Table 1).

Scheme 2

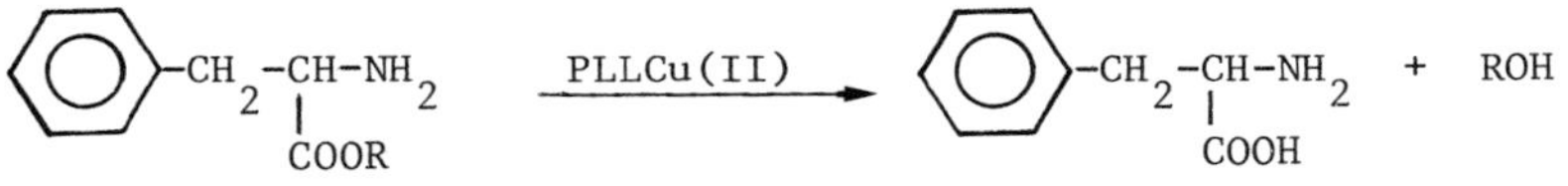

$$V(R=CH_3, C_2H_5)$$

The PLLCu(II) complex is also able to hydrolyze preferentially
D-V isomer, the activity ratio between D and L isomers being 2.41
(Table 1). Circular dichroism data suggested that PLLCu(II) gives
with D-V a complex more stable than with L-V, the difference of
stereoelectivity being ascribed to the different affinity between
the enantiomeric substrate and the catalyst |14|.

The effect of chain conformation on asymmetric induction has been
also observed with heterogeneous catalysts obtained by supporting
metallic palladium on different optically active poly(α-amino acid)s.

Supported metallic palladium on poly(L-leucine), poly(γ-benzyl-L-
glutamate), poly(β-benzyl-L-aspartate) and poly(L-valine) is report
ed by Beamer et al. |15-17| to be catalytically active for the asym-

metric hydrogenation of α–methylcinnamic acid(VI) and α–acetoamido cinnamic acid(VII) to give (R)– or (S)–dihydro–α–methylcinnamic a–cid(VIII) and (S)– or (R)–phenylalanine(IX), respectively(Scheme 3), the optical yield being in the range of 1–6% (Table 2).

Scheme 3

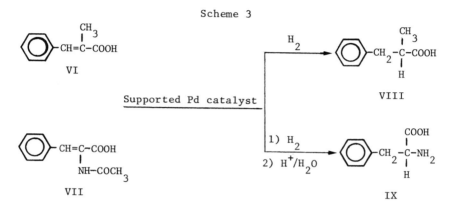

The authors concluded that the asymmetric induction observed was mainly connected with the preferential helical screw sense of the main chain(secondary structure) and not with the configuration of the asymmetric carbon atoms present in the amino acid residues (primary structure) of the optically active polymeric support|16|.

Complexes of ruthenium and rhodium derivatives with poly(L–glu–tamic acid) (PLGA) and its alkaline salts (PLGNa or PLGK) have been recently obtained in our Institute |18,19|. Analytical and infrared data did not allow to suggest a well defined coordinative situation for these systems, probably due to the different binding properties of the polymer ligand. These polymer–bound complexes have been used as catalysts for the asymmetric hydrogenation, in water medium, of ethylacetoacetate to give 3–hydroxy–ethylbutyrate, the optical purity of which being in the range 0.1–2.6% (Table 3). The low asymmetric induction observed seems to indicate that the

TABLE 2

Stereospecific Hydrogenation of α-Methylcinnamic (VI) and α-Acetamidocinnamic (VII) Acids by Palladium Supported on Poly(α-amino acid)s [a]

Polymeric support	Substrate			
	α-Methylcinnamic acid		α-Acetamidocinnamic acid	
	Predominant enantiomer formed in VIII	Optical yield %	Predominant enantiomer formed in IX	Optical yield %
Poly(γ-benzyl-L-glutamate)	R	4.1	S	6.0
Poly(γ-benzyl-L-aspartate)	S	1.4	R	1.0
Poly(L-leucine)	R	1.5	S	5.4
Poly(L-leucine)	R	1.2	S	5.2
Poly(L-valine)	S	0.9	R	4.3

[a] Hydrogen starting pressure : 4.2 atm.

TABLE 3

Asymmetric Hydrogenation in Water of Ethylacetoacetate by Catalytic Systems Based on Poly(L-glutamic acid)/Ruthenium or Rhodium Derivatives [a]

Catalytic system	T (°C)	Conversion %	$|\alpha|_D^{25}$	Optical purity [b] %
			3-hydroxy-ethylbutyrate	
PLGA/RuCl$_3$	80	96.4	+0.037	0.20
PLGK/RuCl$_3$	80	100	+0.024	0.13
PLGA/RhCl$_3$	50	42.5	+0.05	0.27
PLGK/RhCl$_3$	25	18.6	≈0	–
PLGNa/|RuBr$_2$(CO)$_3$|$_2$	100	18.4	+0.48	2.57
PLGA/Ru(CH$_3$COO)$_2$	80	12.8	≈0	–

[a] Hydrogen starting pressure : 80 atm .

[b] Assuming $|\alpha|_D^{25}$ = 18.7 for the optically pure 3-hydroxy-ethylbutyrate.

chiral environment of the polymer chain on the active sites is not
sufficient to affect markedly the stereochemical pathway of the
involved reaction. This fact could be explained taking into account the distance between the metal centers and the polymer backbone.

COMPLEXES OF TRANSITION METAL DERIVATIVES WITH STEREOREGULAR SYN-
THETIC POLYMERS

Atactic polymers have been largely used in the recent years as
macromolecular ligands for anchoring catalytically active metal
complexes, with the aim of obtaining heterogeneous catalysts with
the high activity and selectivity of the corresponding low molecular weight homogeneous systems. On this subject excellent reviews
have been recently reported |20-22|. In general, crosslinked pol-
ymers were used and no attempt was done to examine possible effects
of polymer microstructure on activity, selectivity and stereochemistry of the reactions involved.

As isotactic vinyl polymers are known to assume stereordered
conformations, not only in the crystalline state but also in so-
lution |23,24|, metal complexes bound to these macromolecular ma-
trices could be suitable for controlling the stereochemical pathway of the catalyzed reactions. In this contest macromolecular
complexes with definite structure were prepared by reacting both
isotactic and atactic phosphenated poly(styrene)s such as   poly-
|(p.diphenylphosphino)styrene| |poly(PSt)| and poly|styrene-co-(p.
diphenylphosphino)styrene| |poly(St/PSt)| with $RuBr(C_3H_5)(CO)_3$
|25| (Scheme 4).

All the complexes showed the cis-structure X and were insoluble
in the most common organic solvents. According to the elemental

analysis the percentage of PSt units coordinated to the metal was in the range 85-95%, independently of stereoregularity, molecular weight and content of phosphenated units in the macromolecular ligand.

Scheme 4

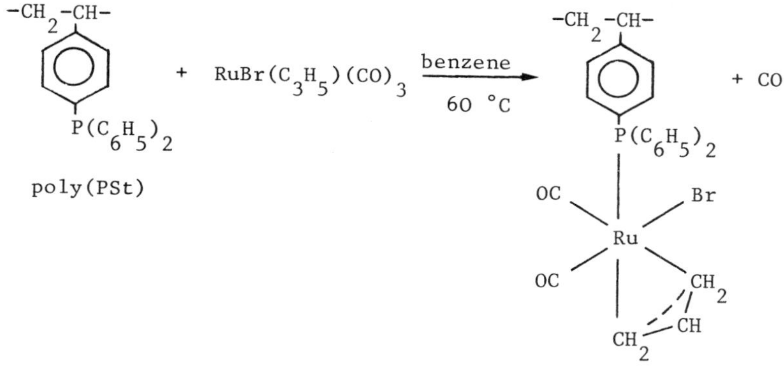

These polymeric complexes were catalytically active for the isomerization of 1-butene in liquid/solid system under nitrogen at 100 °C using toluene as liquid phase |25|. The homogeneous analog with triphenylphosphine displayed higher catalytic activity than the polymeric systems. However, for these last the isomerization rate was strongly affected by the characteristics of the polymer ligand, decreasing with increasing molecular weight and tacticity (Table 4). This behaviour could be explained according to the corresponding decrease of the swelling extent of the polymer matrix, thus giving a lower availability of the active sites to the substrate. A confirmation to this explanation is given by the 1-butene isomerization experiments carried out with the same catalysts in

TABLE 4

Isomerization of 1-butene in Liquid/solid and Gas/solid Systems at 100 °C in the Presence of Catalysts Based on $RuBr(C_3H_5)(CO)_3$ and Phosphenated Poly(styrene)s

PSt units (mol %)	Catalyst Polymer ligand Tacticity	Molecular weight	Ru %	Reaction rate [a] $(mole\ l^{-1}\ h^{-1}\ g\ at.\ Ru^{-1})$ liquid-solid system [b]	gas-solid system [c]
47.0	isotactic	very high[d]	9.37	840	33.6
23.7	atactic	high[e]	7.62	3,000	44.0
100	atactic	low[f]	17.05	5,100	17.5
29.6	atactic	low[g]	11.14	9,200	19.0

[a] Average in the first two hours.

[b] 1-butene: 4.14 M in toluene; 1-butene(mole)/Ru(g at.): 660.

[c] 1-butene: 15.6 mmoles; catalyst: 6.8×10^{-5} g at. Ru .

[d] Isotactic poly(styrene) used as starting material had $\bar{M}_v \approx 1,200,000$.

[e] Atactic poly(styrene) used as starting material had $\bar{M}_v \approx 187,000$.

[f] $|\eta| = 0.28$ dl/g in toluene at 30 °C .

[g] $|\eta| = 0.20$ dl/g in toluene at 30 °C .

gas/solid system (Table 4). In these experiments a large decrease
of catalytic activity was observed with lost of any effect of pol-
ymer ligand microstructure|25|.

A further approach to evidence the role of primary and secondary structure of the macromolecular ligand in determining the catalytic properties of polymer-bound metal complexes is based on ster
eoregular optically active polymers.

Structurally simple macromolecules with definite configurational
and conformational characteristics can be obtained by stereospecif
ic polymerization of chiral vinyl monomers. When the side chain a-
symmetric carbon atom is in the α or β position with respect the
main chain, the macromolecules assume a preferential screw sense
helical conformation|23,24|, thus providing an evidence of confor-
mational analogy with poly(α-amino acid)s and proteins.

Therefore coisotactic copolymers of chiral α-olefins such as
(R)- or (S)-3,7-dimethyl-1-octene(DMO) with styrene were submitted
to the chemical transformations reported in Scheme 5. Thus chiral
stereoregular DMO/(p.diphenylphosphinomethyl)styrene copolymers
|poly(DMO/PMSt)| and DMO/(p.diphenylphosphino)styrene/styrene ter-
polymers |poly(DMO/PSt/St)| having a high conformational homogene-
ity were obtained|26|. Such P-donor polymers were used as ligands
for different ruthenium and rhodium derivatives|27-29|.

The insoluble macromolecular complexes derived from poly(DMO/
PSt/St) or poly(DMO/PMSt) and $RuBr(C_3H_5)(CO)_3$ showed for the metal
complexed units the structures X and XI, respectively.

Starting with different rhodium compounds such as $|Rh(CO)_2Cl|_2$,
$RhH(CO)(PPh_3)_3$ and $RhCl_3 3H_2O$ and the overmentioned optically active
phosphenated polymeric materials, polymer-bound rhodium derivatives
with complexed metal units XII, XIII, XIV, XV, and XVI were obtain
ed, respectively.

Scheme 5

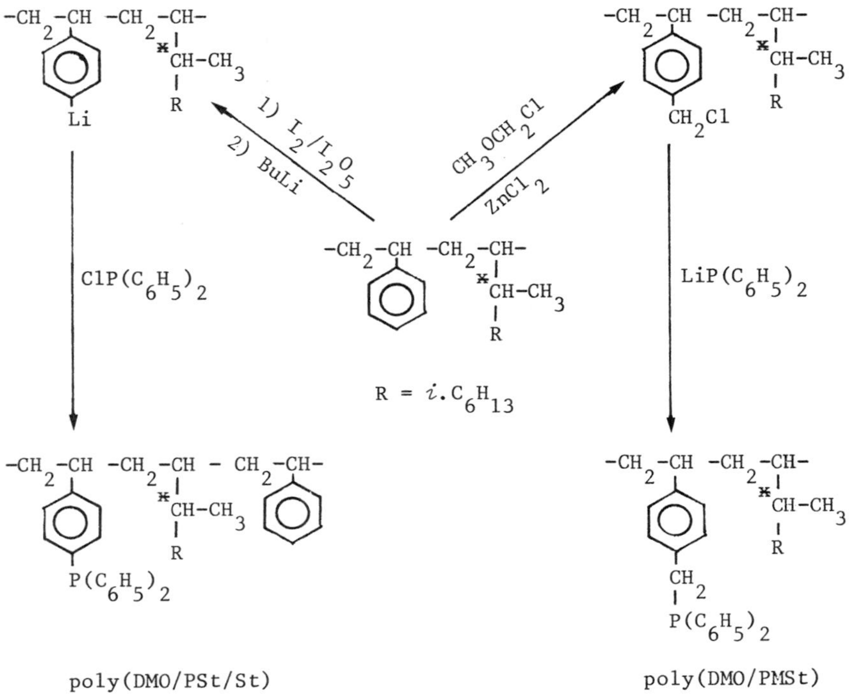

poly(DMO/PSt/St)   poly(DMO/PMSt)

The above complexes were employed as catalysts in reaction involving racemic or prochiral substrates in non aqueous medium (Table 5).

In the isomerization of racemic 4-methyl-1-hexene, carried out at 100 °C using toluene as reaction medium, in the presence of η-allyl ruthenium derivatives bound to the optically active poly(DMO/PSt/St) or poly(DMO/PMSt), a *cis* and *trans*-4-methyl-2-hexene mixture was obtained with a very low optical yield (≈ 0.1%)|27|.

TABLE 5

Ruthenium and Rhodium Complexes Bound to Optically Active Stereoregular Phosphenated Polymers and Their Catalytic Activity in Asymmetric Syntheses

Catalyst Polymer ligand (composition)	Metal complexed unit	Catalyzed reaction	Substrate	Substrate (mole) / Metal (g at.)	Product Type	Optical yield %
Poly(DMO/PSt/St) (0.49/0.27/0.24)	X	Isomerization[a]	(R)(S)-4-methyl-1-hexene	430	4-methyl-2-hexene (*cis*+*trans*)	< 0.1
Poly(DMO/PMSt) (0.52/0.48)	XI	"	"	670	"	< 0.1
Poly(DMO/PSt/St) (0.49/0.27/0.24)	XV	Hydroformylation[b]	Styrene	4,750	Hydratropic aldehyde	< 0.1
Poly(DMO/PSt/St) (0.32/0.28/0.40)	XII	"	"	2,030	"	≃ 0
Poly(DMO/PSt/St) (0.32/0.39/0.29)	XIII	"	"	1,700	"	≃ 0
Poly(DMO/PMSt) (0.52/0.48)	XIV	Hydrosilylation[c]	Acetophenone	210	1-phenyl-ethanol	0.2
Poly(DMO/PSt/St) (0.49/0.18/0.33)	XVI	"	"	170	"	2.5

[a] (R)(S)-4-methyl-1-hexene: 2.5 M in toluene; [b] Styrene: 3.5 M in benzene; [c] Acetophenone: 1 M in benzene, diphenylsilane(mole)/acetophenone(mole): 2.

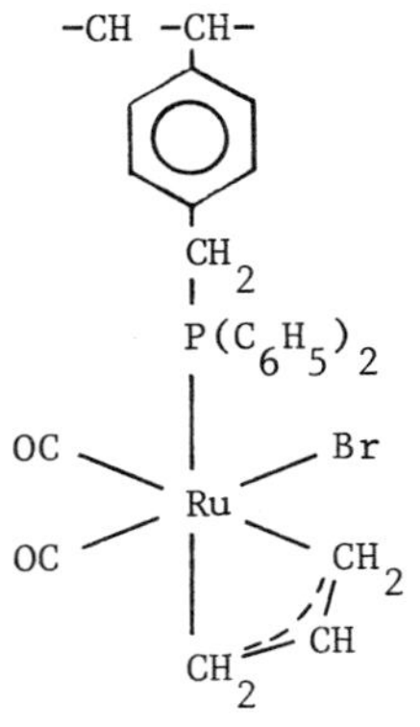

-CH -CH-

CH$_2$
|
P(C$_6$H$_5$)$_2$
|
OC   Br
    \   /
     Ru
    /   \
OC   CH$_2$
          //
          CH
CH$_2$

XI

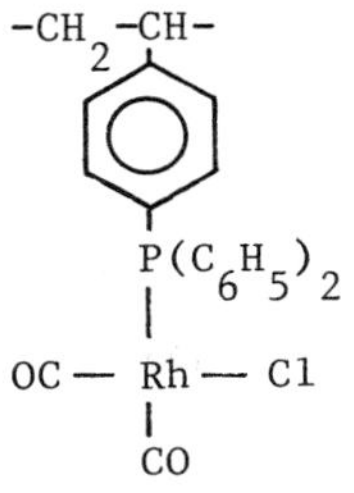

-CH$_2$-CH-

P(C$_6$H$_5$)$_2$
|
OC — Rh — Cl
|
CO

XII

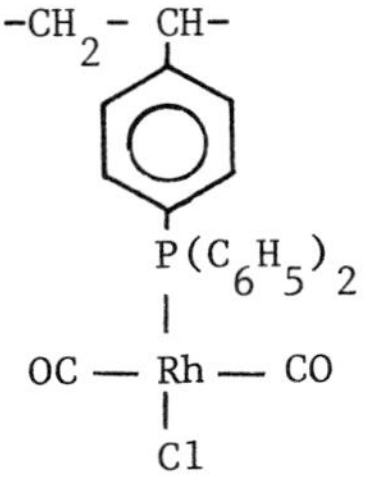

-CH$_2$- CH-

P(C$_6$H$_5$)$_2$
|
OC — Rh — CO
|
Cl

XIII

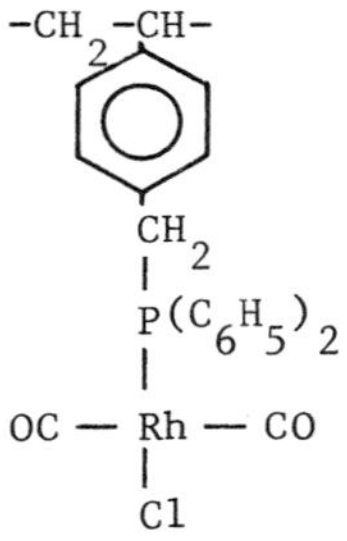

-CH$_2$-CH-

CH$_2$
|
P(C$_6$H$_5$)$_2$
|
OC — Rh — CO
|
Cl

XIV

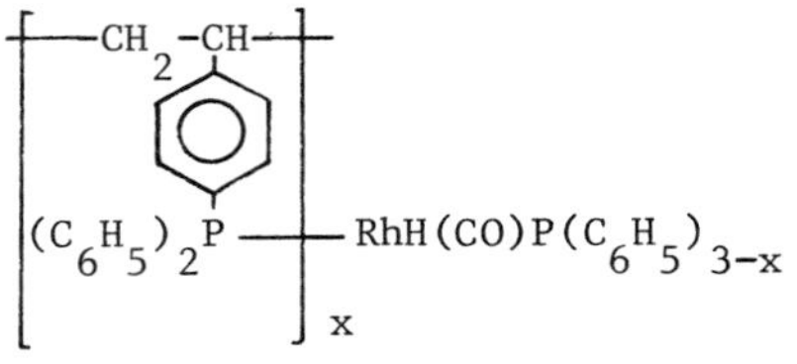

[—CH$_2$-CH—]

(C$_6$H$_5$)$_2$P —— RhH(CO)P(C$_6$H$_5$)$_{3-x}$

x

XV

-CH$_2$-CH-

P(C$_6$H$_5$)$_2$
|
RhCl$_x$

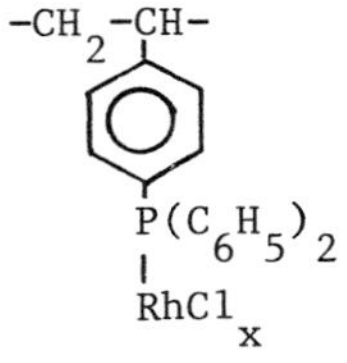

XVI

Asymmetric selectivity of the same order of magnitude was obtained in the hydroformylation of styrene to hydratropic aldehyde at 80 °C and 90 atm of CO/H$_2$ (1/1), using catalysts based on rhodium complexes attached to optically active poly(DMO/PSt/St) having different compositions |29|.

In the reaction of a prochiral ketone such as acetophenone with diphenylsilane at 20 °C in the presence of catalysts obtained from RhCl$_3$·3 H$_2$O or |Rh(CO)$_2$Cl|$_2$ and chiral poly(DMO/PSt/St) and poly-(DMO/PMSt), respectively, optically active 1-phenylethanol was ob-

tained having the same absolute configuration as DMO units and op-
tical purity in the range 0.2-2.5%|27| (Table 5).

These results, even if indicating that the asymmetric polymer
support affects the reaction stereochemistry, also show that the
dissymmetric environment of the polymer main chain is not sufficient
to induce in a large extent the chiral perturbation of the coordination sphere of the metal, probably due to the large distance between polymeric backbone and metal sites in the systems examined
up to now. The low values of optical yield could be also related to
the substrates used, the nature of which might not be suitable for
evidencing local asymmetric environment of the active sites. Accord_
ingly also rhodium complexes with low molecular weight chiral-P
phosphine ligands display low asymmetric induction(1-2%) in the hy_
drogenation of prochiral substrates having similar structure to
those mentioned above|30,31|. In this contest the results obtained
by Hirai and Furuta|32,33| in the homogeneous asymmetric hydrogen-
ation of prochiral ketones by poly(L-methylethylenimine)/RuCl$_3 \cdot$3H$_2$O
in water at pH 5.5, indicate that the substrate may play an important role in determining the stereochemistry of the reaction involved. In fact, in spite of the close vicinity of the metal sites
to the backbone of the chiral stereoregular polymer ligand, only
in the case of bifunctional ketones such as methylacetoacetate and
methyl-i.butenyl-ketone an asymmetric induction, even if in low ex_
tent, occurred (Table 6).

COMPLEXES OF TRANSITION METAL DERIVATIVES WITH ALTERNATING COPOLYMERS

The results described in the previous section suggest the necessity of reducing as far as possible the distance between the macromolecular backbone and the metal atom in order to improve the in-

TABLE 6

Asymmetric Hydrogenation of Prochiral Ketones by Poly(L-methylethylenimine)/Ru(III) Catalytic Systems at 80 °C and 80 atm [a]

Substrate	Ligand/Ru(III) molar ratio	Total yield %	Hydrogenation product Type	Optical yield %
Methylacetoacetate	2.5	54	Methyl(-)(R)-3-hydroxybutyrate	0.5
	10.0	26		5.3
Methyl-i.butenyl-ketone	2.5	32	(-)(R)-4-methyl-2-pentanol	0.4
	5.0	22		0.5
Methyl-i.butyl-ketone	5.0	65	(R)(S)-4-methyl-2-pentanol	0.0

[a] Substrate(mole)/Ru(g at.) in the range 400-465.

fluence of the primary and secondary structure of the polymer lig-
and on activity, selectivity and stereochemistry of the catalytic
systems. A further improvement can be achieved by designing structurally ordered polymeric ligands where each attached metal atom
is at a suitable distance from a peculiar functional group able to
assist either the coordination step of the substrate or its succes
sive transformation. Finally using side chains with different hydro
philic properties it should be possible to modify in a large extent
the interactions between the liquid reaction medium and the polymeric ligand, thus affecting polymer swelling and availability of
the metal sites to the substrate.

Copolymers of ethylene(Et) or alkyl vinyl ethers with maleic
acid(MAc)|34,35|, obtained according to Scheme 6, appeared to pos-
sess the above requirements.

Scheme 6

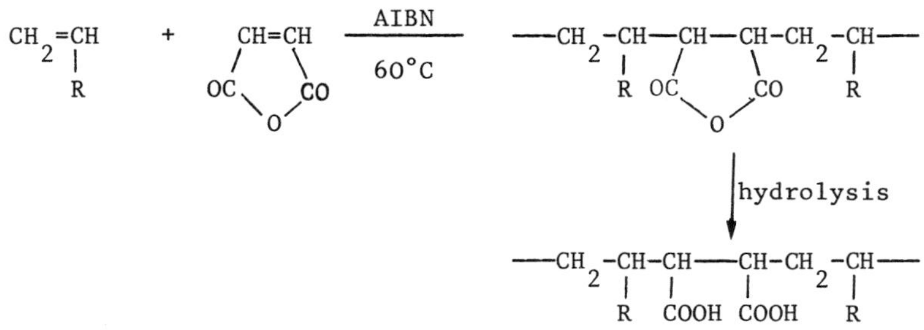

In fact these systems have an alternating structure and the metal
binding units(MAc units *via* carboxylic groups) are inserted between
two counits bearing one R group, the nature of which can be varied
in a large extent. Furthermore the anchoring of the transition met

al derivative to the carboxylic groups guarantees a close vicinity
to the polymer backbone. In this connection alternating copolymers
of MAc with Et |poly(Et/MAc)|, benzyl vinyl ether|poly(BVE/MAc)|
and 2-methylbutyl vinyl ether|poly(BuVE/MAc)| as well as vinyl al-
cohol/BVE/MAc terpolymers|poly(VA/BVE/MAc)| with different content
of VA units, obtained by ether cleavage reaction with dry gaseous
HBr|36| of the corresponding poly(BVE/MAc), were used as ligands to
coordinate ruthenium(II) and rhodium(I) derivatives. By reaction
of |RuCl$_2$(CO)$_3$|$_2$ with the sodium salts of the copolymers and ter-
polymers (Scheme 7), insoluble polymer-bound ruthenium derivatives
containing different amount of sodium and ruthenium carboxylate
groups, were obtained|37|.

Scheme 7

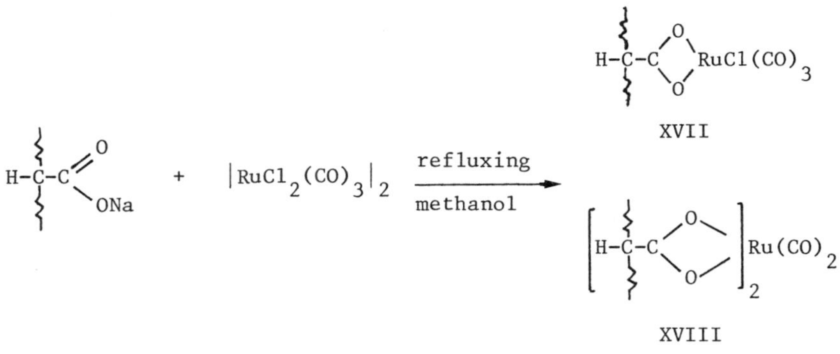

On the basis of elemental analysis and I.R. data, the structures
XVII and XVIII could be assigned to the metal containing moieties.

The above macromolecular ligands have been also allowed to re-
act with RuH$_2$(PPh$_3$)$_4$ and RuH$_2$(CO)(PPh$_3$)$_3$ under the same conditions

as used to prepare low molecular weight carboxylate analogs |38|
(Scheme 8).

Scheme 8

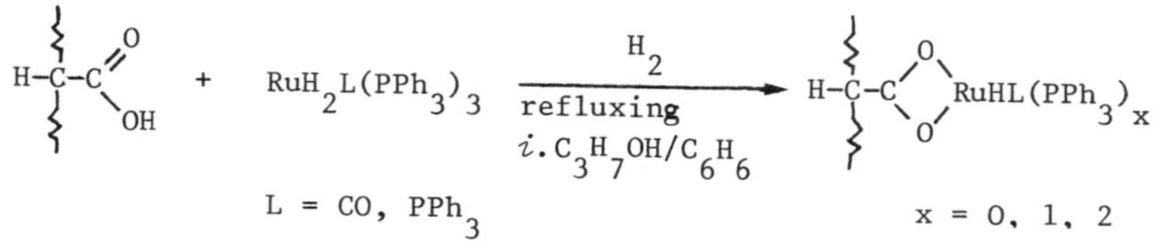

$$H-\overset{|}{\underset{|}{C}}-C\overset{\displaystyle O}{\underset{\displaystyle OH}{<}} \quad + \quad RuH_2L(PPh_3)_3 \quad \xrightarrow[\substack{\text{refluxing} \\ i.\ C_3H_7OH/C_6H_6}]{H_2} \quad H-\overset{|}{\underset{|}{C}}-C\overset{\displaystyle O}{\underset{\displaystyle O}{<}}RuHL(PPh_3)_x$$

$$L = CO,\ PPh_3 \qquad\qquad\qquad x = 0,\ 1,\ 2$$

$$XIX$$

The I.R. spectra indicate for these last insoluble complexes the
presence of a ruthenium-hydrogen bond and of a bidentate coordina-
tion of the carboxylate group (structure XIX) |37, 39|.

It is of interest to note that the P/Ru ratio is in the range
of 1.4-1.8, lower than that theoretically expected (x = 2) and ac-
tually observed in the case of poly(acrylic acid) |poly(AA)|. This
indicates that, on the average, more than one triphenylphosphine
molecule per ruthenium atom is displaced with consequent formation
of "coordinatively unsaturated" metal species. This could be due
both to the presence in MAc units of two carboxylic groups directly bound to adjacent carbon atoms and to the bulkiness of triphenyl
phosphine ligands of the metal derivative |39|.

An analogous route was followed |39| to prepare poly-carboxylate
-triphenylphosphine rhodium(I) complexes (Scheme 9).

In this case the metal complexes obtained have no hydridic nature and the carboxylate group functions as unidentate ligand as
for the corresponding derivatives with low molecular weight car-

boxylic acids |40|. Thus the metal complexed moieties show the structure XX, where x, that is P/Rh ratio, assumes average values in the range 1.5-1.9, according to the presence of "coordinatively unsaturated" metal species also in this case |39|.

Scheme 9

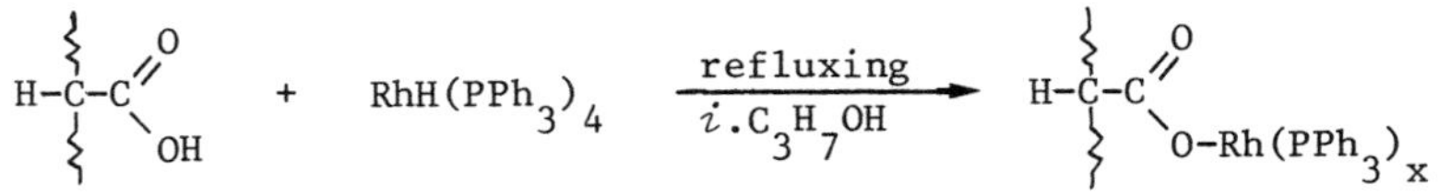

$$x = 1, \ 2, \ 3$$

XX

All the polymeric metal complexes obtained from different ruthen-ium and rhodium derivatives and the overmentioned macromolecular carboxylic ligands have been tested as heterogeneous catalysts for the isomerization and hydrogenation of 1-pentene in the presence of a liquid medium.

The polymer-bound catalysts derived from $|RuCl_2(CO)_3|_2$ and poly-meric sodium carboxylates containing different amount of sodium and rhutenium carboxylate groups (Table 7), isomerize 1-pentene at 100 °C under nitrogen in the presence of toluene and a small amount of ethanol, which is necessary for the formation of active ruthenium-hydride species|37|. The catalytic activity of such systems marked-ly increases with decreasing the content of sodium carboxylate groups . This trend has been related to the contemporary decrease of the swelling extent of the polymeric ligand by the liquid reac-tion medium, with consequent lower availability of the active sites to the substrate. The effect of the polymer ligand on the swelling extent is better evidenced by the large difference in catalytic ac-

TABLE 7

Isomerization of 1-pentene at 100 °C in the Presence of Polycarboxylate Ruthenium(II) Catalysts Obtained from $|RuCl_2(CO)_3|_2$ and Polymeric Sodium Carboxylates [a]

Complex	Polymer ligand (composition)	-COONa total carboxylate groups %	Heterogeneity of the catalysis [b] %	Initial reaction rate ($mmol \ l^{-1} \ h^{-1}$)
1	Poly(AANa)	73	99.9	42
2	Poly(VA/BVE/MAcNa) (0.33/0.17/0.50)	34	93.0	150
3	Poly(Et/MAcNa) (0.5/0.5)	28	67.0	636
4	Poly(VA/BVE/MAcNa) (0.25/0.25/0.50)	27	97.6	1,150
5	Poly(BuVE/MAcNa) (0.5/0.5)	26	95.0	828
6	Poly(AANa)	6	87.0	1,038

[a] 1-pentene: 0.83 M in toluene/ethanol (25/1); 1-pentene(mole)/Ru(g at.): 365.
[b] Evaluated as % of the total ruthenium present in the insoluble residue after the catalytic reaction.

tivity (about one order of magnitude) between the two complexes
(2 and 4 , Table 7) containing about the same content of sodium
carboxylate groups, but different extent of VA hydrophilic counits
|37, 41|.

The systems derived from $RuH_2(PPh_3)_4$ and different polycarboxi-
lic matrices exhibit, in the isomerization and hydrogenation of 1-
pentene at 50 °C, catalytic activity much lower than that of the
closely related homogeneous models |39| (Table 8).

On the contrary, the catalysts obtained from the same polymeric
materials and $RuH_2(CO)(PPh_3)_3$ show a comparable or higher activity
than that of their homogeneous counterparts |39|. This behaviour
has been related to the presence, in the polymeric catalysts, of
"coordinatively unsaturated" metal species to which the substrate
may be easily coordinated without displacement of ligands. Accordingly the polymeric catalyst with the lowest P/Ru ratio (complex
14, Table 9) displays the highest activity (about one order of magnitude with respect to its homogeneous analog 13). A confirmation
to these results derives from the very low catalytic activity of
the system 17, prepared from poly(AA) (Table 9), the metal species
of which are almost coordinatively saturated(P/Ru= 1.9). The higher
activity due to the "coordinative unsaturation" may be related to
the mechanism, generally proposed for isomerization of olefins,
also in the presence of carboxylate ruthenium complexes|38|. Accord
ingly the olefin coordination occurs through displacement of a triphenylphosphine group and/or cleavage of one of the two ruthenium-
oxygen bonds of the bidentate carboxylate ligand (Scheme 10). Generally soluble carboxylate carbonyl triphenylphosphine complexes
obtained from low molecular weight carboxylic acids are less active
than the corresponding carboxylate triphenylphosphine derivatives

TABLE 8

Isomerization and Hydrogenation of 1-pentene at 50 °C in the Presence of Polycarboxylate Ruthenium(II) Hydrido-triphenylphosphine Catalysts Obtained from $RuH_2(PPh_3)_4$ and Their Low Molecular Weight Analogs[a]

Catalytic system (polymer composition)	$\dfrac{P}{Ru}$	Heterogeneity of the catalysis[b] %	Initial reaction rate (mmol l^{-1} h^{-1})	
			Isomerization	Hydrogenation
7 $\|RuH(PPh_3)_3\|_2\|O_2C(CH_2)_2CO_2\|$	3	0	332	232
8 $RuH_2(PPh_3)_4$/poly(Et/MAc) (0.5/0.5)	1.83	97.0	3.5	0.4
9 $RuH(O_2CCH_2OH)(PPh_3)_3$	3	0	224	176
10 $RuH_2(PPh_3)_4$/poly(VA/BVE/MAc) (0.25/0.25/0.50)	1.38	98.0	5	0.4

[a] 1-pentene: 0.83 M in hydrocarbon solvents; 1-pentene(mole)/Ru(g at.): 365; p_{H_2}: 1.06 atm.
[b] See note b, Table 7

TABLE 9

Isomerization and Hydrogenation of 1-pentene at 50 °C with Polycarboxylate Ruthenium(II) Hydrido-tri-phenylphosphine Catalysts Obtained from $RuH_2(CO)(PPh_3)_3$ and Their Low Molecular Weight Analogs[a]

Catalytic system (polymer composition)	$\dfrac{P}{Ru}$	Heterogeneity of the catalysis[b] %	Initial reaction rate ($mmol\ l^{-1}\ h^{-1}$)	
			Isomerization	Hydrogenation
11 $\lvert RuH(CO)(PPh_3)_2 \rvert_2 \lvert O_2C(CH_2)_2CO_2 \rvert$	2	0	38	1.7
12 $RuH_2(CO)(PPh_3)_3$/poly(Et/MAc) (0.5/0.5)	1.38	100	22	0.5
13 $RuH(CO)(O_2CCH_2OH)(PPh_3)_2$	2	0	36	0
14 $RuH_2(CO)(PPh_3)_3$/poly(VA/BVE/MAc) (0.25/0.25/0.50)	1.35	97.0	295	20
15 $RuH_2(CO)(PPh_3)_3$/poly(VA/BVE/MAc) (0.33/0.17/0.50)	1.76	96.0	175	17
16 $RuH(CO)(PPh_3)_2 \lvert O_2CCH(CH_3)_2 \rvert$	2	0	25	0
17 $RuH_2(CO)(PPh_3)_3$/poly(AA)	1.9	95.0	3	0

[a]See note a, Table 8.

[b]See note b, Table 7.

|compare for example the activities of the catalysts 11 and 13 (Table 9) with those of the catalysts 7 and 9 (Table 8), respectively| as the triphenylphosphine displacement is more difficult when a π-acceptor carbonyl group is coordinated to the metal|38|.

Scheme 10

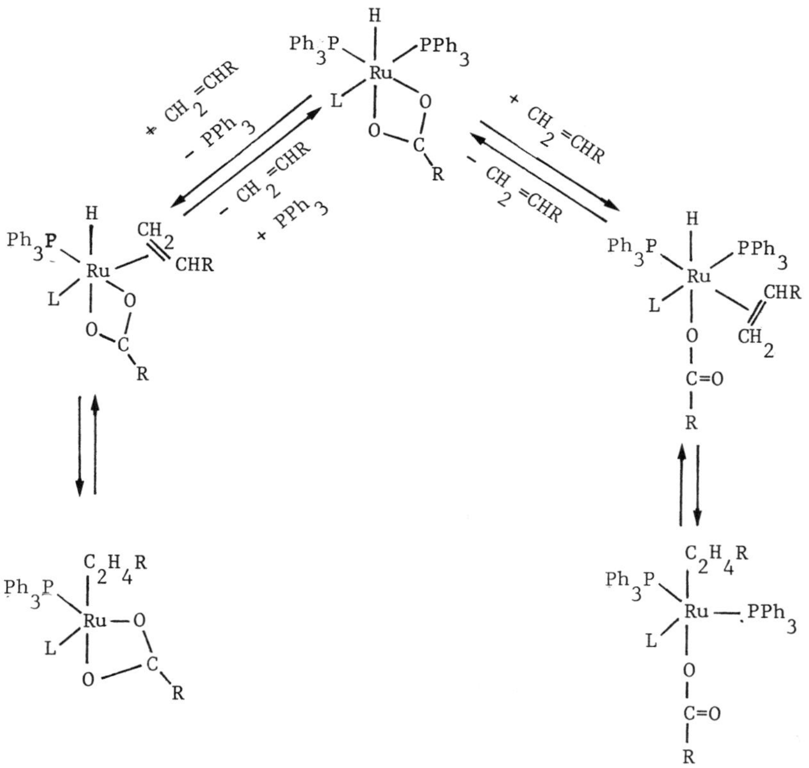

In the case of the polymeric complexes the opposite behaviour is observed, as shown by the catalysts 10 (Table 8) and 14 (Table 9), derived from the same macromolecular ligand and having comparable

P/Ru ratio. According to Scheme 10, this result can be explained
|39,42| considering that, due to the presence of "coordinatively
unsaturated" metal species the triphenylphosphine displacement is
no longer necessary to activate the olefin and thus this step does
not affect the reaction rate. On the other side the carbonyl group
present in the above "coordinatively unsaturated" species favours
the Markownikoff insertion of the olefin which is the active step
for the isomerization process.

Also for the polymeric systems based on rhodium derivatives the
catalytic activity, in the isomerization and hydrogenation of 1-
pentene at 50 °C, was found to be dependent on the P/Rh ratio, connected with the relative amount of "coordinatively unsaturated"
metal species|39|(Table 10).

Recently such species, able to react with O_2 to give paramagnetic $Rh(II)O_2^-$ derivatives, have been detected in the polymeric catalysts by e.s.r. spectroscopy and quantitatively determined. A
linear dependence of the overall initial 1-pentene conversion rate
(isomerization + hydrogenation) on the content of paramagnetic
species was observed and strictly related to the P/Rh ratio|43|,
confirming the model previously proposed.

In the above systems the effect of the swelling extent of the
catalyst on the activity was finally evidenced using solvents having different affinity with the polymeric matrices as reaction medium. Accordingly the overall catalytic activity strongly increases
using 2-propanol in the place of toluene|39|(Table 10).

FINAL REMARKS

The aim of the present review is to call attention of research
people to the enormous potentiality offered by polymer/metal complexes in catalysis provided more attention is given to introduce

TABLE 10

Isomerization and Hydrogenation of 1-pentene at 50 °C with Polycarboxylate Rhodium(I) Catalysts Obtained from $RhH(PPh_3)_4$ and Their Low Molecular Weight Analog [a]

Catalytic system (polymer composition)	$\dfrac{P}{Rh}$	Reaction medium type	ml	Heterogeneity of the catalysis[b] %	Initial reaction rate $(mmol\ l^{-1}\ h^{-1})$ Isomerization	Hydrogenation
$Rh\vert O_2CCH(CH_3)_2\vert (PPh_3)_3$	3	toluene ethanol	24 1	0	482	450
$RhH(PPh_3)_4$/poly(AA)	3	toluene ethanol	24 1	86	53	78
$RhH(PPh_3)_4$/poly(Et/MAc) (0.5/0.5)	1.5	toluene ethanol	24 1	98.5	85	196
$RhH(PPh_3)_4$/poly(VA/BVE/MAc) (0.25/0.25/0.50)	1.9	toluene ethanol	24 1	96	87	121
$RhH(PPh_3)_4$/poly(VA/BVE/MAc) (0.25/0.25/0.50)	1.9	2-propanol ethanol	24 1	92	245	171
$RhH(PPh_3)_4$/poly(BVE/MAc) (0.5/0.5)	1.6	toluene ethanol	24 1	95	65	104
$RhH(PPh_3)_4$/poly(BVE/MAc) (0.5/0.5)	1.6	2-propanol ethanol	24 1	95	240	100

[a] 1-pentene: 0.83 M; 1-pentene(mole)/Rh(g at.): 365; p_{H_2}: 1.06 atm.

[b] See note b, Table 7.

useful characteristics in the polymer ligand. The use of organic
resins to heterogenize homogeneous transition metal complexes is
certainly of interest but reduces the polymer ligand to a very mod-
est role which can be better interpreted, at least from economical
view-point, by inorganic supports. The use of polymers with definite structure, binding the transition metal in a very definite way
to give catalysts working with a known and controllable mechanism,
can justify more work in the area as it may offer peculiar possibility in special cases. The work done up to now in this connection
is certainly not sufficient to disclose all this potentiality and
the examples reported up to now in the literature and described
here indicate lines along which future studies should move. In
spite of the limited results reached we think to have shown how
high selectivity and in particular catalytic activity under very
mild conditions and without sophisticated solvents and atmosphere
can be approached with polymer/metal complexes provided polymer
structure is planned looking at the expected application.

REFERENCES

|1| P. L. Luisi, in Charged & Reactive Polymers: Optically Active
     Polymer, E. Sélégny, Ed., Vol. 5, Reidel, Dordrecht,1979,p.357

|2| R. J. P. Williams, Pure Appl. Chem., 40, 38 (1975).

|3| A. E. Dennard and R. J. P. Williams, Transition Metal Chemis-
     try, R. L. Carlin, Ed., Dekker, New York, Vol. 2, 1966, p. 116.

|4| S. Inone, Adv. Polymer Sci., 21, 78 (1976).

|5| A. Levitzki, I. Pecht, and M. Anbar, Nature, 207, 1386 (1965).

|6| I. Pecht, A. Levitzki, and M. Anbar, J. Am. Chem. Soc., 89,
     1587 (1967).

|7| A. Levitzki, I. Pecht, and A. Berger, J. Am. Chem. Soc., 94, 6844 (1972).

|8| M. Palumbo, A. Cosani, M. Terbojevich, and E. Peggion, Macromolecules, 11, 1271 (1978).

|9| M. Hatano, T. Nozawa, S. Ikeda, and T. Yamamoto, Makromol. Chem., 141, 11 (1971).

|10| T. Nozawa and M. Hatano, Makromol. Chem., 141, 31 (1971).

|11| M. Hatano, T. Nozawa, S. Ikeda, and T. Yamamoto, Makromol. Chem., 141, 1 (1971).

|12| T. Nozawa and M. Hatano, Makromol. Chem., 141, 21 (1971) .

|13| T. Nozawa, Y. Akimoto, and M. Hatano, Makromol. Chem., 158, 21 (1972).

|14| T. Nozawa, Y. Akimoto, and M. Hatano, Makromol. Chem., 161, 289 (1972).

|15| R. L. Beamer, C. S. Fickling, and J. H. Ewing, J. Pharm. Sci., 56, 1029 (1967).

|16| R. L. Beamer, R. H. Belding, and C. S. Fickling, J. Pharm. Sci., 58, 1142 (1969).

|17| R. L. Beamer, R. H. Belding, and C. S. Fickling, J. Pharm. Sci., 58, 1419 (1969).

|18| G. Bini, Thesis, University of Pisa, June 1975.

|19| F. Ciardelli, C. Carlini, and E. Chiellini, Abstracts of XXIVth IUPAC Internat. Symposium on Macromolecules, Jerusalem, 1975, p. 63.

|20| Y. Chauvin, D. Commereuc, and F. Dawans, Progr. Polym. Sci., 5, 95 (1977).

|21| F. R. Hartley and P. N. Vezey, Advances in Organometallic Chemistry, F. G. A. Stone. and R. West, Eds., Academic Press, New York, Vol. 15, 1977, p. 189.

|22| G. Manecke and W. Storck, Angew. Chem. Int. Ed. Engl., 17, 657 (1978).

|23| P. L. Luisi and F. Ciardelli, in Reactivity Mechanism and Structure in Polymer Chemistry, A. D. Jenkins and A. Ledwith, Eds., Wiley, London, 1974, p. 471.

|24| F. Ciardelli, C. Carlini, E. Chiellini, P. Salvadori, L. Lardicci, R. Menicagli, and C. Bertucci, in Preparation and Properties of Stereoregular Polymers, R. W. Lenz and F. Ciardelli, Eds., Reidel, Dordrecht, 1979, p. 353.

|25| C. Carlini, G. Braca, F. Ciardelli, and G. Sbrana, J. Mol. Catal., 2, 379 (1977).

|26| E. Chiellini and C. Carlini, Makromol. Chem., 178, 2545 (1977).

|27| F. Ciardelli, E. Chiellini, C. Carlini, and R. Nocci, A.C.S. Polymer Preprints, 17, 188 (1976).

|28| L. Pardini, Thesis, University of Pisa, 1974.

|29| L. Pacchiarini, Thesis, University of Pisa, 1976.

|30| J. D. Morrison, R. E. Burnett, A. M. Aguiar, C. J. Morrow, and C. Phillips, J. Am. Chem. Soc., 93, 1301 (1971).

|31| P. Bonvicini, A. Levi, G. Modena, and G. Scorrano, J.C.S. Chem. Commun., 1188 (1972).

|32| H. Hirai and T. Furuta, J. Polym. Sci. B, 9, 459 (1971).

|33| H. Hirai and T. Furuta, J. Polym. Sci. B, 9, 729 (1971).

|34| S. Machi, T. Sakai, M. Gotoda, and T. Kagiya, J. Polym. Sci. A-1, 4, 821 (1966).

|35| E. Chiellini, M. Marchetti, C. Villiers, C. Braud, and M. Vert, Europ. Polym. J., 14, 251 (1978).

|36| H. Yuki, K. Hatada, K. Ohta, I. Kinishita, J. Polym. Sci., 7, 1517 (1969).

|37| G. Braca, C. Carlini, F. Ciardelli, and G. Sbrana, Proc. VIth
      Int. Congr. Catalysis, G. C. Bond, P. B. Wells, and F. C.
      Tompkins, Eds., The Chemical Society, London, Vol. 1, 1977,
      p. 528.

|38| G. Sbrana, G. Braca, and E. Giannetti, J. Chem. Soc., Dalton,
      1847 (1976).

|39| G. Sbrana, G. Braca, G. Valentini, G. Pazienza, and A. Alto-
      mare, J. Mol. Catal., 3, 111 (1977/1978).

|40| S. D. Robinson and M. F. Uttley, J. Chem. Soc., Dalton, 1912
      (1973).

|41| G. Braca, F. Ciardelli, G. Sbrana, and G. Valentini, Chim.
      Ind. (Milan), 59, 766 (1977).

|42| G. Braca, G. Sbrana, G. Valentini, and F. Ciardelli, Chim.
      Ind.(Milan), 60, 530 (1978).

|43| G. Braca, G. Sbrana, G. Valentini, A. Colligiani, and C. Pin-
      zino, J. Mol. Catal., 457 (1980).

The Potential of π-Bonded Organometallic Polymers in Catalyst Design

D. W. SLOCUM, B. CONWAY, M. HODGMAN, K. KUCHEL,
M. MORONSKI, R. NOBLE, K. WEBBER and S. DURAJ

Neckers Laboratory
Southern Illinois University
Carbondale, IL 62901

A. SIEGEL

Department of Chemistry
Indiana State University
Terre Haute, IN   47809

and D. A. OWEN

Department of Chemistry
Murray State University
Murray, KY   42071

ABSTRACT

Reductive substitution, the process whereby a hydrocarbon ligand, usually a cyclopentadienyl group, in a transition metal complex is replaced by another unsaturated hydrocarbon with concomitant metal reduction has been utilized to prepare novel complexes of Cr^o, Cr^I, Mn^I, Co^I and Ni^o. Similarly, ligand substitution, defined more generally as replacement of ligand(s) of a metal complex by an unsaturated hydrocarbon but with no change in the oxidation state of the metal, has resulted in the synthesis of a series of tetraphenylborate and substituted

357

tetraphenylborate complexes of Fe^{II}, Mo^I, Cr^I, W^I, and Co^I. These procedures and the complexes derived therefrom are considered models for polymers incorporating a variety of novel features.

INTRODUCTION

Reductive substitution and ligand substitution of metallo-cenes and related compounds has been successful in the synthesis of a wide variety of new and unusual π-complexes. Reductive substitution is defined as the process whereby a hydrocarbon ligand, usually a cyclopentadienyl group, in a transition metal complex is replaced by a new hydrocarbon ligand containing an appropriately different number of π-electrons so as to bring the new complex into conformity with the EAN rule; an additional feature is that formal reduction of the metal atom in the new complex is effected. Ligand substitution is a related process: one hydrocarbon ligand replaces another hydrocarbon ligand or even non hydrocarbon ligand such as CO, halide, etc., but with no accompanying reduction of the metal. As the reducing or displac-ing ligand a neutral unsaturated species, its corresponding radical anion or other suitably structured anion may be used.

The procedures for the syntheses of these complexes can serve as prototypes for the π-complexation of diene and arene polymers by transition metals. It is anticipated that these polymers will avoid certain drawbacks of the more conventional method of attachment of transition metal catalysts to polymers, i.e., phosphine ligation. Phosphine liganding polymers are notorious for their tendency to "leak" their complexed metals. These new polymers will be multiply-bonded to the metal atom, usually by an η^4 or η^6 ligation; said multiple-bonding should significantly decrease the tendency for the metal atoms to "leak". It is also anticipated, based on the model compounds studied, that reductive substitution and ligand substitution will afford several new types of organometallic polymers.

As models for these polymers, a number of high molecular weight dienes and arenes have been used. These include terphenyl and related polyaromatic systems, the tetraphenylborate anion and substituted tetraphenylborate anions, fluorene, carbazole and tetracyclones (tetraphenylcyclopentadienones).

REDUCTIVE SUBSTITUTION

Reductive substitution of cobaltocene with the tetracyclone ligand and substituted tetracyclone ligands has afforded a series of $C_5H_5Co^I(\eta^4\text{-diene})$ complexes (equation 1). This reaction is envisioned

(1)

as proceeding <u>via</u> an $\eta^5 \rightarrow \eta^1$ rearrangement prior to complete dissociation and reductive displacement of the cyclopentadienyl ligand (equation 2). Our thinking is

(Generalized, Diene)   (2)

that all the reductive substitutions described in this section proceed <u>via</u> this or a related mechanism. Substituted complexes of this class that have been prepared are bis-substituted at the p-positions of the 3- and 4- position phenyl groups (Table I).

TABLE I

$(\eta^5\text{-Cyclopentadienyl})(\eta^4\text{-tetracyclone})$cobalt Compounds

Prepared by the Reductive Substitution of Cobaltocene[a,b]

Para 3- and 4- position substituent	% Yield	m.p. (°C)
-H	37	325–327
-OCH$_3$	22.9	238–239
-CH$_3$	11.5	241–242
-Cl	30.5	299–300

[a]Reactions run in refluxing diethylbenzene or reactants simply melted together with no solvent present.

[b]All compounds are red crystals from CH$_2$Cl$_2$/petroleum ether.

Studies of the electronic influence of these remote substituents on the metal atom using ^{13}C nmr[1] and mass spectroscopy are in progress, but preliminary indications are that these remote substituents do not have much of an effect.

Direductive bis-substitution of nickelocene (equation 3) with tetracyclone has afforded bis-tetracyclone nickel, an extremely intractable material. A second model derivative that has been prepared by this technique is bis-duroquinone nickel, but at present the yield is extremely low. Attempts to prepare a mixed complex of these two ligands have failed.

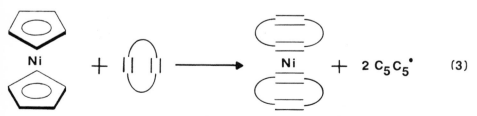

$$\text{Diēne} = \text{Tetracyclone or Duroquinone}$$

Manganocene and chromocene upon reductive substitution have afforded a series of polyphenyl derivatives of Mn^I, Cr^I, and Cr^O. This process is illustrated with terphenyl radical anion functioning as the reducing ligand (equation 4).

$$M = Mn, Cr \tag{4}$$

Terphenylcyclopentadienylmanganese is an air-stable, high melting solid whereas the analogous chromium derivative, being only a 17-electron system, is not air stable. Ir and nmr evidence indicate that complexation has taken place at the central ring of the terphenyl moiety. Other arenecyclopentadienylmanganese complexes that have been prepared are listed in Table II, although

TABLE II

$(\eta^5-C_5H_5)Mn(\eta^6-ArH)$ Derivatives

Prepared by Reductive Substitution of Manganocene

Arene	m.p.(°C dec)	% Yield	Color
$C_6H_5-C_6H_5$	116	30	orange
$C_6H_4-1,2-(C_6H_5)_2$	oil	30	red
$C_6H_4-1,3-(C_6H_5)_2$	106	35	red
$C_6H_4-1,4-(C_6H_5)_2$	113	61	orange
$C_6H_3-1,3,5-(C_6H_5)_3$	147	20	red
$C_6H_1-1,2,3,4,5-(C_6H_5)_5$	158-162	25	red
$C_6H_5C(C_6H_5)_6=C(C_6H_5)_2$	76	20	dk. red

the evidence is not as strong that the indicated arene ring is the
one π-bonded to the metal atom.

An 18-electron bis-terphenylchromium derivative has resulted
from direductive bis-substitution of chromocene where the ratio of
radical anion to metallocene was 2:1.   This result coupled with
the results of direductive bis-substitution in the nickel series
(equation 3) suggests that this process may prove useful in crosslinking of certain types of polymers.

LIGAND SUBSTITUTION

Ligand   substitution   is   a   process   whereby   a   hydrocarbon
ligand   or σ-type ligands such as CO or halogen are displaced by
a new hydrocarbon ligand but without accompanying reduction of the
metal atom of the complex.   Several notable sequences incorporating this process have been carried out.

Cyclopentadienylirondicarbonyl iodide upon treatment with the
bulky, negatively charged tetraphenylborate anion produced a novel
zwitterionic organometallic complex (equation 5).   In like manner

a series of phenyl-substituted tetraphenylborates have been used
to synthesize a series of correspondingly substituted zwitterionic
complexes (Table III).
Modification of this route has allowed preparation of not only the
parent complex but also the methylcyclopentadienyl derivative
(equation 6).

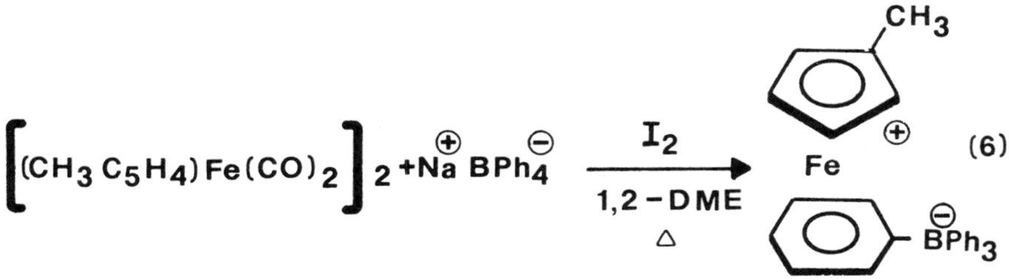

$$
\left[(CH_3 C_5 H_4) Fe(CO)_2 \right]_2 + Na^{\oplus} BPh_4^{\ominus} \xrightarrow[\substack{1,2-DME \\ \Delta}]{I_2}
$$

(6)

Ir, mass spectral, ^1H and ^{13}C nmr data support the assignment of the unusual zwitterionic structure in these complexes.

Reductive substitution of the parent iron zwitterionic complex with manganocene gave a bimetallic compound, illustrating the fact that two contiguous rings can be complexed (equation 7). This result also suggests

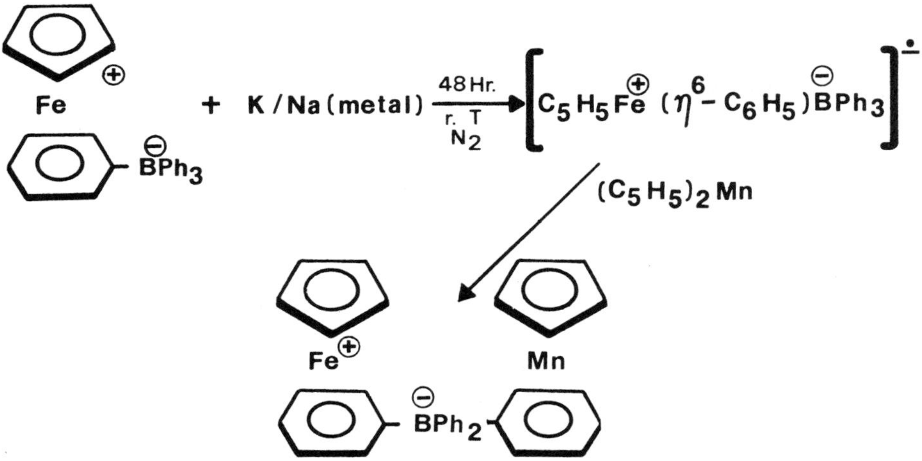

$$
+ \; K/Na\,(metal) \xrightarrow[\substack{r.\,T \\ N_2}]{48\,Hr.} \left[C_5 H_5 \overset{\oplus}{Fe} \, (\eta^6 - C_6 H_5) \overset{\ominus}{BPh_3} \right]^{\cdot -}
$$

(7)

$(C_5 H_5)_2 Mn$

that more than one transition metal may be incorporated into a suitable polymer with this technique.

Group VI hexacarbonyls have been found to be susceptible to displacement by the bulky, negatively charged tetraphenylborate ion. Equation 8 illustrates the synthesis of a series of arene metal tricarbonyls.

TABLE III

$(\eta^5\text{-Cyclopentadienyl})(\eta^6\text{-tetraarylborate})$ iron Complexes
Prepared by Ligand Substitution of
Cyclopentadienylirondicarbonyl iodide

Aryl substituent	% Yield	m.p.(°C dec)
C_6H_5-	65	275-278
$p-CH_3C_6H_4-$	60	241-242
$m-CH_3C_6H_4-$	56	190-192
$p-CH_3OC_6H_4-$	18	198-200

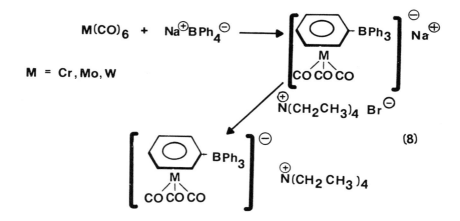

(8)

These molecules were isolated and characterized as their tetraethylammonium salts. In addition to the usual ir and nmr characterization of these compounds, an x-ray structure of the molybdenum member of the series has been completed[2].

It was desirable to demonstrate that at least one of these model compounds could function as a homogeneous catalyst. To this end the most soluble of the zwitterionic complexes was examined. This was prepared by a method analogous to the synthesis of the iron compound (equation 5), i.e., cycloheptatrienyl molybdenumdi-carbonyl iodide was treated with sodium tetraphenylborate (equation 9). In addition

$$(9)$$

to ir, ^{1}H and ^{13}C nmr data for this complex, and x-ray structural analysis has been obtained [2]. Its visible spectrum was consistent with a significant amount of low-lying electron density on the metal atom while the x-ray diffraction study revealed that the molybdenum atom might well possess additional coordination sites. The anticipated result of these observations, namely, that this complex would promote reduction of olefins under mild conditions, has now been realized [3].

Because of the interesting and potentially significant properties of this molybdenum complex, a series of analogues have been synthesized utilizing substituted tetraphenylborates. The fully characterized members of this series are recorded in Table IV.

A tetraphenylborate derivative of cobalt had recently been prepared. Bis-(norbornadienecobalt dicarbonyl) was prepared by heating dicobaltoctacarbonyl with two equivalents of norborna-diene. This intermediate was treated with sodium tetraphenyl-borate (equation 10)

$$(10)$$

thereby producing a 45% yield of the zwitterionic cobalt complex. This compound is the least soluble of the tetraphenylborate complexes; as such UV-visible and nmr spectra could not be obtained.

Lastly, an intriguing series of complexes containing polynuclear aromatic rings has been discovered[4]. These are the group VI complexes of carbazole and fluorene. Preparation of the tricarbonyl and tetraphenylborate members of these two series are illustrated in equations 11 and 12, respectively.

X = N, CH ; M = Cr, Mo, W   Isolated as $N(CH_2CH_3)_4$ Salts

Isolated as $N(CH_2CH_3)_4$ Salts

TABLE IV

$(\eta^7-Cycloheptatrienyl)(\eta^6-tetraphenylborate)M^{VI}$ Complexes

Metal (M)	Aryl Substituent	Yield (%)	m.p.(°C dec)
Mo	H	83	318–320
Mo	p–CH$_3$	11	267
Mo	m–CH$_3$	51	283–284
Cr	H	--	221–224
W	H	--	180

Thus complexes containing Ni^o, Co^I, Fe^{II}, Mn^I, Cr^o, Cr^I, Mo^o and W^o have been prepared. Most of these materials possess high thermal and air stability. One of these complexes has been demonstrated to function as a homogeneous catalyst for the reduction of olefins. All of these results are encouraging in our efforts to incorporate these same ideas into the manufacture of transition metal complexed polymers.

ACKNOWLEDGEMENTS

Technical assistance of Jeff Janos and Iwona Herod, Southern Illinois University, and R. Lin, Indiana State University, was helpful. Financial support by the Coal Extraction and Utilization Research Center, Southern Illinois University - Carbondale, was invaluable. DAO is grateful to the Committee on Institution Studies and Research (C.I.S.R.) at Murray State for two grants; 1978-1979, 1979-1980. Appreciation is expressed to Marilyn Flannery and Sandra Kosecki for typing the manuscript and to Wally Covert and Vince Flotta for illustration of the equations.

REFERENCES

(1)   In collaboration with K. Moedritzer and R. E. Miller,
      Monsanto Agricultural Products Co.

(2)   M. B. Hossain and D. van der Helm, Inorg. Chem., 17, 2893
      (1978).

(3)   D. A. Owen, A. Siegel, R. Lin, D. W. Slocum, B. Conway,
      M. Moronski and S. Duraj, Annals of the New York Academy of
      Sciences, 333, 90 (1980).

(4)   Group VI carbonyl of fluorene have concurrently been studied
      elsewhere, cf. A. N. Nesmeyanov, N. A. Ustynyuk, L. G.
      Makarova, S. Andre, Yu. A. Ustynyuk, L. N. Novikova and Yu.
      N. Luzikov, J. Organometal. Chem., 154, 45 (1978); A. N.
      Nesmeyanov, N. A. Ustynyuk, L. N. Novikova, T. N. Rybina, Yu.
      A. Ustynyuk, Yu. F. Oprunenko and O. I. Trifonova, J.
      Organometal. Chem., 184, 63 (1980).

Rate Accelerating Polymeric Cofactors in the Homogeneous Hydrogenation of Alkenes

D. E. Bergbreiter and M. S. Bursten
Department of Chemistry
Texas A&M University
College Station, TX 77843

ABSTRACT

The novel use of polymeric reagents as selective phosphine absorbers in a homogeneous catalysis is presented. Silver(I) p-toluene sulfonate (A) is shown to absorb tertiary phosphines from organic solution and to restore PPh_3-inhibited homogeneous hydrogenation of alkenes by $RhClL_3$ ($L=PPh_3$) to a non-inhibited rate. Treatment of $RhClL_3$ with ethylene in presence of A followed by addition of a second alkene and removal of ethylene under reduced pressure leads to an increased rate of hydrogenation for several alkenes.

INTRODUCTION

Dissociation of a nonvolatile ligand has become accepted as essential to alkene hydrogenations catalyzed by homogeneous transition metal compounds containing triphenylphosphine.[1] This

$$1 \qquad X_m ML_n \;\rightleftharpoons\; X_m ML_{n-1} \;+\; L \qquad L=PPh_3$$

dissociation was first postulated by Wilkinson and coworkers in 1966 to explain the low solution molecular weights obtained for

$RhClL_3$ (L=PPh_3), an active alkene hydrogenation catalyst.[2]
Though subsequent studies indicated dissociation is less extensive than was originally claimed [3], the postulate that the active catalyst is a coordinatively unsaturated species formed by
dissociation of a PPh_3 ligand is generally regarded as correct.
[4,5]

Although dissociation of PPh_3 is not the rate-determining
step for hydrogenations catalyzed by $RhClL_3$ [6], if equation 1
could be shifted to the right, faster rates of alkene hydrogenation might be achieved. This rightward shift could be accomplished by removal of PPh_3 from solution, thus generating more of
the active form of the catalyst.

In the past several years attempts at selective removal of
PPh_3 from a homogeneous solution have been made. Reverse osmosis
has been effective but requires pressures up to 100 atmospheres.
[7] Complexation of PPh_3 with soluble Lewis acids has also been
attempted [8]; although side reactions were observed, no rate enhancement attributable to a shift of equation 1 was observed.
Thus, although some attempts have been made, a rate enhancement
in homogeneous hydrogenation by a rightward shift of equation 1
has not been demonstrated previously.

Herein we report the use of functionalized polymers to absorb
free PPh_3 from homogeneous solutions. Although the use of supported homogeneous catalysts has become chemically fashionable in
the last decade, polymer functionalization to selectively absorb
a non-volatile ligand is a novel application of polymers to homogeneous catalysis. Such a polymer might be expected to effect a
significant rate enhancement in a system which dissociates a lig-
and to form the catalytically active species. A supported Lewis
acid should interact with free Lewis base, thus shifting equilibrium 1 in the desirable rightward direction. Side reactions,
observed by Shriver [8] when he used soluble Lewis acids to shift
this equilibrium, should be eliminated.

A polymeric cofactor for ligand absorption in homogeneous
catalysis must meet several criteria. It must not be catalyti-

cally active itself nor absorb the homogeneous catalyst. It should be an efficient absorber of non-volatile ligand--in this case, PPh_3. Preferably, it should also be prepared from readily available precursors, and be air-stable.

EXPERIMENTAL

All solutions were handled under nitrogen, argon, hydrogen, or ethylene, using standard Schlenk techniques.[10] Gases were used without further purification. Toluene, diethyl ether, tetrahydrofuran, and pentane were freshly distilled under nitrogen from sodium benzophenone ketal. Ethanol and methanol were purged with a strong flow of nitrogen for one hour. 1-Hexene, 1-octene, cyclohexene, styrene, and norbornadiene were passed through neutral alumina to remove stabilizers and peroxides, carefully degassed, and then purged with nitrogen for 1/2 hour.

$RhClL_3$ ($L=PPh_3$), $[RhCl(C_8H_{14})]_2$, and $PEtPh_2$ were obtained from Strem Chemicals and used as received. PPh_3 was obtained from Fluka. $RhCl(PEtPh_2)_3$ [11] was prepared by literature methods and identified by nmr, ir and melting point. Macroreticular ion exchange resin, polystyrene sulfonic acid, (Amberlyst 15), was purchased from Chemalog and extracted for 48 hours in a Soxhlet extractor with dimethylformamide before use in order to remove kaolins and surfactants present. Conventional flow techniques were used for exchange of metal ions.[12] After exchange, the polymer was dried under vacuum for 24 h. Silver substitution was 1.0 mmol Ag^+/g polymer (analyzed as AgCl). Proton nmr spectra were taken on a Varian T-60 nmr; ^{31}P spectra on a Varian XL 200 MHz nmr spectrometer. Infrared spectra were run on a Perkin-Elmer 297 ir spectrometer. Melting points were determined in a Thomas/Hoover melting point apparatus and were uncorrected.

Adsorbtion of Phosphines by Polymeric Reagents

A solution of PPh_3, 0.04 $\underline{M}$ in THF or acetone, was prepared.
20 mL of this solution was injected into a flask containing 2 g
(1.0 mmol metal ion/g polymer) of metal ion exchanged polystyrene
sulfonate beads and covered with a serum cap. 0.05 mL aliquots
of this solution were withdrawn after 1 and 12 hours and analyzed
for PPh_3 concentration by ^1H and ^{31}P nmr; p-xylene and $OPPh_3$ were
used as internal standards for ^1H and ^{31}P nmr respectively. Sodium polystyrene sulfonate and polystyrene sulfonic acid were
both used in control experiments, and showed no detectable PPh_3
absorption at these concentrations. A similar experiment was run
using a THF solution 0.04 $\underline{M}$ in styrene to determine whether the
alkene were being absorbed by either silver(I) polystyrene sulfonate ($\underline{A}$) or polystyrene sulfonic acid. In neither case was any
change in styrene concentration observed.

Hydrogenations

Hydrogenations were run at 25.0 ± 0.1°C in a 100 mL three-
necked flask equipped with a solid addition tube and a gas inlet
tube covered with a serum cap. A 1 3/8" egg-shaped magnetic stir-
bar provided vigorous mixing. The flask was connected to a hydrogenation apparatus consisting of a two-way stopcock to vacuum and
a three-way stopcock to the H_2 source and to a 50 mL buret. A
leveling bulb was connected to the bottom of the buret with rubber tubing and filled with dibutyl pthalate. Constant atmospheric pressure was maintained throughout the reaction by adjus-
ting this leveling bulb as the reaction proceeded.

In order to ascertain the effects of $\underline{A}$, typical hydrogenations were run in the three following ways: 1) polymer beads were
added at the start of (or during) a hydrogenation; 2) polymer beads
were present as the catalyst dissolved under H_2, and were stirred
with the solution of the transition metal catalyst under H_2 before
the addition of alkene; or 3) procedure 2 was followed, but under
C_2H_4 atmosphere, which was removed before hydrogenation commenced.

A typical hydrogenation using procedure 1 was as follows: The catalyst (10-60 mmol) was placed in the 100 mL 3-necked flask and the polymer beads (0.2-0.4 g) were put in the solid addition tube. The apparatus was carefully assembled, evacuated and flushed with H_2 three times, and then held under vacuum for 10 min. The apparatus was flushed with H_2 and evacuated three more times, then filled with H_2 and 10 mL of solvent (toluene, unless otherwise noted) was added. The solution was carefully degassed at reduced pressure and filled with H_2 three times and then stirred vigorously under H_2 until the catalyst was fully dissolved. At this point, stirring was stopped and alkene (2-20 mmol) was injected. When polymeric beads were to be added at the beginning of a hydrogenation, they were added at this point by turning the solid addition tube. Hydrogenation was initiated by resumption of stirring. Alternatively, polymeric beads were added after hydrogenation had been initiated.

Hydrogenations according to procedure 2 were run in essentially the same manner, except that the catalyst and polymeric beads were both placed in the hydrogenation vessel as the reaction was set up. The solid addition tube was not used; a glass stopper was inserted in its place. All other procedures were identical to those used in procedure 1.

The hydrogenations according to procedure 3 were run after utilizing an ethylene pretreatment. As in procedure 2, catalyst and polymer beads were placed in the hydrogenation vessel. The third neck was fitted with a gas inlet valve connected to the C_2H_4 source. The hydrogenation apparatus was evacuated and filled with C_2H_4 three times. Toluene (10 mL) was then added, and carefully degassed and flushed with C_2H_4 three times. The solution was allowed to stir under C_2H_4 for 900 s at which time the alkene to be hydrogenated (2-20 mmol) was injected. After an additional 100 s the apparatus was carefully evacuated and H_2 was introduced. The apparatus was carefully evacuated and vigorously stirred for 15 s; the apparatus was then carefully evacuated and flushed with H_2. After two more cycles consisting solely of flushing with H_2

and careful evacuation, the apparatus was filled with H_2 and hydrogenation begun by initiation of stirring.

When C_2H_4 was the alkene to be hydrogenated, the apparatus was carefully evacuated then filled with H_2 three times, after the catalyst had been stirred with polymer beads under C_2H_4 atmosphere for 1000 s, i.e., the solution was not stirred under reduced pressure.

Hydrogenation rates were calculated by plotting the consumption of H_2 vs. time and fitting a straight line to the region where the rate was nearly constant; this region encompassed at least the first 50–500 s after introduction of alkene. Rates are reported both as mmol H_2/s and mmol H_2/[catalyst],$\underline{M}$·min.

In order to determine whether any catalytic activity resided upon the polymer beads, some hydrogenation experiments were stopped before alkene had been consumed. The polymer beads were allowed to settle and the supernatant was transferred via syringe to another H_2-filled hydrogenation apparatus and hydrogenation resumed. The polymeric beads were stirred two times with toluene (15 mL) which was then syringed off. Fresh toluene (10 mL) and alkene (2–20 mmol) were injected and hydrogenation was initiated. No catalytic activity was observed to reside on these beads.

RESULTS AND DISCUSSION

Initial attempts to meet the criteria for an effective ligand absorbing polymer were focused on using polyethylene functionalized with transition metal organometallics.[13] Some PPh_3 was indeed absorbed by $FeCp(CO)_2$-polyethylene, but low functionalization (10^{-2} mmol Fe/g polymer) and low thermal and air stability caused abandonment of this functionalized polymer.[14] Alternatively, macroreticular ion exchange resin, polystyrene sulfonic acid (Amberlyst 15) was exchanged with various metal cations. These metal-ion containing polymers were checked for triphenylphosphine absorbtion by observing the change in PPh_3 concentration in THF or acetone solution by ^{1}H nmr. Results of these experiments are summarized in Table I. Silver(I) polystyrene sulfonate (A) was clearly the most efficient phosphine absorber, and this polymer was used in subsequent studies.

To further elucidate the behavior of A, absorption of PPh_3 was observed by ^{31}P nmr. Use of this technique permits low phosphine concentrations similar to those expected in a catalytic reaction to be observed. At 4×10^{-2} M, the amount of PPh_3 remaining in solution was 44% of the original amount after 12 min and 17% after 46 min; at 106 min the amount of PPh_3 was still 11% of the original amount and a small signal due to $OPPh_3$ had appeared. This is not surprising as this reaction was run in air. As the amount of PPh_3 remaining in solution is decreasing much less than exponentially, this suggests an equilibrium may ultimately be attained (equation 2). The low concentration of PPh_3 remaining accounts for the nonobservance of a signal in the ^{1}H nmr. The interval observed

$$2 \qquad PPh_3 \quad + \quad A \rightleftharpoons A-PPh_3$$

for the disappearance of most of the PPh_3 is consistent with the length of the induction period in PPh_3-inhibited hydrogenations (vida infra).

Once the absorption of PPh_3 by A had been verified, attempts to use A as a rate accelerating cofactor in homogeneous catalysis were initiated. $RhClL_3$ was chosen as it is among the most-studied homogeneous hydrogenation catalysts.[1,5]

TABLE 1. Absorbtion of PPh_3 by Metal-ion
Substituted Polystyrene sulfonate

metal ion	%PPh_3 absorbed from 0.04 M THF solution (12 h)
Co^{+3}	0
Co^{+2}	10
Ni^{+2}	10
Cu^{+2}	50
Ag^{+}	100 (1 h)

Initial results on hydrogenation of 1-octene, cyclohexene, and styrene proved somewhat disappointing in that the rate of hydrogenation, when $\underline{A}$ was added at the initiation of a hydrogenation (Procedure 1) was essentially unaffected. However, the efficiency of the polymeric reagent as a PPh_3 absorber in the presence of $RhClL_3$ was easily demonstrated by rapid formation of $[RhClL_2]_2$ when a solution of $RhClL_3$ was placed in contact with $\underline{A}$ under nitrogen. Dimer formation was observed both by ^{31}P nmr and by the formation of many small pale red crystals in the previously homogeneous solution. A control solution of $RhClL_3$ in absence of $\underline{A}$ showed no indication of dimer formation over a 3 h interval. The dimer is much less soluble than the monomer [2], and exhibits only 1/10 the catalytic activity of the monomer.[15] This greatly reduced activity was indeed observed when $RhClL_3$ was dissolved under N_2 in presence of $\underline{A}$ and subsequently used as a hydrogenation catalyst.

The lack of dimer formation in a catalytically active solution is largely attributable to the affinity of $RhClL_2$ for H_2.[16] Indeed, a solution prepared by dissolving $RhClL_3$ under H_2, in presence of $\underline{A}$ (procedure 2) displayed no loss in catalytic activity. H_2RhClL_3 is believed to dissociate to a much smaller degree than $RhClL_3$, and has a concomitantly smaller tendency to dimerize.[1,15]

Evidence of the effectiveness of $\underline{A}$ in absorbing PPh_3 from a catalytically active solution is illustrated in Figure 1. When $\underline{A}$ was added during a hydrogenation which was inhibited by excess PPh_3 the rate increased to that observed in the absence of any excess ligand. Thus $\underline{A}$ absorbs free PPh_3 from a catalytically active solution but does not activate H_2RhClL_3 during a catalysis.[16]

The difference between this ligand-absorbing polymer and its monomeric analog is apparent. Silver(I)$\underline{p}$-toluene sulfonate ($\underline{B}$) is light sensitive, decomposing over two days; samples of the polymeric analog ($\underline{A}$) show no sign of decomposition or loss of activity after six months. When $\underline{B}$ was added to a catalytically active solution of $RhClL_3$, H_2, and styrene, the reaction quickly stopped and a black insoluble precipitate formed. This may be a silver-rhodium

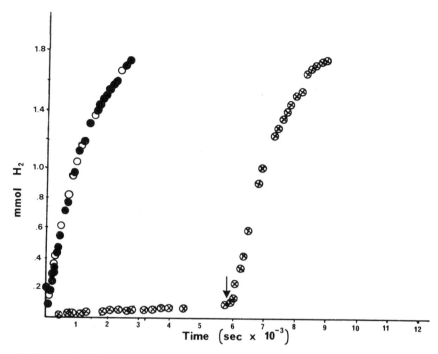

FIGURE 1. Hydrogenation of styrene (0.11 $\underline{M}$) catalyzed by 2.3 x 10^{-3} $\underline{M}$ RhClL$_3$ in toluene at 25°C in the absence of both $\underline{A}$ and any excess PPh$_3$ (0), in the presence of $\underline{A}$ (●), and in the presence of excess PPh$_3$ (1.6 x 10^{-2} $\underline{M}$ (⊗)). The effect of rate acceleration in the latter experiments reflects addition of $\underline{A}$ at 5900 s (arrow in Figure).

complex of the type reported by Vrieze.[17] When excess PPh$_3$ was present, the addition of $\underline{B}$ gave the same result as addition of $\underline{A}$. As [$\underline{B}$] $\cong$ [PPh$_3$] in this case, apparently $\underline{B}$ reacts with PPh$_3$ rather than with RhClL$_3$.

Competition for the dissociated PPh$_3$ ligand appeared to be the reason for lack of any enhanced hydrogenation rate in presence of $\underline{A}$. Therefore, displacement of a PPh$_3$ ligand prior to catalysis might lead to enhanced hydrogenation activity. Fortunately, RhClL$_3$ which is only a poor hydrogenation catalyst for ethylene [18], reversibly forms a π-complex with ethylene; this is stable in solution only under ethylene atmosphere.[19]   (equation 3) $RhClL_3$ was

$$3 \quad RhClL_3 \; + \; C_2H_4 \rightleftharpoons (C_2H_4)RhClL_2 \; + \; L \quad K_{eq} \cong 0.4 \; [19]$$

dissolved under ethylene atmosphere in presence of A; after 900 s alkene was added and after 1000 s ethylene was carefully removed. This solution hydrogenated various alkenes from 1.3 to 12 times faster than a solution which underwent ethylene pretreatment in absence of A, or a $RhClL_3$ solution which contained no polymeric reagent and did not undergo ethylene pretreatment, i.e., an unmodified hydrogenation.   (These results are displayed in Table 2.) Apparently, ethylene displaces PPh_3 in $RhClL_3$, allowing PPh_3 to be absorbed by A while preventing dimerization of $RhClL_2$.   Addition of alkene and removal of ethylene allow formation of a $RhClL_2$-(alkene) complex; once hydrogenation commenced, equilibrium 1 has effectively been shifted to the right and hydrogenation occurs at a faster rate.   The rate enhancement observed is greatest for those alkenes which form strong complexes with $RhClL_3$ and are slowly hydrogenated, consistent with formation of a higher concentration of $RhClL_2$(alkene) in solution than would be found in absence of ethylene pretreatment in presence of A.

TABLE 2.   Hydrogenation of Alkenes by $RhClL_3$
after Ethylene Pretreatment

alkene	[alkene],M	rate in presence of A[a]	rate in absence of A[a]	rate enhancement
1-hexene	0.35	23.9	12.2	1.96
cyclohexene	0.29	18.1	13.4	1.35
norbornene	1.03	36.6	21.3	1.72
norbornadiene	1.09	4.23	0.35	12.1
ethylene	--	4.4	0.74	5.8
styrene	0.21	60.	60.	1.0

[a]rate expressed as mmol hydrogen consumed/ $[RhClL_3] \cdot min$.

A comparison to work reported by Wilkinson in 1968 [11] is instructive. The catalyst $RhClL_n$ (n=2,3) was formed in situ by combination of varying ratios of PPh_3 and $[Rh(cod)Cl]_2$ with alkene under H_2. These reactions were run at rhodium and alkene concentrations comparable to those used in this work; a rate enhancement of 1.4 for hydrogenation of 1-hexene was observed for n=2 as opposed to n=3. This compares with the twofold rate enhancement seen in this work following ethylene activation in presence of $\underline{A}$, as opposed to an unmodified hydrogenation using $RhClL_3$. The consistency of these results suggests that the concentration of $RhClL_2$ is indeed increased when ethylene pretreatment is employed in presence of $\underline{A}$.

In order to test the effectiveness of ethylene pretreatment for complexes of the general formula $RhClL_3$, hydrogenations using $RhCl(PEtPh_2)_3$ were also investigated. All results with this compound were virtually identical to those found for $RhCl(PPh_3)_3$, except that ethylene pretreatment deactivated the catalyst. No ethylene complex with $RhCl(PEtPh_2)_3$ is known, so this is not entirely unexpected.

The effectiveness of a polymeric phosphine absorber as a rate-accelerating cofactor in homogeneous catalysis has been demonstrated. Application of this reagent to other phosphine-containing homogeneous transition metal catalysts is currently under investigation.

ACKNOWLEDGEMENTS

Generous support of this research was provided by the Department of Energy, the Office of Naval Research, and the Center for Energy and Mineral Resources at Texas A&M University.

REFERENCES

[1] James, B. R. "Homogeneous Hydrogenation"; John Wiley & Sons: New York, 1973.

[2] Osborn, J. A.; Jardine, F. H.; Young, J. F.; Wilkinson, G. J. Chem. Soc.(A) 1966, 1711-1732.

[3]   Lehman, D. D.; Shriver, D. F.; Wharf, I. J. Chem. Soc., Chem.
      Commun. 1979, 1486.   Arai, H.; Halpern, J. ibid 1971,
      1571-2. Eaton, D. R.; Stuart, S. R. J. Amer. Chem. Soc. 1968,
      90, 4170-2.

[4]   Hoffman, P. R.; Caulton, K. G. J. Amer. Chem. Soc. 1975, 97,
      4221-8.

[5]   Dolcetti, G.; Hoffman, N. W. Inorg. Chim. Act. 1974, 9, 269-
      303. Harmon, R. E.; Gupta, S. K.; Brown, D. J. Chem. Rev.
      1973, 73, 21-52.

[6]   Halpern, J; Okamoto, T.; Zakhariev, A. J. Molec. Catal. 1976,
      2, 65-8 and references therein.

[7]   Gosser, L. W.; Knoth, W. H.; Parshall, G. W. J. Molec. Catal.
      1977, 2, 253-263. Gosser, L. W.; Knoth, W. H.; Parshall, G. W.
      J. Amer. Chem. Soc. 1973, 95, 3436-7.

[8]   Strauss, S. H.; Shriver, D. F. Inorg. Chem. 1978, 17, 3069-
      3074. Porter, R. A.; Shriver, D. F. J. Organomet. Chem.
      1975, 90, 41-7. Hidai, M.; Kuse, T.; Hidita, T.; Uchida, Y.;
      Misono, A. Tetrahedron Lett. 1970, 1715-6.

[9]   Chaurm, Y.; Commereuc, C.; Dawans, F. Prog. Polym. Sci. 1977,
      5, 95-226.

[10] Shriver, D. H. "Manipulation of Air-Sensitive Compounds";
      New York: McGraw-Hill, 1969.

[11] Moltelatici, S.; van der Ent, A.; Osborn, J. A.; Wilkinson,
      G. J. Chem. Soc. (A) 1968, 1054-8.

[12] Dorfner, K. "Ion Exchangers: Properties and Applications";
      Ann Arbor Science Publishers, Inc.: An Arbor, Mich., 1973;
      99-115.

[13] Rasmussen, J. R.; Stedronsky, E. R.; Whitesides, G. M. J.
      Amer. Chem. Soc. 1977, 99, 4736-4745. Rasmussen, J. R.;
      Bergbreiter, D. E.; Whitesides, G. M. J. Amer. Chem. Soc.
      1977, 99, 4746-4756.

[14] A report on decarbonylation of metal carbonyls by $RhCl(PPh_3)_3$
      led to abandonment of any metal carbonyl as a phosphine-
      acceptor: Varshavsky, Yu. S.; Shestakova, E. P.; Kiseleva,
      N. V.; Cherkasova, T. G.; Buzina, N. A.; Bresler, L. S.;
      Kormer, V. A. J. Organomet. Chem. 1979, 170, 81-83.

[15] Halpern, J.; Wong, C. S. J. Chem. Soc., Chem. Commun. 1973,
      629-630. Meakin, P.; Jesson, J. P.; Tolman, C. A. J. Amer.
      Chem. Soc. 1972, 94, 3240-3242.

[16] k for $RhClL_3 \rightleftharpoons RhClL_3H_2 \cong 2 \times 10^4 mol^- dm^3$ [17] and Rousseau,
      C.; Evard, M.; Petit, F. J. Molec. Catal. 1977/8, 3, 309-
      324.

[17] Kuyper, J.; Vrieze, K. J. Organomet. Chem. 1976, 107, 129–138.

[18] Ohtani, Y.; Yamagishi, A.; Fujumoto, M. Bull. Chem. Soc. Jpn. 1979, 52, 2149–50.

[19] Tolman, C. A.; Meakin, P. S.; Lindmer, D. L.; Jesson, J. P. J. Amer. Chem. Soc. 1974, 96, 2762–2774. Ohtani, Y.; Yamagishi, A.; Fujimoto, M. Bull. Chem. Soc. Jpn. 1979, 52, 69–72; 1979, 52, 1537–8; 1979, 52, 3437–8; 1977, 50, 1453–9.

POLYPHOSPHAZENES

Polydichlorophosphazene Polymerization Studies

GARY L. HAGNAUER

Polymer Research Division
U.S. Army Materials and Mechanics Research Center
Watertown, Massachusetts 02172

ABSTRACT

Polydichlorophosphazene $(NPCl_2)_x$ is unique as a synthetic precursor
for poly(organo)phosphazenes [1-6] and presents special problems
in polymerization studies because of branching, crosslinking,
and cyclization reactions and in characterization due to the
polymer's hydrolytic instability. This paper reviews the
literature and discusses problem areas and recent advances
relating to polydichlorophosphazene polymerization. Melt,
solution, and irradiation polymerization reactions and mechanisms
are discussed. Previously unpublished results are presented,
and current research efforts and areas for future investigation
are considered.

BACKGROUND

In 1897, H. N. Stokes [7] first reported high temperature
reactions of cyclic chlorophosphazenes forming an insoluble
"inorganic rubber" which was later recognized as being crosslinked
polydichlorophosphazene. Early workers such as Schenck and Römer
[8], Ficquelmont [9], and Schmitz-Dumont [10,11] investigated the
polymerization reaction; but the first detailed studies were not
made until the 1950s by Patat and coworkers [12-15]. Since then,
techniques for high temperature, melt and solution polymerizations
and for irradiation polymerization have been developed; and the

effects of temperature, pressure, catalysts, etc. have been
studied [16-37]. Also, a number of polymerization kinetics
studies have been reported for both catalyzed and uncatalyzed
reactions [18-20, 22, 27]. The most significant advance relating
to the development of technologically useful polymers was made in
1965 when Allcock and coworkers developed methods for preparing
soluble, open-chain polydichlorophosphazene [1, 2]. Unlike the
crosslinked or insoluble polymer, linear or branched polydichlorophosphazene is soluble in a variety of solvents (e.g., benzene,
chloroform, and tetrahydrofuran) and therefore is amenable to
subsequent substitution reactions where the chlorine atoms are
replaced with organic nucleophiles to yield stable, high molecular
weight poly(organo)phosphazenes.

HIGH TEMPERATURE, MELT POLYMERIZATION

The high temperature, melt polymerization reaction of hexa-
chlorocyclotriphosphazene (I) to polydichlorophosphazene has been
most widely studied. Trimer (I) is the principal product obtained
in the synthesis of chlorocyclophosphazenes [8, 38-40]. The cyclic
trimer (I) is a white, crystalline (mp 112-114°C) solid that
sublimes in vacuum and is soluble in organic solvents. Thermal
polymerization proceeds at a faster rate and at lower temperatures
with the trimer (I) than with higher molecular weight cyclic
homologs [18, 25]. Also, melt polymerization has been preferred
because solution polymerization generally is slower and tends to
yield insoluble or low molecular weight products [12, 19, 36].
Melt polymerization reactions are usually run in sealed, evacuated
glass tubes. Below 230°C, the uncatalyzed polymerization is quite
slow and may not proceed at all if the trimer is extremely pure
[18, 29]. Between 230° and 350°C, the rate of polymerization
increases with temperature [18]. With increasing polymerization
time and as the temperature is raised, the rate of crosslink
formation increases. Crosslinking is unpredictable and is greatly
enhanced by the presence of impurities. Also, as the temperature

is increased, depolymerization reactions occur such that at about 600°C, ring–polymer equilibrium reactions prevail and the polymer degrades to low molecular weight cyclics $[NPCl_2]_{3-7}$ [10, 14, 15, 25]. Finally, there is a large number of materials which behave as polymerization catalysts or accelerators and promote polymerization at temperatures as low as 200°C. Additives or impurities which are able to extract chloride ions from phosphorus tend to accelerate the rate of polymerization [18-20, 22, 25, 27, 29, 30, 32-34]. Unfortunately, such catalysts also tend to promote crosslinking [18-20, 27, 30, 33].

To obtain soluble, high molecular weight polymer, the polymerization is best run using highly pure trimer (I) at temperatures between 240 and 255°C in clean, glass tubes sealed under vacuum (0.005 to 0.010 mm Hg). Procedures for purifying polymerization grade trimer have been published [2, 16]. The highly pure commercial cyclic trimer/tetramer mixture, Phosnic 390 [41], may be used without further purification [37]. As polymerization proceeds, the contents of the polymerization tube become increasingly more viscous. When the reaction mixture becomes so viscous that little or no flow occurs, usually within 1-7 days, the reaction should be terminated by reducing the temperature below 200°C. Sealed polymerization tubes may be kept at ambient temperature for years without effecting further reaction. Depending on trimer (I) purity, polymerization temperature, and reaction time, polymer yields up to 75% may be obtained. However, if polymerization is allowed to continue beyond the time at which flow ceases, excessive branching and crosslinking may occur.

Recently the uncatalyzed, high temperature, melt polymerization [Eq. (1)] was investigated in considerable detail [42]. Two
batches of trimer (I) were obtained from different manufacturers
and purified extensively. Purified trimer (I) from each batch was
placed in pyrex tubes, sealed under vacuum, and polymerized in an
aluminum block oven at 250°C. Tubes were removed from the oven
periodically to terminate polymerization at various stages; and
the reaction products were converted into hydrolytically stable
trifluoroethoxyphosphazenes by nucleophilic displacement of chlorine
on the phosphorus atoms.

$$\underset{\underset{Cl}{|}}{\overset{\overset{Cl}{|}}{(N = P)_x}} \xrightarrow[\text{benzene-tetrahydrofuran}]{\text{NaOCH}_2\text{CF}_3} \underset{\underset{OCH_2CF_3}{|}}{\overset{\overset{OCH_2CF_3}{|}}{(N = P)_x}} + 2x\text{NaCl} \qquad (2)$$

Mild reaction conditions were used to preserve the polyphosphazene chain backbone structure and solution precipitation techniques
were used to separate the polymer from unreacted trimer [42, 43].
Polymer yields were determined gravimetrically. The polymer was
then thoroughly characterized using liquid size exclusion chromatography, viscometry, membrane osmometry and light scattering.
Polymer yield is plotted versus polymerization time in figure 1.
The polymerization behavior of the two trimers, designated A and
B, is quite different. After 7 days, polymer prepared from trimer
A contains 2% crosslinked product; whereas no gel is evident in
polymers prepared from trimer B. No induction period is evident.
Polymerization starts almost immediately; however there are great
differences in the rates of polymerization of trimers A and B.
The polymerizations autoaccelerate after 1 day and 6 days for
trimers B and A, respectively. Changes in the intrinsic viscosity
[η] of the stabilized polymer correspond to changes in polymer
yield with polymerization time [Fig. (2)]. Molecular weight tends
to increase with polymerization time and, as shown in figure 3,
significant changes in the molecular weight distribution take place.
The polyphosphazene has a bimodal molecular weight distribution

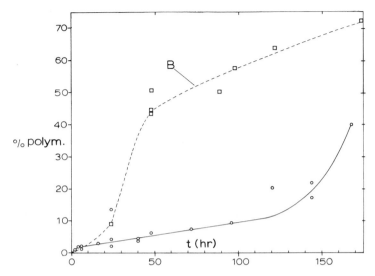

Figure 1. % polymerization of trimer A (——) and trimer B (----)
versus polymerization time at 250°C.

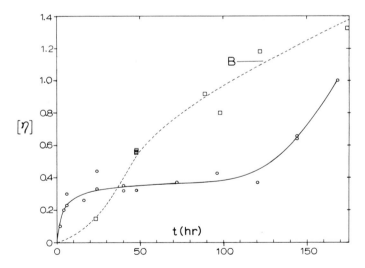

Figure 2. Intrinsic viscosity of polymer from trimer A(——) and
trimer B(----) versus polymerization time.

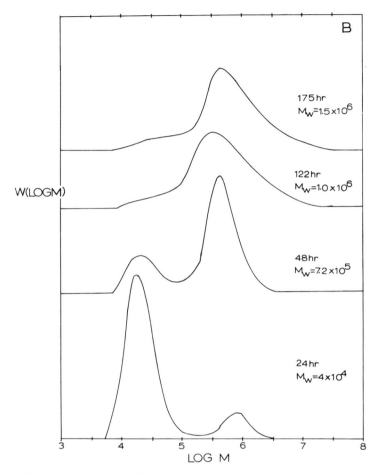

Figure 3. Weight differential molecular weight distributions
of poly[bis(trifluoroethoxy)phosphazene] products from the
polymerization of trimer B.

with peak maxima at 2×10^4 and at 1×10^6 g/mol. In the early
stages of polymerization, the main product is low molecular weight
polymer; but as polymerization proceeds, increasingly larger
amounts of high molecular weight polymer are produced. The
molecular weight distributions of polymers prepared from trimer A
are also bimodal but not as broad as the B polymers. Finally, the

analyses suggest that polymers prepared from trimer A are
relatively linear during the early stages of polymerization; but
after 6 days, branching and crosslinking reactions set in. All the
polymers prepared from trimer B are branched to some extent but
there are no insoluble products, suggesting the absence of a high
degree of crosslinking.

The quality of trimer (I) has a profound effect not only on
the rate of polymerization but also on the nature of the reaction
products [37]. For example, different batches of commercial trimer
[41], designated IJ-3 and IL-22, and of highly pure trimer PN-1
were melt polymerized in sealed, evacuated glass tubes at 250°C.
Special techniques were developed for the separation and dilute
solution characterization of the dichlorophosphazene reaction
products. The commercial trimer samples contained 9% cyclic
tetramer, octachlorocyclotetraphosphazene; however, in both cases,
the trimer/tetramer mixtures were free of apparent impurities.
Polymerization times, yields, and characterization data are shown
in Table 1. Intrinsic viscosity [η], absolute number- and weight-
average molecular weights, and the z-average radius of gyration
$\left\langle s^2 \right\rangle_z^{\frac{1}{2}}$ were measured for each polymer in dilute toluene solution
at 25°C. Liquid size exclusion chromatography was used to analyze
nonpolymeric components in the reaction mixtures and to characterize molecular weight distributions of the polymers. It is noted
that the cyclic tetramer concentration does not change and that
no insoluble products are formed. Seemingly, the tetramer plays
no active role in polymerization; however, there may be an indirect
benefit in that the tetramer tends to keep the viscosity of the
reaction medium from increasing too rapidly as polymerization
progresses and thereby reduce the likelihood of crosslinking.
The commercial samples give comparable polymer yields to the high
purity trimer but with shorter polymerization times.
Also, side reactions occur during polymerization producing high
and intermediate molecular weight oligomers, particularly with
the commercial samples. The intermediate MW oligomers are cyclic

TABLE 1.   Analysis of Polymerization Products

SAMPLE	PN-1	IJ-3	IL-22
Polymerization Time (hours)	100	45	21.5
% Polymer	30.8	26.0	31.4
% High MW oligomers	1.8	12.4	4.0
% Intermediate MW oligomers	2.4	18.4	7.8
% Yield	35.0	56.8	43.2
% Cyclic tetramer	0	9.0	9.0
$[\eta]$ (dl/g)	1.35	0.94	1.68
$\bar{M}_w$ (10^{-5})	24.5	8.96	18.1
$\bar{M}_n$ (10^{-5})	4.30	3.24	4.15
$\bar{M}_w/\bar{M}_n$	5.7	2.77	4.36
$\langle s^2 \rangle_z^{\frac{1}{2}}$ (A)	874	570	774

with molecular weights up to 1000g/mol and the high MW oligomers
are cyclic and linear species ranging between 1000 and 8000 g/mol.
The polymers have high molecular weights and broad molecular weight
distributions; and the $[\eta]$ and $\langle s^2 \rangle_z^{\frac{1}{2}}$ values suggest that there is
little, if any, polymer chain branching. Upon comparison, there
may be a relationship between the somewhat lower molecular weight
and narrower distribution of sample IJ-3 and the relatively high
concentration of high and intermediate MW oligomers formed during
its polymerization.

The presence of trace impurities may explain why different
batches of trimer (I) purified in an identical manner or obtained
from the same manufacturer should polymerize at different rates
and form products with different structures. For example, small
amounts of water (0.02 to $\sim$0.1 mol %) increase the rate of polymerization; while larger amounts of water (>0.2 mol %) tend to retard
polymerization [33]. Trace amounts of phosphorus pentachloride
(<0.10 mol %), a reagent used in the synthesis of chlorophospha-

zenes, tend to retard polymerization and lower polymer molecular weights [33]. Hydrogen chloride, formed when chlorophosphazenes react with water, in trace amounts (<0.083 mol %) also reduces the rate of polymerization and lowers molecular weight [33]. Indeed slight differences in the surface treatment and' volume of the glass polymerization tubes or in the amount of air and oxygen trapped in the trimer may have a catalytic effect on polymerization [29, 32, 42].

A variety of compounds are reported to catalyze the melt polymerization of the cyclic trimer (I) and tetramer. Alcohols, ethers, ketones, esters, carboxylic acids, peroxides, nitromethane, and metals, like tin, zinc and sodium, enhance the rate of polymerization at 200°–235°C when added in catalytic amounts [18–20, 25, 27]. At higher temperatures water [27, 33, 34], sulfur [30], oxygen [29], and the surface treatment of the walls of glass polymerization tubes [22, 32] have a catalytic effect on polymerization. In the past, most attempts to catalyze melt polymerizations resulted in the formation of products which were partly or wholly insoluble and therefore unsuitable as precursors for the synthesis of poly(organo)phosphazenes. However recently, soluble, high molecular weight polymers have been prepared in catalyzed, melt polymerizations by using highly pure chlorocyclophosphazenes and limiting conversions to about 50%. For example, at temperatures between 200° and 300°C, very low concentrations of water (0.005–0.10 mol %) catalyze polymerization [33, 44]. Catalysis of trimer (I) polymerization at 231°C with a Lewis acid (0.024 mol % Et_3Al_2Cl)$_3$yields 78% polymer after 6 hours reaction [45]. The polymer has a high molecular weight as indicated by its intrinsic viscosity ($[\eta] = 1.46$ dl/g) with only 0.79% insoluble gel [45]. Other Lewis acids, such as $AlCl_3$, $EtAlCl_2$, and Et_2AlOEt, also have proven to be effective catalysts producing soluble polymer with little or no gel and polymerization times ranging from 60 hours at 175°C to 1.5 hours at 245°C [46–48]. Melt polymerization with Ziegler catalysts containing $TiCl_3$ and Et_3Al yielded 50% polymer after 12 hours at 220°C [49].

SOLUTION POLYMERIZATION

Solution polymerization techniques have been applied in
attempts to prepare high molecular weight polydichlorophosphazene
[12, 13, 19, 36]. The potential advantages of solution polymerization include lower viscosity and better control of polymer chain
structure as compared with the melt polymerization. Conceptually,
it should be possible to completely polymerize the trimer and
obtain high molecular weight, fully soluble polydichlorophosphazene.
Initiation, termination, chain transfer and branching reactions
should be controllable to the extent that catalysts and solvents
in conventional solution polymerizations are effective. In practice,
however, solution polymerizations have not been so successful. At
the high temperatures required for polymerization (200–300°C) many
solvents react with the chlorophosphazenes to form oily or
insoluble residues [12, 13, 19, 36]. Trimer (I) polymerizations
at 270–300°C with carbon tetrachloride as the solvent and 5–30%
solution concentrations resulted in higher yields than obtained
by melt polymerizations run at similar temperatures and times;
but, although polymer fractions with molecular weights as high as
130000 were analyzed, less than half of the products were soluble
[12]. For trimer (I) polymerizations at 210°C, only crosslinked
products are obtained in cyclohexane; mixtures of soluble and
crosslinked products are formed in chlorobenzene with the highest
yields of soluble polymer (ca. 40%) and the least amount of crosslinked product (ca. 5%) produced in the 40–60% concentration range;
and soluble, low molecular weight (<10000 g/mol) products are
formed in benzene over the 20–80% concentration range [36].
Soluble fractions from the chlorobenzene polymerization have
molecular weights of 2.6 X 10^6 g/mol which is similar to values
of melt polymerized polymers and, upon comparison with results
using benzene and cyclohexane as solvents, suggests an effect
relating to the polarity or electrophilicity of the solvent [36].

Trimer (I) solution polymerizations at 210°C with benzene as
the solvent are catalyzed by benzoic acid, ethyl ether, and ethyl
alcohol but yield only soluble, low molecular weight (ca. 1300

g/mol) products [19]. A more promising polymerization uses
polyphosphoric acid (1%) as a catalyst and is run at 200°C using
50% trimer (I) solutions with 1,2,4-trichlorobenzene [50]. The
polymerization yields soluble polymer which is amenable for use in
substitution reactions but has a somewhat lower molecular weight
and broader molecular weight distribution than polymers obtained
by melt polymerization [5]. Metal or quaternary ammonium salts of
carboxylic, sulfonic, picric, or phosphorus acids are also effective
catalysts for solution polymerization of chlorocyclophosphazenes
in 1,2,4-trichlorobenzene [51]. For example, a stirred solution
consisting of 1 part cyclic trimer (I), 1 part trichlorobenzene
and 0.01 part $(CH_3SO_3)_2$ Hg and blanketed with nitrogen at 217°C
for 3 hours produces soluble, high molecular weight polymer [51].
There is also evidence that the presence of certain fluorine
compounds, sulfur, or sulfur-donors prevent gelation in the
solution polymerization of chlorocyclophosphazenes at temperatures
in the range 140-225°C when solvents are used that dissolve the
chlorocyclophosphazenes and have dielectric constants $\varepsilon \geq 2$ [52, 53].
An amazing 94% yield of soluble polymer and no insoluble gel is
obtained when a 50% solution of cyclic trimer (I) and 0.5 mol %
sulfur in tetralin is heated for 70 hours at 205°C [52]. In the
absence of sulfur, insoluble products with concentrations as high
as 15% are found [52].

SOLID STATE POLYMERIZATION

The cyclic trimer (I) polymerizes in the solid state when
irradiated with 50kV X-rays [23]. The rate of polymerization
increases with temperature to a maximum of 0.8% per hour at the
melting point of the trimer and drops to zero above the melting
temperature [23]. The presence of water, oxygen, solvent, and
even cyclic tetramer decreases the polymerization rate and the
polymerization appears to terminate after about 10% conversion
[23, 26]. Finally, it is noted that the cyclic tetramer does not
polymerize in the solid state [23].

Properties of polymer prepared by solid state irradiation are
identical to those of crosslinked polydichlorophosphazene prepared
by high temperature, melt polymerization. The polymer is insoluble,
exhibits elastomer behavior, and crystallizes when stretched to
give an x-ray diffraction spectrum identical to the melt polymerized
polymer [23, 26].

POLYMERIZATION MECHANISM

The mechanism of the polydichlorophosphazene polymerization
is not known with certainty. Attempts to study the polymerization
kinetics have been plagued not only with experimental difficulties
due to the high polymerization temperatures and hydrolytic
instability of dichlorophosphazenes, but also with the problem of
reproducibility caused by trace impurities. Additionally, there
are problems of interpretation where the reaction mixture includes
cyclic and linear oligomers, soluble linear and branched polymers,
and insoluble, crosslinked products. For uncatalyzed, melt
polymerization, second-order polymerization kinetics are reported
with activation energies of 42, 51.3 and 57 kcal/mol [12, 22, 27].
However rate constants determined at 250°C differ by almost an
order of magnitude, probably as a consequence of trace impurities.
The actual rate constant may be appreciably smaller than reported
if impurities are entirely eliminated [29]. Catalyzed, melt
polymerizations follow first-order kinetics and are reported to
have induction periods which decrease with increasing temperature
[18-20, 27]. Other investigators find no indication of an
induction period for either catalyzed or uncatalyzed polymerization [12, 22, 29, 33, 34, 42]. For melt polymerizations catalyzed
with benzoic acid, the activation energy is 24.3 kcal/mol in the
range 200-220°C and the rate increases exponentially with catalyst
concentration [27]. Finally, it is noted that insoluble products
were formed in most studies and recent findings indicate that the
polymerization is too complex to analyze as simple first- or
second-order kinetics [37, 42].

Polydichlorophosphazene polymerization reactions are exothermic [10, 11, 24] and the extent to which the polymerization is exothermic decreases with increasing size of the cyclic oligomer [25]. This is also indicated by the fact that the rate of polymerization of the cyclic tetramer is much slower than for the trimer [18]. As mentioned previously, the rate of polymerization increases with increasing temperature [12,18, 19, 22, 26, 27, 29] and with certain impurities or catalysts present [18-20, 22, 25, 27, 29, 30, 32-34, 44-51]. Other compounds lower the rate of polymerization and reduce polymer molecular weight [33]. Increasing temperature and the presence of impurities or catalysts also increase the rate of crosslinking [2, 18-20, 27, 30, 33]; while other compounds inhibit crosslink formation [52, 53]. Compounds, such as phosphorus pentachloride, cause polymer chain degradation [33] and may promote side reactions during polymerization of the cyclic trimer to produce high molecular weight cyclic and linear oligomeric chlorophosphazenes [37, 54]. At high polymerization temperatures, especially above 300°C, polymer chain depolymerization reactions occur and consideration of ring-polymer equilibria becomes important [16, 55]. Increasing pressure between 10 and 70 kbars favors the formation of polymer; however the rate of polymerization decreases as the pressure is raised [24]. This suggests that the rate determining polymerization step has a transition state that occupies a larger volume than the reactants and argues in favor of a ring-opening chain propagation mechanism. Of course, there are ample spectral, chemical and physical property data supporting that polydichlorophosphazene is an open-chain polymer and not a series of linked cyclic units [3, 16].

Since high molecular weight polymer is formed at the very onset of polymerization [12, 42], there is little doubt that a chain growth polymerization is involved; however there are questions concerning the exact nature of the active center for kinetic chain growth. Free radical mechanisms have been proposed for both high temperature, melt and solution polymerizations

[12-14, 22, 30]. Of the compounds that are sufficiently stable
to be used as free radical initiators at elevated temperatures,
sulfur has a powerful catalytic effect at 250°C [30] and oxygen
appears to be an active initiator above 300°C [29] for melt
polymerizations. In fact the exceedingly slow polymerization of
highly pure trimer at 300°C may be a thermal polymerization
involving free radical formation and initiation of the trimer [29].
However most evidence suggests, at least for temperatures below
300°C, that a free radical mechanism is unlikely. Melt polymerization of trimer is not initiated at elevated temperatures by ultraviolet irradiation [18], X-rays [17], or accelerated electrons
[56]. A different mechanism involving bond interchange may explain
why irradiation with X-rays induces trimer polymerization in the
solid state [23, 26]. And finally, during melt polymerization at
250°C, no free radical species are detected by electron spin
resonance spectroscopy [57].

Most evidence favors an ionic mechanism for polymerization.
Compounds which typically catalyze ionic polymerizations are also
observed to accelerate polymerization of the trimer [18-20, 25,
27, 33, 44-51]. It is also noted that the reaction of a 5:1 molar
mixture of cyclic trimer and phosphorus pentachloride at 250-**300°C**
proceeds by an ionic mechanism to form low molecular weight,
linear products of the type $[Cl(PCl_2 = N)_{3x}PCl_3]PCl_6$ [28]. And
during melt polymerization of the cyclic trimer, both the conductance and dielectric constant of the reaction mixture increase
rapidly as the temperature is raised from 203° to 253°C and
continue to increase with time at 253°C [57]. When the temperature
is lowered from 253° to 202°C, the reaction mixture's dielectric
constant value remains high which can be attributed to the presence
of polymer, and the fact that part of the conductance is reversible
is indicative of ionization equilibria [57]. Conductance data
indicates that ionization of the trimer is the primary initiation
step and that ionization occurs concurrently with polymerization.
Finally, an ionic mechanism is consistent with solvent effects
reported in the uncatalyzed, solution polymerization of trimer

[36]; i.e., to produce high molecular weight polymer the dielectric constant of the solvent must be high. In an ionic polymerization a high dielectric medium is effective in shielding or solvating the active center from its counterion to achieve higher rates of polymerization and high molecular weights, whereas free radical reactions generally are insensitive to their environment.

Two types of initiation processes are proposed [57] involving (a) heterolytic cleavage of a phosphorus-nitrogen bond or (b) dissociation of a chloride ion from phosphorus.

$$(3)$$

Mechanism (a) is unfavored since the cyclic trimer $[NP(OC_6H_5)_2]_3$, which has a phosphorus-nitrogen bond strength similar to that of the chlorophosphazene trimer (I), neither polymerizes nor shows large irreversible increases in conductance or dielectric constant over the temperature range 120-350°C [57]. Mechanism (b) is consistent with the conductance behavior of the trimer and the low conductance of $[NP(OC_6H_5)_2]_3$ at 350°C, since the phosphorus-oxygen bond in $[NP(OC_6H_5)_2]_3$ is quite stable to ionic reactions [16]. The chloride ionization mechanism (b) is also consistent with the facts that catalysts for the polymerization are generally those which also are able to extract chloride ions from phosphorus [18-20, 25, 27, 32-34, 44-51] and that hydrogen chloride is found in reaction tubes after polymerization [18, 20, 27, 32-34]. Furthermore, the decrease in the rate of polymerization and reduction in polymer molecular weight upon adding small amounts of hydrogen chloride during the melt polymerization of

trimer [33] agrees with mechanism (b) in Eq. (3) and a mechanism
for terminating chain propagation.

   The high conductivity of the reaction mixture during
polymerization argues against the chloride ion being the
initiating species [57]. Because of the greater electropositivity
of phosphorus, the electrophilic center of the initiating species
is probably located on phosphorus rather than nitrogen and propagation can be envisioned as the electrophilic phosphorus center
attacking another ring and inducing ring-opening.

$$(4)$$

Eq. (4) is in accord with conductance and dielectric data and with
the observed pressure dependence of the polymerization rate [24]
which suggests that the rate determining step has a transition
state occuping a larger molar volume than the sum of the reactant
volumes. The high molecular weight, linear polymer found at low
conversions implies that the active center for propagation is
quite stable [42]. In fact the uncatalyzed, melt polymerization
of highly pure trimer at 250°C has many of the characteristics of
a "living" polymer polymerization [42]. If initiation were
controlled and side reactions could be avoided, high molecular
weight, linear polymer having a narrow molecular weight distribution probably could be realized.

   Eq. (3) and (4) suggest mechanisms for polymer chain
branching and crosslinking. Ionization of a chloride ion from
chlorophosphazene units within the polymer chain could generate
an active center for cationic chain propagation and thereby cause
branching [Eq. (5)] or the active center may attack a nitrogen
atom on the backbone of another chain resulting in crosslinking
[Eq. (6)].

$$\underset{\underset{\overset{||}{\underset{\zeta}{N}}}{Cl-P-Cl}}{} \underset{+Cl^-}{\overset{-Cl^-}{\rightleftharpoons}} \underset{\underset{\overset{||}{\underset{\zeta}{N}}}{Cl-P^+}}{} \overset{[NPCl_2]_3}{\longrightarrow} \underset{\underset{\overset{||}{\underset{\zeta}{N}}}{Cl-}\ P-(N=P)_2^{Cl\ Cl}\!\!\!-N=\overset{Cl}{\underset{Cl}{P^+}}}{} \quad (5)$$

$$\underset{\underset{\overset{||}{\underset{\zeta}{N}}}{Cl-}\ P^+}{} \quad \underset{\underset{\overset{||}{\underset{\zeta}{N}}}{:N}}{\overset{P^{\diagdown Cl}_{\diagdown Cl}}{}} Cl^- \longrightarrow \underset{\underset{\overset{||}{\underset{\zeta}{N}}}{Cl-}\ P-N}{} \overset{P^{\diagdown Cl}_{\diagdown Cl}}{} \quad (6)$$

An active chain end also may produce a branch unit by attacking
a backbone nitrogen atom [Eq. (7)].

$$\sim\!(N=\underset{Cl}{\overset{Cl}{P}})_x\!-N=\underset{Cl}{\overset{Cl}{P^+}}\ Cl^- \ :N\!\!\overset{\overset{\displaystyle P\diagup Cl}{\displaystyle \quad\diagdown Cl}}{} \longrightarrow \sim\!(N=\underset{Cl}{\overset{Cl}{P}})_x\!-N=\underset{Cl}{\overset{Cl}{P}}\!-N\!\overset{\overset{\displaystyle P\diagup Cl}{\displaystyle \quad\diagdown Cl}}{} \quad (7)$$

A phosphorus atom in the polymer chain backbone is probably
not highly prone to ionization when one considers that, in
comparison with the cyclic trimer, the molecular structure and
chemical behavior of an $NPCl_2$ chain unit and higher cyclic
chlorophosphazenes $[N = PCl_2]_n$ (n = 4, 5, 6, ...) are quite
similar and that the higher cyclics polymerize slowly, if at all,
at temperatures below 300°C. Indeed, for the uncatalyzed, melt
polymerization of a cyclic trimer/tetramer mixture at 250°C, the
tetramer concentration does not change with polymer yields as high
as 50% [37, 54] demonstrating a resistance both to ionization and
to attack by an active center. Under the same reaction conditions
polymer with no apparent polymer chain branching or crosslinking
is produced [37, 54]. One may conjecture that units within the
polymer chain behave similar to the tetramer and inherently are
more resistant than the cyclic trimer to ionization and attack by
an active center. Perhaps this resistance reflects a greater
degree of shielding of both the phosphorus to inhibit an active
center and the nitrogen for protection against electrophilic
attack in higher cyclics and chain units. It is also noted that

the active center generated by the equilibrium in Eq. (5) is less
electrophilic than the active end in Eq. (4) and therefore
probably less likely to attack a nitrogen atom.

An increase in ionization, as shown in Eq. (3b) and (5), and
a reduction is steric barriers may explain both the higher rate
of polymerization and the increase in branch or crosslink formation
with increasing temperature above 250°C. Autoacceleration effects,
as noted in the section High Temperature, Melt Polymerization,
may also be interpreted in terms of the ionic mechanisms. Autoacceleration occurs at 10-15% conversion of trimer [Fig. (1)]
and is accompanied by a rapid rise in molecular weight. In the
early stages of melt polymerization, the viscosity and dielectric
constant ($\varepsilon \sim 2.5$) of the reaction mixture are relatively low; but
as polymerization proceeds, the mixture becomes more viscous and
its dielectric constant increases ($\varepsilon \longrightarrow 3.29$) [57]. Envisioning
melt polymerization as an ionic, solution polymerization where
the solvent medium is changing from trimer to trimer plus
increasing amounts of polymer, a state of reaction (ca. 10-15%
conversion) is achieved where, because of increases in the
dielectric constant, the solvent effectively shields the active
center from its counterion resulting in an increased rate of
polymerization. Low molecular weight products formed early in
the reaction, where the dielectric constant is lowest, may be a
consequence of poor dissociation of the ion pair. Viscosity of
the reaction medium may also be important in that charge transfer
or perhaps chain backbiting mechanisms are more effective for
terminating chain growth in the early stages of polymerization
when the molecules are most mobile. The enhanced rate of
polymerization of trimer with added tetramer [37] may be the
result of the tetramer being more polarizable than the trimer
and persumably, therefore, raising the dielectric constant during
polymerization.

The dramatic increase in gel formation at about 70% conversion
of trimer during melt polymerization [20] is indicative of a
change in mechanism. At high conversions where viscosity is so

high that flow essentially ceases, the trimer no longer effectively
shields adjacent polymer chains but chain units have enough
mobility that branching or crosslinking mechanisms [Eq. (6) and
(7)] become important. Indeed the reactions probably involve
charge transfer to adjacent chain units to perpetuate the active
center. As 100% conversion is approached, the rate of
polymerization naturally decreases but ionic centers may remain
active and continue to increase the degree of crosslinking.

Mechanisms involving depolymerization also should be
considered. In the melt polymerization of highly pure trimer at
250°C, small amounts of cyclic homologues $[NPCl_2]_x$ (x>3) are formed
at long polymerization times and slowly increase in concentration
with time [42]. In view of ring-polymer equilibria reported at
higher temperatures [10, 14, 15, 25], cyclic formation is
understandable and probably involves a backbiting mechanism
[Eq. (8)].

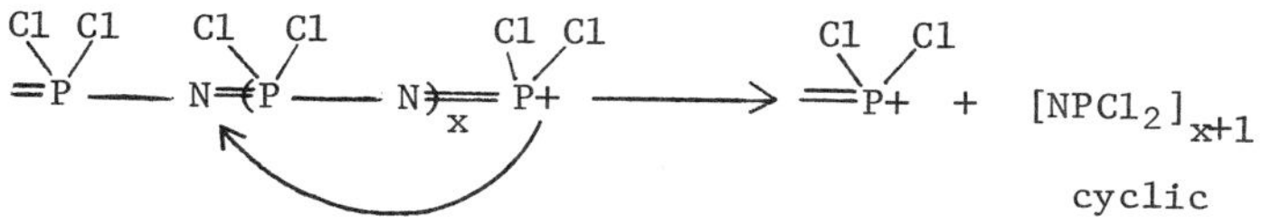

$$\equiv P \overset{Cl}{\underset{Cl}{}} — N\equiv P \overset{Cl}{\underset{Cl}{}} — N \overset{}{\underset{x}{}} P+ \overset{Cl}{\underset{Cl}{}} \longrightarrow \equiv P+ \overset{Cl}{\underset{Cl}{}} + [NPCl_2]_{x+1} \qquad (8)$$

cyclic

Intermolecular reactions involving ring-ring or ring-chain inter-
conversion also may explain the formation of higher cyclic
homoloques [35].

Backbiting and branching reactions [Eq. (7)] have been cited
as possible termination mechanisms. Trace impurities could also
terminate chain growth. Melt polymerizations are terminated by
lowering the temperature which suggests a mechanism involving
association of ion pairs [Eq. (9)].

$$\sim N \equiv \overset{Cl}{\underset{Cl}{P+}} \qquad Cl^- \longrightarrow \sim N \equiv PCl_3 \qquad (9)$$

For typical melt polymerizations with conversions less than 70%,
no additional polymerization or crosslinking occurs if the polymer-

ization mixtures are kept at room temperature. However if the
residual trimer is removed from the polymer, crosslinking
reactions are indicated in that the solubility of the trimer
decreases with time. Experiments suggest that the crosslink units
are different from P-O-P type crosslinks which are due to
hydrolysis [55]. Speculation is that ionization and crosslinking
[Eq. (6)] are possible at room temperature.

The preceding discussion assumes that thermal ionization is
an effective initiation mechanism for polymerization. There is
some question whether polymerization would occur at 250°C using
a totally pure trimer in an inert environment [29, 33]. Minute
amounts of impurities acting as catalysts may initiate
polymerization and dictate the course of polymerization [Fig. (1)].
Mechanism (b) in Eq. (3) may explain the catalysis by compounds
which facillitate removal of chlorine from phosphorus; however
the actual mechanism is probably more complex. Since Lewis acid
catalysts are particularly effective in promoting polymerization
of trimer [45-51], perhaps a different cationic mechanism should
be considered for initiation. In cationic systems polymerization
is commonly initiated by Bronsted acids (water, alcohols, mineral
acids, etc); but when the Bronsted acids do not initiate sufficiently, a Lewis acid (e.g., halides of aluminum, iron, tin,
titanium and boron) can serve as an effective cocatalyst
[Eq. (10)]. Mechanism (d) is similar to (b) in Eq. (3) and would
explain the generation of hydrogen chloride observed in many
catalyzed polymerizations. In mechanism (e) the proton attaches
itself to a nitrogen atom to form an intermediate structure which
ring-opens producing a positive center for chain propagation.
Considering that the nitrogen atom is more electronegative than
phosphorus, mechanism (f) may occur but does not involve the
generation of hydrogen chloride. Mechanism (g) is a possible
intermediate route for mechanism (d). An intermediate similar to
the one in mechanism (g) has been proposed to explain catalysis
by the surfaces of glass polymerization tubes [32].

$$AlCl_3 \; + \; HN \longrightarrow H^+ \, [AlCl_3X]^-$$

$$H^+[AlCl_3X]^- + \text{(trimer)} \xrightarrow[-[AlCl_3X]^-]{d} \text{(cation)} \; + \; HCl \qquad (10)$$

Catalysis by water also may involve an oxophosphazane intermediate [33]. Hydrolysis of the trimer by catalytic amounts of water (0.005–0.10 mol%) may produce an oxophosphazane unit which is thermally unstable and may cause the ring to open and generate active species for either anionic or cationic chain propagation [Eq. (11)].

$$\text{(trimer)} \xrightarrow[-HCl]{H_2O} \text{(OH-trimer)} \rightleftharpoons \text{(oxophosphazane)} \qquad (11)$$

$$+P(N=P)_2\text{-}NH^- \qquad \text{or} \qquad P(N\text{---}P)_2\text{-}N\text{---}H$$

COMMENTS

The production and development of technologically useful poly(organo)phosphazenes depend directly on optimizing the polydichlorophosphazene polymerization reaction. Promising developments have been made in the areas of high temperature, melt and solution polymerization. Advances are being made in gaining a better understanding of the polymerization mechanism(s). However there are still major problem areas which must be surmounted for polyphosphazenes to be commercially viable. Polymerization run times are often too long and temperatures are too high for safe and efficient handling and cost effectiveness. Highly pure trimer must be used and conversions must be limited to prevent crosslink formation. The polymerization behavior from batch-to-batch of trimer is not reproducible and polymer chain structure cannot be controlled with certainty. And finally, there are problems in handling hydrolytically unstable polydichloro-phosphazene and controlling or preventing changes in polymer chain structure that may occur during nucleophilic substitution [5]. More basic research is required, especially in developing catalysts and chain transfer agents for lowering polymerization temperatures, obtaining higher conversions, preventing crosslink and cyclic oligomer formation, and controlling polymer molecular weight, molecular weight distribution, and branching. Up to the present, efforts to expand polymer production have relied upon scaled-up laboratory polymerization conditions. Consideration should be given to innovations required for the pilot plant and full scale production of polydichlorophosphazene.

REFERENCES

[1] H. R. Allcock and R. L. Kugel, J. Am. Chem. Soc., 87, 4216 (1965).
[2] H. R. Allcock, R. L. Kugel, and K. J. Valan, Inorg. Chem., 5, 1709 (1966).
[3] R. E. Singler, N. S. Schneider, and G. L. Hagnauer, Polymer Eng. & Sci., 15, 321 (1975).
[4] H. R. Allcock, Chem. Technol., 5, 552 (1975).
[5] R. E. Singler and G. L. Hagnauer, Organometallic Polymers, C. E. Carraher, J. E. Sheats, C. U. Pittman, Ed., Academic Press, New York, 1978, p. 257.
[6] R. E. Singler, G. L. Hagnauer, and N. S. Schneider, Polymer News, 5, 9 (1978).
[7] H. N. Stokes, Am. Chem. J., 19, 782 (1897).
[8] R. Schenck and G. Römer, Chem. Ber., 57B, 1343 (1924).

[9]   A. M. de Ficquelmont, C. R. Acad. Sci., 204, 689, 867 (1937).
[10] O. Schmitz-Dumont, Z. Electrochem., 45, 651 (1939).
[11] O. Schmitz-Dumont, Angew. Chem., 52, 498 (1939).
[12] F. Patat and F. Kollinsky, Makromol. Chem., 6, 292 (1951).
[13] F. Patat, Angew. Chem., 65, 173 (1953).
[14] F. Patat and K. Frombling, Monatsh. Chem., 86, 718 (1955).
[15] F. Patat and P. Derst, Angew. Chem., 71, 105 (1959).
[16] H. R. Allcock, Phosphorus Nitrogen Compounds, Academic
      Press, New York, 1972.
[17] T. R. Manley, Nature (London), 184, 899 (1959).
[18] J. O. Konecny and C. M. Douglas, J. Polym. Sci., 36, 195
      (1959).
[19] J. O. Konecny, C. M. Douglas, and M. Y. Gray, J. Polym.
      Sci., 42, 383 (1960).
[20] F. G. R. Gimblett, Polymer, 1, 418 (1960).
[21] F. G. R. Gimblett, Plast. Inst., Trans. 28, 65 (1960).
[22] D. Chakrabartty and B. N. Ghosh, J. Polym. Sci., 62, 5130
      (1962).
[23] V. Caglioti, D. Cordischi, and A. Mele, Nature (London),
      195, 491 (1962).
[24] J. R. Soulen and M. S. Silverman, J. Polym. Sci., Part A
      1, 823 (1963).
[25] J. K. Jacques, M. L. Mole, and N. L. Paddock, J. Chem. Soc.,
      London p. 2112 (1965).
[26] D. Cordischi, A. D. Site, A. Mele, and P. Porta,
      J. Macromol. Chem., 1, 219 (1966).
[27] J. R. MacCallum and A. Werninck, J. Polym. Sci., Part A-1
      5, 3061 (1967).
[28] E. F. Moran, J. Inorg. Nucl. Chem., 10, 1405 (1968).
[29] R. O. Colclough and G. Gee, J. Polym. Sci., Part C 16,
      3639 (1968).
[30] J. R. MacCallum and J. Tanner, J. Polymer. Sci., Part B
      7, 743 (1969).
[31] G. Allen, C. J. Lewis, and S. M. Todd, Polymer, 11, 31
      (1970).
[32] J. Emsley and P. B. Udy, Polymer, 13, 593 (1972).
[33] H. R. Allcock, J. E. Gardner, and K. M. Smeltz,
      Macromolecules, 8, 36 (1975).
[34] V. V. Korshak, S. V. Vinogradova, D. R. Tur, N. N. Kasarova,
      L. I. Komarova, and L. M. Gil'man, Acta Polym., 30, 245
      (1979).
[35] V. V. Kireev, V. V. Korshak, G. I. Mitropol'skaya, and
      W. Sulkowski, Vysokomol. Soedin., Ser. A 21, 100 (1979).
[36] J. Retuert, S. Ponce, and J. P. Quijada, Polym. Bull., 1,
      653 (1979).
[37] G. L. Hagnauer, Recent Advances in Size Exclusion
      Chromatography, T. Provder, Ed., ACS Symposium Monograph,
      Washington, D.C., to be published in 1980.
[38] L. G. Lund, N. L. Paddock, J. E. Proctor, and H. T. Searle,
      J. Chem. Soc., London p. 2542 (1960).

[39] M. Becke-Goehring and E. Fluck, Angew. Chem., 74, 382
      (1962).
[40] M. Becke-Goehring and W. Lehr, Z. Anorg. Alleg. Chem., 327,
      128 (1964).
[41] Nippon Fine Chemical Co., Ltd., Inabata and Co., Ltd.,
      51, Junkeimachi 2-chrome Minami-Ku, Osaka, Japan.
[42] G. L. Hagnauer, to be published.
[43] G. L. Hagnauer and R. E. Singler, ACS Org. Coat. Plast.
      Chem., 41, 88 (1979)
[44] H. R. Allcock, K. M. Smeltz, and J. E. Gardner, (Firestone
      Tire & Rubber Co.) U.S. Patent 3,937,790 (1976).
[45] Firestone Tire & Rubber Co., Neth. Appl. 76 09,253
      (1977).
[46] D. L. Snyder, M. L. Stayer, and J. W. Kang, (Firestone Tire
      & Rubber Co.) U.S. Patent Appl. 606,802 (1975).
[47] R. L. Dieck and A. B. Magnusson, (Armstrong Cork Co.)
      U.S. Patent Appl. 731,745 (1976).
[48] R. E. Dieck, T. B. Garrett, and A. B. Magnusson (Armstrong
      Cork Co.) U.S. Patent Appl. 747,626 (1976).
[49] D. L. Snyder, J. W. Kang, and J. W. Fieldhouse, (Firestone
      Tire & Rubber Co.) U.S. Patent Appl. 898,007 (1978).
[50] J. E. Thompson, J. W. Wittman, and K. A. Reynard,
      Horizons, Inc., Cleveland, Ohio, NASA Contract NAS9-14717,
      April 1976 (N76-27424).
[51] K. A. Reynard and A. H. Gerber, (Horizons Inc.) U.S. Patent
      Appl. 474,055 (1974).
[52] Firestone Tire & Rubber co., Neth. Patent Appl. 78 05,383
      (1979).
[53] A. F. Halasa and J. E. Hall (Firestone Tire & Rubber Co.)
      Ger. Offen. 2,820,082 (1979).
[54] G. L. Hagnauer and T. N. Koulouris, to be published.
[55] H. R. Allcock, J. Macromol. Sci., Rev. Macromol. Chem. 4,
      3 (1970).
[56] M. W. Spindler and R. L. Vale, Makromol. Chem., 43, 231
      (1961).
[57] H. R. Allcock and R. J. Best, Can. J. Chem., 42, 447 (1964).

Poly(difluorophosphazene): A New Intermediate for the Synthesis of Poly(organophosphazenes)

T. L. Evans and H. R. Allcock

Department of Chemistry
The Pennsylvania State University
University Park, Pennsylvania 16802

ABSTRACT

Although many poly(organophosphazenes) have been synthesized, new preparative pathways are needed, especially for polymers that contain alkyl side groups. A new development involves the use of poly(difluorophosphazene), $(NPF_2)_n$, instead of poly-(dichlorophosphazene), $(NPCl_2)_n$, as a substrate for reactions with organometallic reagents. This approach has allowed the preparation of a new class of poly(organophosphazenes) that possess substituent groups linked to the skeleton through direct phosphorus-carbon bonds. The synthesis of uncrosslinked poly-(difluorophosphazene) and its reactions with alkoxides and amines are also reviewed.

INTRODUCTION

Poly(organophosphazenes) are an unusual and structurally diverse group of inorganic-organic macromolecules that provide valuable chemical and physical differences from conventional organic polymer systems [1, 2, 3]. They are normally prepared by substitution reactions carried out with the highly reactive polymeric intermediate, poly(dichlorophosphazene),

409

$(NPCl_2)_n$.   The well-known interactions of $(NPCl_2)_n$ with sodium
alkoxides, sodium aryloxides, primary amines, or secondary
amines have been employed to prepare over 100 polymeric
derivatives (Scheme 1).   Some of these are of interest as new
structural materials or as carrier molecules for chemotherapeutic
agents or transition metal species.

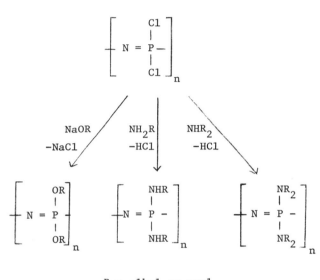

R = alkyl or aryl

n = 10,000

Scheme 1

A major objective in our laboratory over the past 5 years
has been to extend Scheme 1 to include the interactions of poly-
(dichlorophosphazene) with organometallic reagents with a view to
the preparation of poly(alkyl- or arylphosphazenes),
$(NPR_2)_n$.   However, the reactions of $(NPCl_2)_n$ with organolithium
or organomagnesium reagents involve concurrent chain scission
and substitution reactions.   Thus, it has not yet been possible
to prepare polymers from $(NPCl_2)_n$ that possess significant

amounts of aryl or alkyl pendent groups without encountering a
concurrent molecular weight decline [4].

We have now been able to solve this problem by a new
approach which involves the use of poly(difluorophosphazene),
$(NPF_2)_n$, in place of $(NPCl_2)_n$ [5] as a polymeric reaction
substrate.   Initially, little was known about the reactions of
poly(difluorophosphazene) with simple nucleophiles such as amines
or alkoxides.   Indeed, until recently, a soluble form of
$(NPF_2)_n$ had not been prepared.   Here, we review three aspects
of the chemistry of poly(difluorophosphazene):   (1) a synthesis
route to the preparation of an organic solvent-soluble
modification of $(NPF_2)_n$;   (2) the reactions of $(NPF_2)_n$ with
sodium alkoxides, aryloxides, and amines;   (3) the reactions of
$(NPF_2)_n$ with organolithium and organomagnesium reagents.

Preparation and Properties of Poly(difluorophosphazene)

A synthesis route to crosslinked poly(difluorophosphazene),
$(NPF_2)_n$, has been known since 1958.   However, the crosslinked
form is unsuitable for substitution reactions [6].   The first
organic-solvent-soluble version of $(NPF_2)_n$ was prepared in our
laboratory in 1974 (Scheme 2).

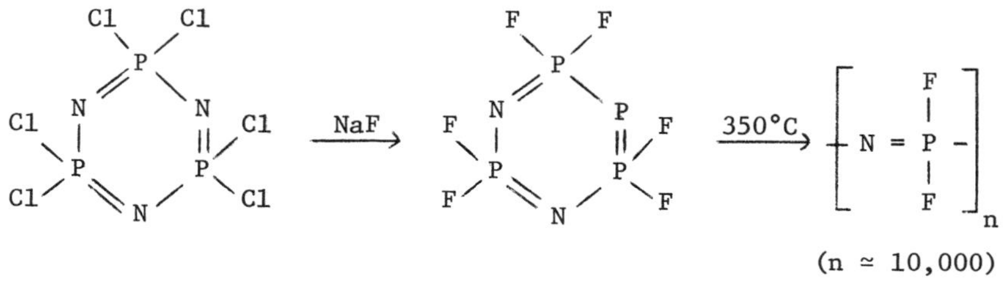

$$(n \simeq 10,000)$$

Scheme 2

This was accomplished by the use of a methodology developed
approximately 10 years earlier for the isolation of uncrosslinked
poly(dichlorophosphazene).   In that earlier work it was found

that the crosslinking of poly(dichlorophosphazene) occurs only
after 70% or more of the $(NPCl_2)_3$ has been converted to high
polymer.   If the polymerization is terminated before this
stage, organic solvent-soluble polymer can be isolated [7].
The polymerization of $(NPF_2)_3$ to $(NPF_2)_n$ also passes through
these same stages, from unchanged molten trimer, through
mixtures of trimer with uncrosslinked $(NPF_2)_n$, to crosslinked,
insoluble $(NPF_2)_n$.   Thus, the isolation of poly(difluorophosphazene) in a form that is suitable for substitution
experiments requires that the polymerization be terminated
during the second stage (70% conversion was the upper limit for
the isolation of soluble polymer) [8].

The mechanism of polymerization of trimeric halophosphazenes
to poly(halophosphazenes) has not yet been firmly established.
However, a ^{31}P NMR study of the conversion of $(NPF_2)_3$ to $(NPF_2)_n$
has suggested that no appreciable equilibration between $(NPF_2)_3$
and $(NPF_2)_4$ is involved in the polymerization process (Figure 1).

Poly(difluorophosphazene) is a white or colorless elastomeric
compound that is hydrolytically unstable and is soluble only in
perfluorinated solvents.   The unusual solubility behavior of
$(NPF_2)_n$ presented some complications when substitution reactions
were attempted, because the nucleophiles employed in these
present studies are soluble in etheric solvents, such as tetrahydrofuran, but are insoluble in perfluorinated solvents.   Thus,
the substitution reactions described here were carried out in
heterophase reaction media.

The Reactions of Poly(difluorophosphazene)

1. _Information available from model system studies._   The
substitution reactions of polymers are often complex.   Thus, it
is useful to carry out initial mechanistic studies with small
molecule model compounds and to utilize the information obtained
as a predictive technique for the high polymer reactions [9].
We have employed this method successfully to anticipate the
reactions of poly(difluorophosphazene) from studies of the

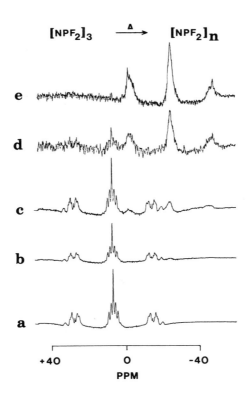

FIG 1. ^{31}P NMR spectra obtained at different stages
during the polymerization of $(NPF_2)_3$ to $(NPF_2)_n$.   Spectra a, b,
c, d, and e are from samples in which the degrees of conversion
to the polymer were 1.5%, 5.0%, 15.6%, 63%, and 94%,
respectively.   The spectra illustrate the disappearance of
$(NPF_2)_3$ and the appearance of $(NPF_2)_n$ without the accompanying
formation of other species.   The spectra were obtained for
samples in perfluorobutyltetrahydrofuran solvent and were
referenced to an 85% aqueous H_3PO_4 external reference.

cyclic trimeric and tetrameric model compounds $(NPF_2)_3$ and
$(NPF_2)_4$.   The results of these model studies are as follows.

The reactions of $(NPF_2)_4$ with sodium trifluoroethoxide or
sodium phenoxide resulted in the formation of $[NP(OCH_2CF_3)_2]_4$
or $[NP(OC_6H_5)_2]_4$.   However, the interactions of $(NPF_2)_4$ with
primary or secondary amines, such NH_2CH_3, $NH_2C_4H_9$, or
$NH(CH_3)_2$, yielded exclusively non-geminally substituted
aminofluorophosphazenes.   When the same reaction conditions
were employed with $(NPCl_2)_3$, complete amination was accomplished.
For example, the reaction of $(NPF_2)_4$ with an excess of n-butylamine yielded $[N_4P_4(NHC_4H_9)_4F_4]$ (Scheme 3), but no compounds
that possessed higher degrees of amination could be isolated
(even at reaction temperatures as high as 60°C).   This same
substitution pattern was observed for the cyclic trimeric
compound $(NPF_2)_3$, although these reactions were slower than for
$(NPF_2)_4$.   However, use of the more reactive reagent, $LiN(CH_3)_2$
for reactions with $(NPF_2)_4$ did yield cyclic compounds such as
$[N_4P_4(N(CH_3)_2)_7F]$ and $[NP(N(CH_3)_2)_2]_4$ [10].

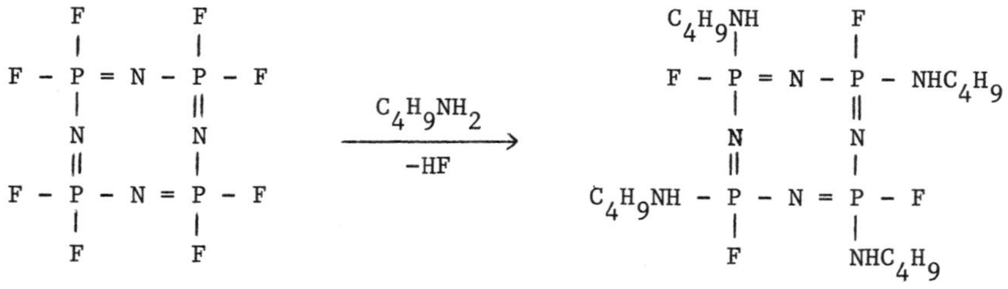

Scheme 3

The reaction mechanism for the amination of fluorocyclophosphazenes is still not completely understood.   However, the
inability of compounds such as $N_4P_4(NHR)_4F_4$ and $N_3P_3(NHR)_3F_3$
to undergo further   amination is probably a consequence of three
factors: (1) Electron donation takes place from a pendent amino
substituent to a phosphorus atom.   This deactivates that

phosphorus atom to further reaction.   The presence of the highly electronegative fluorine atom at that site probably enhances electron donation by the amino group compared with the situation when a chlorine atom is present.   (2) The low reactivity of P-F bonds to substitution reactions may contribute to this effect. (3) The weak nucleophilic character of amines toward phosphazene phosphorus atoms reduces the overall reactivity.   For these reasons, units such as [P(F)(NHR)] are more deactivated than [P(Cl)(NHR)] to further aminolysis, and very reactive nitrogen nucleophiles (e.g. $LiN(CH_3)_2$) must be employed before all the P-F bonds can be replaced by amino pendent groups.

Investigations of the reactions of the fluorocyclophosphazenes $(NPF_2)_3$ or $(NPF_2)_4$ with organometallic reagents such as methyllithium, n-butyllithium, phenyllithium or methylmagnesium chloride have been carried out in several laboratories [11, 12, 13, 14].   It has been found that the fluorocyclophosphazenes are significantly more resistant to ring degradation reactions than are chlorocyclophosphazenes.

2. The reactions of $(NPF_2)_n$ with sodium alkoxides, aryloxides, or amines.   The reactions of high polymeric $(NPF_2)_n$ with sodium trifluoroethoxide or sodium phenoxide are considerably slower than those of $(NPCl_2)_n$ with the same reagents, but complete fluorine replacement can be effected to yield $[NP(OCH_2CF_3)]_n$ or $[NP(OC_6H_5)_2]_n$.   The interactions of $(NPF_2)_n$ with amines such as NH_2CH_3, $NH_2C_4H_9$, $NH_2C_6H_5$ or $NH(CH_3)_2$ resulted exclusively in the formation of products in which each phosphorus atom possessed one fluorine and one amino substituent [8].   The similarity to the cyclic model systems is striking.   These results are illustrated in Scheme 4 for the interactions of $(NPF_2)_n$ with n-butylamine. With $(NPCl_2)_n$, a non-geminal amination pattern has been detected in only one case, and this involved the interactions of $(NPCl_2)_n$ with the bulky diethylamine.   The substitution pattern for the reaction of $(NPF_2)_n$ with amines is probably not entirely a consequence of steric inhibition because amines with small steric dimensions (e.g. methylamine) also yield fluoro-amino phosphazenes.

In a practical sense, the partial amination of $(NPF_2)_n$ is a complicating factor because the residual P-F bonds impart hydrolytic instability to the products.   However, it also provides a synthetic advantage for the preparation of new polyphosphazene structures.   For example, the reactions of polymers such as $[NP(NHC_4H_9)(F)]_n$ with sodium trifluoroethoxide opens up new synthetic pathways to the formation of polymeric species of the type, $[NP(NHC_4H_9)(OCH_2CF_3)]_n$ (Scheme 4).   Mixed substituent phosphazenes with the substituent groups arrayed in a regular fashion along the backbone should possess properties that are different from those with the substituents oriented randomly.

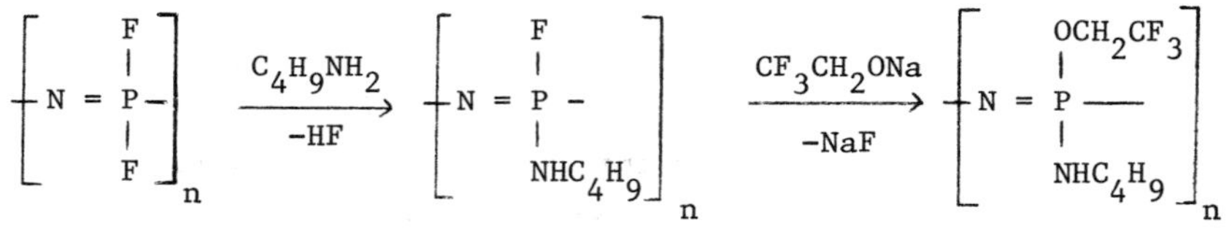

$$\left[\begin{array}{c} F \\ | \\ -N = P- \\ | \\ F \end{array} \right]_n \xrightarrow[-HF]{C_4H_9NH_2} \left[\begin{array}{c} F \\ | \\ -N = P - \\ | \\ NHC_4H_9 \end{array} \right]_n \xrightarrow[-NaF]{CF_3CH_2ONa} \left[\begin{array}{c} OCH_2CF_3 \\ | \\ -N = P \underline{\hspace{1em}} \\ | \\ NHC_4H_9 \end{array} \right]_n$$

Scheme 4

3. Reactions of $(NPF_2)_n$ with Phenyllithium.   Although poly(difluorophosphazene) is a useful intermediate for the synthesis of alkoxy, aryloxy, or aminophosphazene high polymers, its principle utility is as a substrate for reactions with organometallic reagents to yield species with side groups linked to the skeleton through carbon-phosphorus bonds.   Chain cleavage reactions are far less facile when poly(difluorophosphazene) is used as a substrate than when poly(dichlorophosphazene) is employed.   This conclusion is based on the studies described in the following sections.

The interactions of $(NPF_2)_n$ with phenyllithium were carried out as illustrated in Scheme 5.   (Sodium trifluoroethoxide was used in most cases as a second nucleophile to ensure complete fluorine replacement).   These reactions yielded the first examples of phosphazene high polymers that possess significant

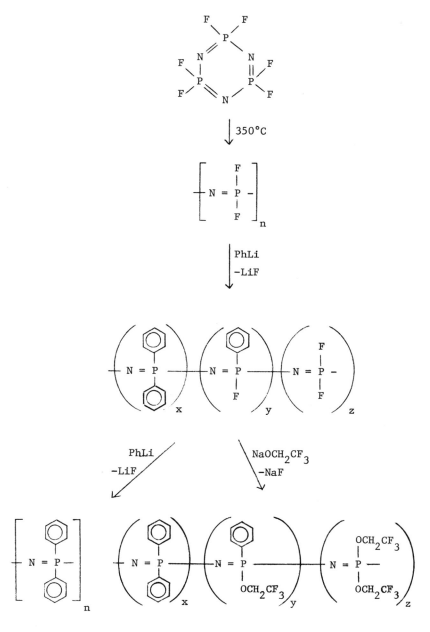

Scheme 5

percentages of side groups attached to the skeleton through
direct P-C bonds [15].   The polymers isolated, their
molecular weights and their glass transition temperatures (T_g)
are listed in Table 1.

The fully phenylated derivatives had lower molecular weights
than the mixed-substituent phenyl-trifluoroethoxy polymers.   In
fact, it was found that approximately 70% of the fluorine atoms
could be replaced by phenyl groups before a molecular weight
decline was detected.   These results are in marked contrast to
the rapid molecular weight loss that accompanies substitution
during the interactions of phenyllithium with $(NPCl_2)_n$ [4]
(Figure 2).

An interesting feature of the reaction of $(NPF_2)_n$ with
phenyllithium is the dramatic decrease in reaction rate that
occurs after approximately 70% of the fluorine atoms of $(NPF_2)_n$
have been replaced by phenyl groups.   Chain cleavage is
encountered during this period of slow substitution.   A study
of the mechanism of chain cleavage is currently under
investigation in our laboratory.

Another interesting feature of the phenylation process is
the tendency for phenyl substitution to proceed predominantly,
although not exclusively, by a geminal mechanism.   This implies
that phosphorus atoms that possess one fluorine atom and one
phenyl group are more reactive to arylation than are phosphorus
atoms that possess two fluorine atoms.   This substitution
pattern is in contrast to the amination reactions of $(NPF_2)_n$
described in the previous section.   These differences could
reflect the higher reactivity of the organolithium reagent than
the amines, and possibly the existence of different mechanisms
for the two substitution processes.

A preliminary study of the thermal behavior of the
derivative $[NP(C_6H_5)_{1.24}(OCH_2CF_3)_{0.76}]_n$ was carried out.   It
showed that the presence of the phenyl groups on a polyphosphazene chain increases the stability at 300°C compared with the

TABLE 1. Properties of Poly(alkyl- or arylphosphazenes)

Polymer	GPC M.W.[a]	T_g (°C)[b]
$[NP(C_6H_5)_{0.38}(OCH_2CF_3)_{1.62}]_n$ [c]	$> 1 \times 10^6$	+7
$[NP(C_6H_5)_{0.64}(OCH_2CF_3)_{1.36}]_n$ [c]	$> 1 \times 10^6$	+25
$[NP(C_6H_5)_{0.96}(OCH_2CF_3)_{1.04}]_n$ [c]	$> 1 \times 10^6$	+45
$[NP(C_6H_5)_{1.24}(OCH_2CF_3)_{0.76}]_n$ [c]	$> 1 \times 10^6$	+60
$[NP(C_6H_5)_2]_n$ [c]	5.0×10^4	+70
$[NP(CH_3)_{1.8}(OCH_2CF_3)_{0.2}]_n$ [d]	crosslinked	–
$[NP(CH_3)_{1.0}(OCH_2CF_3)_{1.0}]_n$ [d]	crosslinked	–
$[NP(C_4H_9)_{0.98}(OCH_2CF_3)_{1.02}]_n$ [e]	crosslinked	–
$[NP(C_4H_9)_{1.5}(OCH_2CF_3)_{0.5}]_n$ [e]	crosslinked	–
$[NP(C_4H_9)_2]_n$ [f]	5.0×10^4	–45
$[NP(C_2H_5)_{1.8}(OCH_2CF_3)_{0.2}]_n$ [g]	6.0×10^5	–60

[a] GPC Molecular Weights are relative to polystyrene standards.

[b] T_g values were obtained by the torsional pendulum method.

[c] Obtained from the interactions of $(NPF_2)_n$ and phenyllithium (at 25°C) followed by treatment with sodium trifluoroethoxide.

[d] Obtained from the interactions of $(NPF_2)_n$ and methyllithium (THF; –60°C) followed by treatment with sodium trifluoroethoxide.

[e] Obtained from the interactions of $(NPF_2)_n$ and n-butyllithium (THF; –60°C) followed by treatment with sodium trifluoroethoxide.

[f] Obtained from the interactions of $(NPF_2)_n$ and an excess of n-butyllithium (benzene; +2°C).

[g] Obtained from the interactions of $(NPF_2)_n$ and diethylmagnesium (THF; 25°C) and then sodium trifluoroethoxide treatment.

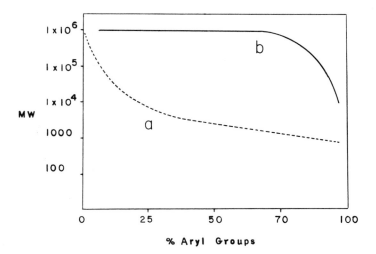

FIG. 2.   Comparisons of the variation in GPC average
molecular weight for $[NP(C_6H_5)_x(OCH_2CF_3)_y]_n$ versus the
percentage of phenyl groups attached to the backbone.   The
broken curve represents the behavior of the system when
$(NPCl_2)_n$ is used as a reaction substrate.   The solid line
illustrates the behavior with $(NPF_2)_n$.

homopolymer, $[NP(OCH_2CF_3)_2]_n$.   Phenyl pendent groups could
increase the thermal stability of a polyphosphazene in two ways.
First, the low reactivity of a P-C bond as compared with P-O or
P-N linkages could reduce the likelihood of side group reactions
at high temperatures.   Second, the steric bulk of the phenyl
group might block the unzipping or backbiting processes that
lead to the formation of low molecular weight oligomers.

4. <u>Reactions of $(NPF_2)_n$ with Alkyllithium Reagents</u>.   The
interactions of $(NPF_2)_n$ with methyllithium or <u>n</u>-butyllithium are
more complicated than those with phenyllithium.   With the use of
these reagents, <u>crosslinked</u> products were obtained.   The cross-

linking process was detected over a wide range of reaction temperatures (for methyllithium from 25°C to -70°C, and for n-butyllithium from 0°C to -70°C).   Furthermore, with butyllithium, only low molecular weight products were isolated when the reaction temperature exceeded 0°C.   In fact, no reaction conditions were found that did not involve either chain cleavage or crosslinking [16].

However, it was found that the number of crosslinks per chain decreased as the reaction temperature was lowered.   When very low reaction temperatures (-50°C to -70°C) were employed for n-butyllithium or methyllithium, lightly crosslinked products were isolated that possessed high percentages of alkyl pendent groups.   These polymers were sufficiently swelled by the reaction media to allow complete fluorine atom replacement to take place with sodium trifluoroethoxide.   Some of the polymers prepared in this manner are listed in Table 1.

The crosslinking reaction observed with the use of alkyllithium reagents and $(NPF_2)_n$ probably involves two steps: first, proton abstraction from pendent alkyl groups by the alkyllithium compound;   second, a coupling reaction between the pendent alkyllithium substituent from one polymer chain and a phosphorus-fluorine bond from a second polymer chain.   This reaction sequence is illustrated in Scheme 6 for the interactions of $(NPF_2)_n$ with methyllithium.

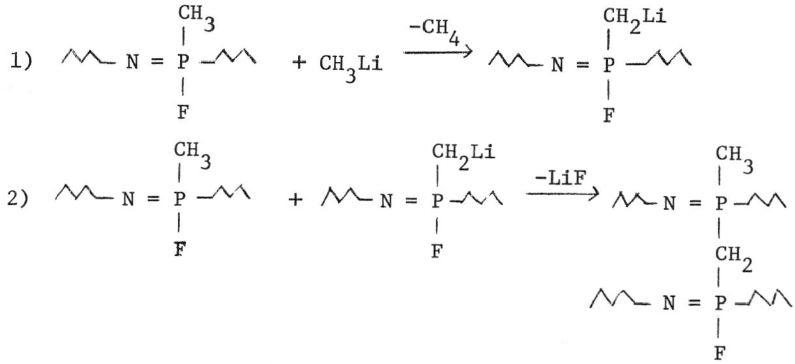

Scheme 6

5. Reactions of $(NPF_2)_n$ with Organomagnesium Compounds.
The interactions of $(NPF_2)_n$ with organomagnesium reagents, such
as RMgX or MgR_2 yielded mainly crosslinked species.   The
quantity of organic solvent-soluble products that were
isolated from these interactions never exceeded 10%.   Even
phenylmagnesium bromide or diphenylmagnesium reacted with
$(NPF_2)_n$ to yield crosslinked products.   It seems unlikely that
proton abstractions from phenyl groups could occur at an
appreciable rate at 25°C.   Thus, another mechanism for
crosslinking, not described by Scheme 6, must be operative with
these reagents.   Additional research with small molecule model
compounds is needed before the mechanistic basis for these
reactions can be understood.

ACKNOWLEDGMENT

This work was supported by the Office of Naval Research.

REFERENCES

[1]   H. R. Allcock, Angew. Chem. (Internat. Ed. English), 16, 147
      (1977).

[2]   S. H. Rose, J. Polymer Sci., B7, 837 (1965).

[3]   R. E. Singler, N. S. Schneider, and G. L. Hagnauer, Polymer
      Eng. Sci., 321, 15 (1975).

[4]   H. R. Allcock and C. T.-W. Chu, Macromolecules, 12, 551
      (1979).

[5]   H. R. Allcock, D. B. Patterson, and T. L. Evans, J. Am. Chem.
      Soc., 99, 6095 (1977).

[6]   F. Seel and J. Langer, Z. Anorg. Allgem. Chem., 295, 316
      (1958).

[7]   H. R. Allcock and R. L. Kugel, J. Am. Chem. Soc., 87, 4212
      (1965).

[8]   H. R. Allcock, D. B. Patterson, and T. L. Evans, Macro-
      molecules, 12, 172 (1979).

[9]   H. R. Allcock, Acc. Chem. Res., 12, 351 (1979).

[10] T. L. Evans and H. R. Allcock, Inorg. Chem., 18, 2342 (1979).

[11] C. W. Allen and T. Moeller, Inorg. Chem., 19, 2177 (1968).

[12] T. Moeller, A. Failli, and F. Y. Tsang, Inorg. Nucl. Chem.
      Letters, 1, 49 (1965).

[13] T. N. Ranganathan, S. M. Todd, and N. L. Paddock, Inorg.
      Chem., 12, 316 (1973).

[14] H. R. Allcock, R. J. Ritchie, and P. R. Suszko, unpublished
      results.

[15] H. R. Allcock, T. L. Evans, and D. B. Patterson, Macro-
      molecules, 13, 201 (1980).

[16] H. R. Allcock, T. L. Evans, D. B. Patterson, and P. R.
      Suszko, submitted for publication.

Synthesis of Poly(dialkylphosphazenes) from N-Silylphosphinimines

Robert H. Neilson and Patty Wisian-Neilson

Department of Chemistry
Texas Christian University
Fort Worth, Texas   76129

ABSTRACT

Many N-silylphosphinimines $Me_3SiN=P(X)RR'$ undergo facile thermal decomposition with elimination of substituted silanes Me_3SiX and formation of cyclic or polymeric phosphazenes $(RR'PN)_n$. The process which is a new, general synthesis of phosphazenes has been used to prepare poly(dimethylphosphazene), $(Me_2PN)_n$, in nearly quantitative yield from $Me_3SiN=P(OCH_2CF_3)Me_2$. Convenient synthetic routes to the necessary silicon-nitrogen-phosphorus precursors are described and the results of their decomposition reactions are reported.

INTRODUCTION

Inorganic polymers with a backbone of alternating phosphorus and nitrogen atoms $(\overset{|}{\underset{|}{P}}=N)_n$ are known as polyphosphazenes [1]. Many such polymers with a variety of substituents at phosphorus have been prepared and they generally exhibit many useful properties including low temperature flexibility, resistance to chemical attack, flame retardancy, stability to UV and

425

visible radiation, and moderate to high thermal
stability.

Although numerous polyphosphazenes have been
prepared, the only method that has proven generally
useful for their synthesis is that developed by Allcock and coworkers [1,2]. This procedure involves
the initial preparation of poly(dihalophosphazene)
(halogen = F,Cl) from the cyclic trimers (eqs 1 and 2)
and subsequent nucleophilic displacement of the halogens along the chain (eq 3). In each case, the
substituents at phosphorus must be introduced after

$$nPCl_5 + nNH_4Cl \xrightarrow{-4nHCl} \frac{1}{n}(Cl_2PN)_n \qquad (1)$$
$$n = 3,4$$

$$(X_2PN)_3 \xrightarrow{\Delta} (X_2PN)_n \qquad (2)$$
$$X = Cl,F \qquad n = 15,000$$

$$(X_2PN)_n \begin{cases} \xrightarrow{NaOR} [(RO)_2PN]_n \\ \xrightarrow{RNH_2} [(RNH)_2PN]_n \end{cases} \qquad (3)$$

polymerization since the substituted cyclic phosphazenes do not polymerize [3]. Until recently, no
general method has been reported which allows the
incorporation of the desired substituents before
polymerization.

A common feature of the known phosphazene polymers is that the organic substituents are bonded to

phosphorus through oxygen or nitrogen thereby providing pathways for decomposition or depolymerization on heating above about 200 °C. It has long been speculated that directly P-C bonded alkyl or aryl groups might enhance the thermal and chemical stability of the polymers and give rise to interesting physical properties [3,4]. Furthermore, the alkyl substituted polyphosphazenes $(R_2PN)_n$ would be isoelectronic with the silicone polymers $(R_2SiO)_n$ which are an extremely versatile and useful class of polymers. The published attempts to prepare the directly P-C bonded polyphosphazenes have been largely unsuccessful. Treatment of poly(dihalophosphazenes) with organometallic reagents (e.g. RMgX or RLi) results in either incomplete substitution under mild conditions or chain cleavage and and molecular weight reduction under more vigorous conditions [4,5].

In this context, we report here a new, general, and direct method for the synthesis of polyphosphazenes. The new synthesis is based on the premise that suitably constructed N-silylphosphinimines will eliminate substituted silanes to form cyclic and/or polymeric phosphazenes (eq 4). This method has sev-

$$Me_3SiN=\underset{\underset{R'}{|}}{\overset{\overset{R}{|}}{P}}-X \longrightarrow Me_3SiX + \frac{1}{n}\{\underset{\underset{R'}{|}}{\overset{\overset{R}{|}}{P}}=N\}_n \qquad (4)$$

eral potential advantages, the most important of which is the ability to incorporate the desired phosphorus substituents directly into the starting Si-N-P com-

pound. Therefore, by designing the appropriate
precursors, it should be possible to widely vary the
pattern of substituents in the phosphazene products.
It is this procedure which has resulted in the successful preparation of the first fully P-C bonded
poly(dialkylphosphazene), $(Me_2PN)_n$ [6].

SILICON-NITROGEN-PHOSPHORUS COMPOUNDS

Studies of the chemistry and stereochemistry of
silicon-nitrogen-phosphorus compounds [7-13] have led
us into the phosphazene area in the following manner.
In an attempt to prepare the first example of an acy-
clic pentacoordinate phosphorus compound containing a
silylamino substituent, lithium bis(trimethylsilyl)-
amide was treated with some substituted fluorophosphoranes [13]. The desired silylaminophosphoranes,
however, were not isolated. Instead, the reaction
(eq 5) proceeded with elimination of fluorotrimethyl-
silane as well as lithium fluoride to produce a new
series of P-fluoro-N-silylphosphinimines. Furthermore,

$$RR'PF_3 + (Me_3Si)_2NLi \xrightarrow{-LiF} \begin{bmatrix} Me_3Si & & F & R \\ & N-P & \\ Me_3Si & & F & R' \end{bmatrix}$$

R	R'
Ph	F
Me$_2$N	F
Me	F
Ph	Ph

$$\xrightarrow{-Me_3SiF} \quad (5)$$

$$Me_3SiN=P-F$$
with R above and R' below P.

it was soon established that these phosphinimines
would readily undergo further fluorosilane elimination
(eq 6) to form phosphazenes [13]. Closer examination
of these phosphazenes has indicated that only smaller

$$Me_3SiN=\overset{\overset{\displaystyle R}{|}}{\underset{\underset{\displaystyle R'}{|}}{P}}-F \longrightarrow \frac{1}{n}\overset{\overset{\displaystyle R}{|}}{\underset{\underset{\displaystyle R'}{|}}{[P=N]}}_n + Me_3SiF \qquad (6)$$

R	Ph	Ph	Me	Me	Me_2N
R'	F	Ph	F	Me	F

cyclic compounds (i.e., trimer and tetramer) are obtained. Nonetheless, these decompositions appeared
to be a novel synthesis of phosphazene rings, especially non-geminally substituted compounds in which
each phosphorus bears two different substituents.

These preliminary results led us to speculate on
the generality of such silane eliminations and, hence,
to propose the general phosphazene synthesis outlined
above (eq 4). A most obvious variable in this new
method is the effect of changing the leaving group (X)
in the precursor N-silylphosphinimines. With this in
mind we have prepared a number of new "suitably constructed" N-silylphosphinimines with a variety of
leaving groups, and a mixture of alkyl, aryl, and
alkoxy substituents at phosphorus.

The synthesis of the new N-silylphosphinimines
was contingent upon the availability of relatively
large amounts of another type of Si-N-P compound, the
bis(trimethylsilyl)aminophosphines, $(Me_3Si)_2NPRR'$.

Initially, the synthesis of such compounds involved
reactions of $(Me_3Si)_2NLi$ with chlorophosphines R_2PCl
[12]. The preparation and handling of the latter reagents, however, is often tedious, especially where
R = alkyl. These problems have been avoided by our
development of a convenient, "one-pot" synthesis
(eq 7) which uses readily available starting materials [11]. The silylaminophosphines can be prepared

$$(Me_3Si)_2NH \xrightarrow[Et_2O]{n-BuLi} (Me_3Si)_2NLi \xrightarrow[-78\,°C]{PCl_3} (Me_3Si)_2NPCl_2$$

$$(Me_3Si)_2NPR_2 \xleftarrow[0\,°C]{2RMgBr} \qquad (7)$$

$$R = Me, Et$$

in molar quantities and are isolated in 70-75% yields
by vacuum distillation.

One of the first appropriately substituted N-
silylphosphinimines was obtained in a related study
of the oxidation of silylaminophosphines [7,12]. When
$(Me_3Si)_2NPMe_2$ is oxidized by either oxygen or tert-
butyl(trimethylsilyl) peroxide, the structurally
rearranged siloxyphosphinimine $Me_3SiN=P(OSiMe_3)Me_2$ is
isolated in high yield (eq 8). Indeed, the lability

$$(Me_3Si)_2NPMe_2 \xrightarrow{[O]} \left[\begin{array}{c} Me_3Si \\ \diagdown \\ \diagup \quad N-PMe_2 \\ Me_3Si \end{array} \begin{array}{c} O \\ \| \end{array} \right]$$

$$\underset{\underset{Me}{|}}{\overset{\underset{Me}{|}}{Me_3SiN=P}}-OSiMe_3 \longleftarrow \qquad (8)$$

of silyl groups toward intramolecular migrations is a
dominant feature of the chemistry of such compounds
[7,9,11,12].

A much more important development in the synthesis of potential phosphazene precursors was the preparation of a new series of P-bromo substituted
phosphinimines [14]. This was accomplished by the
simple reaction of bromine with bis(trimethylsilyl)-
aminophosphines in benzene (eq 9). Unlike trialkyl-
and triarylphosphines which react with bromine to give

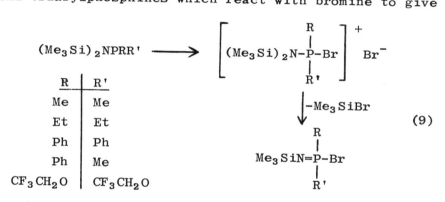

$$(Me_3Si)_2NPRR' \longrightarrow \left[(Me_3Si)_2N-\overset{\overset{R}{|}}{\underset{\underset{R'}{|}}{P}}-Br \right]^+ Br^-$$

R	R'
Me	Me
Et	Et
Ph	Ph
Ph	Me
CF$_3$CH$_2$O	CF$_3$CH$_2$O

$$\downarrow -Me_3SiBr$$

$$Me_3SiN=\overset{\overset{R}{|}}{\underset{\underset{R'}{|}}{P}}-Br \qquad (9)$$

phosphonium salts $[R_3PBr]^+Br^-$, the analogous silyla-
minophosphonium salts are not isolated. Instead,
Me$_3$SiBr is eliminated, giving high yields of the
P-bromo-N-silylphosphinimines. These compounds are
hydrolytically unstable liquids which are usually
sufficiently volatile to permit purification by vacuum
distillation. They have been fully characterized by
^{1}H, ^{13}C, and ^{31}P NMR spectroscopy and elemental
analysis.

Not only are these P-bromo compounds potential
phosphazene precursors themselves, but they also can

be derivatized at the P-Br bond, thereby facilitating
the introduction of other functional groups [14].   The
P-bromo-N-silylphosphinimines react with dimethylamine
and with alcohols in the presence of Et_3N (to scavenge
HBr) giving the amino and alkoxy substituted phosphinimines in good yields (eqs 10-12).   These products are

$$
\begin{array}{ll}
& \overset{\displaystyle R}{\underset{\displaystyle R'}{Me_3SiN=P-NMe_2}} \quad (10) \\
\xrightarrow{\ Me_2NH\ } & \\
\end{array}
$$

$$
\overset{\displaystyle R}{\underset{\displaystyle R'}{Me_3SiN=P-Br}} \xrightarrow[Et_3N]{MeOH} \overset{\displaystyle R}{\underset{\displaystyle R'}{Me_3SiN=P-OMe}} \quad (11)
$$

$$
\xrightarrow[Et_3N]{CF_3CH_2OH} \overset{\displaystyle R}{\underset{\displaystyle R'}{Me_3SiN=P-OCH_2CF_3}} \quad (12)
$$

colorless liquids which can be purified by vacuum
distillation and which have been characterized by NMR
spectroscopy and elemental analysis.

   These simple reactions have, therefore, produced
a new series of potential phosphazene precursors
$Me_3SiN=P(X)RR'$ (eq 4) with a variety of leaving groups
($X = OSiMe_3$, Br, NMe_2, OMe, and OCH_2CF_3).   Moreover,
these are all easily prepared in synthetically useful
quantities, while quantities of the original P-fluoro-
N-silylphosphinimines are often limited by difficulties in the synthesis and handling of the starting
fluorophosphoranes (eq 5).

PHOSPHAZENE SYNTHESIS

Cyclic Phosphazenes

The thermal stability of each of the new N-silylphosphinimines has been investigated. Typically, the compounds were sealed in evacuated heavy-walled glass ampoules and heated either in an oil bath or an oven. At relatively high temperatures of 200° to 250 °C, the compounds $Me_3SiN=P(X)R_2$, where X is NMe_2, $OSiMe_3$, or OMe and R is Et or Me, showed no signs of any decomposition even after several days. In each case none of the expected silane Me_3SiX was formed and the phosphinimines were recovered unchanged.

The P-bromo-N-silylphosphinimines, however, are much less stable and readily eliminate Me_3SiBr on heating under a variety of conditions. For example, when $Me_3SiN=P(Br)Me_2$ is heated in refluxing benzene for 5 days, a ^{31}P NMR spectrum of the white solids remaining after removal of benzene and Me_3SiBr indicate that nearly equal amounts of tetrameric (δ 27.1) and pentameric (δ21.5) $(Me_2PN)_n$ [15] are formed. On heating a neat sample of the same compound in a sealed ampoule at 190 °C for 2 days the principle ^{31}P NMR signal corresponds to tetramer with only traces of trimer and pentamer. On heating in a sealed ampoule at 250 °C for one day, slightly larger proportions of trimer and pentamer relative to tetramer are observed.

Some related P-bromo-N-silylphosphinimines $Me_3SiN=P(Br)RR'$ (R = R' = Me; R = R' = Et; R = R' = Ph; and R = Me, R' = Ph) also smoothly eliminate Me_3SiBr on heating in sealed glass ampoules. In all cases

only the smaller cyclic products are obtained and, indeed, this appears to be an effective new synthetic
route to such compounds.

Polymeric Phosphazenes

While the N-silylphosphinimines where the potential leaving group is F, Br, NMe_2, $OSiMe_3$ or OMe are
either stable compounds or give exclusively cyclic
phosphazenes when heated, markedly different results
are obtained when the leaving group is trifluoroethoxy,
CF_3CH_2O. On investigating the thermal stability of
$Me_3SiN=P(OCH_2CF_3)Me_2$, we find that $Me_3SiOCH_2CF_3$ is
eliminated from the compound under a variety of conditions. In each case, the product is a solid material which analyzes for Me_2PN but bears no physical
resemblance to the cyclic dimethylphosphazenes $(Me_2PN)_n$
(n = 3,4,5). Although this decomposition occurs slowly at room temperature or under a neat reflux at
atmospheric pressure, best results are obtained by
heating the phosphinimine in a sealed glass ampoule
at 190 °C for 2 days. Removal of the volatile silane
under vacuum leaves an off-white solid which, unlike
the trimer and tetramer, is quite insoluble in water
but is highly soluble in CH_2Cl_2 or $CHCl_3$. The ^{31}P NMR
spectrum contains a sharp singlet at $\delta 8.26$ (in $CDCl_3$,
relative to H_3PO_4) which is far upfield from that of
the cyclic dimethylphosphazenes. Based on elemental
analysis, which confirms the molecular formula as
Me_2PN, the reaction yield is virtually quantitative.
A weight average molecular weight $(\overline{M}_w)$ of 50,000 was
determined by light scattering techniques. This
value corresponds to an average chain length of about

660 monomer units and although this is significantly
less than the typical 10,000-15,000 units reported for
polyphosphazenes with alkoxy, amino, or halo substituents, it is, nevertheless, the first example of a
well-characterized poly(dialkylphosphazene) with
directly P-C bonded substituents.

Other physical data for poly(dimethylphosphazene)
include a glass transition temperature (T_g) of -42 °C
a melting point (T_m) of 158 °C, and an intrinsic
viscosity, $[\eta]$, of 0.44 dl/g (in $CHCl_3$ at 25 °C). When
a CH_2Cl_2 solution of $(Me_2PN)_n$ is allowed to evaporate
slowly, a thin film of the polymer is formed. Pour-
ing a CH_2Cl_2 solution into hexane results in precipitation of a white powder-like form of the polymer and
serves as a means of purification.

Prepared in this manner, poly(dimethylphosphazene) is superficially similar to some lower molecular
weight (3,500-12,500) products reported by Sisler and
coworkers [16]. Their compounds, however, were obtained in low yields from the thermolysis of
$Me_2P(NH_2)_2Cl$ after a tedious preparation and purification. The reproducibility of their results has also
been questioned [3]. Our method, on the other hand,
utilizes easily prepared starting materials, affords
excellent yields, and gives considerably higher molecular weights for $(Me_2PN)_n$.

The elimination of $Me_3SiOCH_2CF_3$ from N-silylphosphinimines appears to be fairly general. Flindt and
Rose [17] have reported that this silane elimination
from $Me_3SiN=P(OCH_2CF_3)_3$ gives low molecular weight
oligomers $(\overline{M}_n \sim 10,000)$ of $[(CF_3CH_2O)_2PN]_n$. Our preliminary results indicate that $Me_3SiOCH_2CF_3$ will also

eliminate from some other N-silylphosphinimines.
Although $Me_3SiN=P(OCH_2CF_3)Ph_2$ is stable on heating for
prolonged periods at 250 °C, both $Me_3SiN=P(OCH_2CF_3)RR'$
(R = R' = Et; and R = Me, R' = Ph) show quantitative
formation of $Me_3SiOCH_2CF_3$. The solid product of the
decomposition of the diethyl compound gives a satisfactory elemental analysis for Et_2PN, but its insolubility in all common solvents has thus far prevented
further characterization. The phenyl/methyl substituted compound decomposes to form a sticky solid, the
^{31}P NMR spectrum of which consists of a sharp singlet
at $\delta 1.67$. Like $(Me_2PN)_n$, this is significantly upfield
from the signals of the small ring compounds ($\delta \sim 27$).

Another interesting system under investigation in
our laboratory is a polymeric phosphazene with a mixture of Me and Et substituents at phosphorus. By
allowing $(Me_3Si)_2NPCl_2$ (eq 7) to react with equimolar
amounts of MeMgBr and EtMgBr, we obtain 1:2:1 proportions of the phosphines $(Me_3Si)_2N-PMe_2$, $-PMeEt$, and
$-PEt_2$. Subsequent reactions as described previously
provide mixtures of $Me_3SiN=P(OCH_2CF_3)RR'$ which decompose readily to give phosphazene products characterized
by 3 broad signals in the ^{31}P NMR spectrum at $\delta 7.8$, 13,
and 18 in approximately 1:2:1 proportions. Presumably
these correspond to the PMe_2, PMeEt, and PEt_2 centers,
respectively, distributed in a random fashion along the
phosphazene chain. Further characterization of these
materials is underway.

CONCLUDING REMARKS

The research described herein demonstrates that
the proposed phosphazene synthesis (eq 4) is, in fact,

a generally useful one. Through the elimination of
Me_3SiBr from N-silylphosphinimines, high yields of
smaller cyclic phosphazenes are readily obtained. More
significantly, the elimination of $Me_3SiOCH_2CF_3$ from
other N-silylphosphinimines is a general approach to
good yields of polymeric phosphazenes, many of which
have evaded synthesis by other methods. This synthetic route is particularly applicable to the preparation
of polyphosphazenes with directly P-C bonded substituents such as $(Me_2PN)_n$. The physical properties and
potential applications of $(Me_2PN)_n$, particularly in
view of its electronic similarity to poly(dimethylsiloxane) $(Me_2SiO)_n$, certainly are worthy of more detailed study. Other features of our synthesis which
need to be further explored include the variation of
substituents at phosphorus, methods of increasing
molecular weight, and the possibility of crosslinking
the polymer chains.

A number of obvious questions involving mechanistic aspects of the new synthetic method remain to be
answered. Since the cyclic phosphazenes $(Me_2PN)_{3,4}$ do
not polymerize [3] even under extreme conditions, it
appears that the $Me_3SiOCH_2CF_3$ elimination reaction must
bypass the cyclics, leading to speculation about the
formation of the Me_2PN monomer. Another interesting
but puzzling fact is that Me_3SiBr elimination gives
only cyclic phosphazenes while $Me_3SiOCH_2CF_3$ elimination yields exclusively polymeric products.

Finally, it seems likely that the development of
this new synthesis of polyphosphazenes, including some
much-sought-after compounds, will serve to further
stimulate the continued growth of phosphazene chemis-

438 NEILSON AND WISIAN-NEILSON

try and, perhaps, to help bring it closer to the realm
of commercial application.

ACKNOWLEDGMENTS

The authors thank the U.S. Army Research Office,
and the donors of the Petroleum Research Fund, administered by the American Chemical Society, for generous
financial support. The characterization data ($\overline{M}_w$, T_g,
etc. ...) for poly(dimethylphosphazene) was kindly
provided by Dr. Gary Hagnauer of the Army Materials
and Mechanics Research Center.

REFERENCES

[1] H.R. Allcock, Phosphorus-Nitrogen Compounds,
     Academic Press, New York, 1972.

[2] H.R. Allcock, Angew. Chem. Int. Ed. Engl., 16,
     147 (1977).

[3] H.R. Allcock and D.B. Patterson, Inorg. Chem.,
     16, 197 (1977).

[4] H.R. Allcock, D.B. Patterson, and T.L. Evans,
     J. Am. Chem. Soc., 99, 6095 (1977).

[5] H.R. Allcock and C.T.-W. Chu, Macromolecules,
     12, 551 (1979).

[6] P. Wisian-Neilson and R.H. Neilson, J. Am. Chem.
     Soc., 102, 2848 (1980).

[7] R.H. Neilson, P. Wisian-Neilson, and J.C. Wilburn,
     Inorg. Chem., 19, 413 (1980).

[8] R.H. Neilson and D.W. Goebel, J. Chem. Soc. Chem.
     Commun., 769 (1979).

[9] J.C. Wilburn. P. Wisian-Neilson, and R.H. Neilson,
     Inorg. Chem., 18, 1429 (1979).

[10] R.H. Neilson and W.A. Kusterbeck, J. Organomet. Chem., 166, 309 (1979).

[11] J.C. Wilburn and R.H. Neilson, Inorg. Chem., 18, 347 (1979).

[12] J.C. Wilburn and R.H. Neilson, Inorg. Chem., 16, 2519 (1977).

[13] P. Wisian-Neilson, R.H. Neilson, and A.H. Cowley, Inorg. Chem., 16, 1460 (1977).

[14] P. Wisian-Neilson and R.H. Neilson, Inorg. Chem., in press (June, 1980)

[15] H.T. Searle, J. Dyson, T.N. Ranganathan, and N.L. Paddock, J. Chem. Soc., Dalton Trans., 203 (1975).

[16] H.H. Sisler, S.E. Frazier, R.G. Rice, and M.G. Sanchez, Inorg. Chem., 5, 327 (1966).

[17] E.-P. Flindt and H. Rose, Z. Anorg. Allg. Chem., 428, 204 (1977).

Index